KINEMATICS

Edited by **Efren Gorrostieta Hurtado**

Kinematics
http://dx.doi.org/10.5772/intechopen.68259
Edited by Efren Gorrostieta Hurtado

Contributors

Laith Sahawneh, Randal Beard, Ondrej Hock, Jozef Šedo, Milan Čoh, Norbert Piotrowski, Adam Barylski, Takashi Harada, Giuliani Garbi, Francisco José Grandinett, Panchanand Jha, Bibhuti Bhusan Biswal, Edgar Alonso Martínez García, Jorge Reyes, Rafael Torres-Córdoba, Ricardo Rodriguez Jorge, Liang Yang, Jizhong Xiao, Bin Li, Zhaoming Liu, Holger Korte, Sven Stuppe, Jan-Hendrik Wesuls, Tsutomu Takagi, Ana Cristina Majarena Bello, Javier Conte, Jorge Santolaria, Raquel Acero, Bashar El-Khasawneh, Abdur Rosyid, Anas Alazzam

Published by InTech
Janeza Trdine 9, 51000 Rijeka, Croatia

Publishing Process Manager Dajana Pemac
Technical Editor SPi Global
Cover InTech Design team

First published December, 2017
Printed in Croatia
Legal deposit, Croatia: National and University Library in Zagreb

Additional hard copies can be obtained from orders@intechopen.com

Kinematics, Edited by Efren Gorrostieta Hurtado
p. cm.
Print ISBN 978-953-51-3687-3
Online ISBN 978-953-51-3688-0

For EU product safety concerns:
IN TECH d.o.o., Prolaz Marije Krucifikse Kozulić 3, 51000 Rijeka, Croatia,
info@intechopen.com or visit our website at intechopen.com.

Meet the editor

Dr. Eng. Efren Gorrostieta is a professor at the Engineering Faculty of the Autonomous University of Queretaro, Mexico. He studied Electronic Engineering in Western Institute of Technology and Higher Education (ITESO) and received a Master degree in Science in Control and Automation and a PhD degree in Mechatronics. He was a cofounder of the Mechatronics Mexican Association, the president of the IEEE Queretaro Section, and the chair of the IEEE Queretaro Computational Intelligence Chapter. He has given lectures in control systems and robotics at different universities and has been a chair, reviewer, and editor in several national/international congresses related to robotics, automation, and artificial intelligence. He has several publications in conferences and journals in the field.

Contents

Preface

The study of the movements of bodies and particles in space is one of the most important tasks in the development of different areas; as an example of this issue is the robotics field where the kinematics occupies an important role in its development, which is complemented by the dynamics.

Given the importance of the subject in different areas of knowledge, there is a constant search for new technology and methods that allow us to understand and facilitate its study and in the same way to contribute with new techniques and ideas in the growth of this area. This is the reason why we can find research groups dedicated to the development of this area. The present work presents a compilation of the efforts made by researchers in several institutions at different parts of the world, and in this way, we can find the evolution in recent years, like applications in navigation systems and new methods for analysis, which are important in the proposal of new paradigms. Also, there are applications in mobile robotic path planning, robot manipulators, parallel robots, development of control systems and kinematic modeling to understand the behavior of the human body, applications in the processes of measurements in different stages of the processes, optimization of the parameters that allows better performance in mechanical systems, and so on. Moreover, we can find works that make use of some artificial intelligence techniques allowing a new approach to the kinematics among other things. Finally, there are works in different areas that allow to see the use of scope and importance of the subject.

This area also offers us an opportunity to grow using and combining many other techniques such as quaternions, differential mechanics, artificial intelligence applications, and dynamic programming, to name a few, while at the same time bringing us the idea on how to develop new technologies.

I appreciate the contribution and professional work of each of the researchers who contributed by sharing their work, development, and ideas that allowed us to present this book today and share all this information with readers and the rest of the scientific community.

Dr. Eng. Efren Gorrostieta
Autonomous University of Queretaro,
Faculty of Engineering, Mechatronics Laboratory,
Cerro de las Campanas S/N,
Queretaro, Qro,
Mexico

Chapter 1

Kinematic Performance Measures and Optimization of Parallel Kinematics Manipulators: A Brief Review

Abdur Rosyid, Bashar El-Khasawneh and Anas Alazzam

Additional information is available at the end of the chapter

http://dx.doi.org/10.5772/intechopen.71406

Abstract

This chapter covers a number of kinematic performance indices that are instrumental in designing parallel kinematics manipulators. These indices can be used selectively based on manipulator requirements and functionality. This would provide the very practical tool for designers to approach their needs in a very comprehensive fashion. Nevertheless, most applications require a more composite set of requirements that makes optimizing performance more challenging. The later part of this chapter will discuss single-objective and multi-objectives optimization that could handle certain performance indices or a combination of them. A brief description of most common techniques in the literature will be provided.

Keywords: parallel kinematics manipulator, kinematic performance measures, optimization, workspace, Jacobian-based performance measures, stiffness, accuracy

1. Introduction

Serial kinematics mechanisms (SKMs) have been widely used for different applications. Although SKMs have many advantages, such serial mechanisms have many drawbacks such as low stiffness, accumulating pose error, low agility, low payload-to-weight ratio, and complicated inverse kinematics. Hence, to overcome these drawbacks, parallel kinematics mechanisms (PKMs) are used particularly for more demanding tasks such as high-speed and high-precision applications. In spite of their many advantages, the PKMs in general also have some drawbacks such as smaller workspace, complicated forward kinematics, and singularity issue. To alleviate these drawbacks, optimization with various techniques is commonly conducted to improve their drawbacks while maintaining their advantages. In terms of the number of objectives being optimized, the optimization can be either single-objective or multi-objective. In most cases, there are more than one objectives required to be optimized. Furthermore, some objectives quite frequently are conflicting each other. For example, most PKMs usually require not only larger workspace but also stiffer structure with lower mass. In fact, enlarging the workspace usually requires longer links which results in the reduction of the stiffness and the increase of mass. In the multi-objective optimization, different objectives might be picked based on the priority of the objectives which depends on the application.

Therefore, in this chapter, a comprehensive review of a number of performance indices are defined and presented which are relevant for different applications. This is followed by a review of the optimization techniques used to design different systems to satisfy certain objective or multiple objectives. This is extremely important given the nonlinearity of the parallel link manipulator systems and the conflicting nature of the different performance indices that could be counter intuitive to optimize by trial and error and hence, mathematical schemes would be the solution.

2. Performance measures

There are quite many measures or indices to indicate the performance of a robot. Patel and Sobh [1] classified them into either local or global, kinematic or dynamic, and intrinsic or extrinsic. The local measures are dependent on the robot configuration, whereas the global measures evaluate the robot performance across the whole workspace. The global performance index (GPI) can be obtained by integrating the local performance index P over the workspace W as given by Eq. (1). If the local performance index cannot be expressed analytically, discrete integration as given by Eq. (2) can be used. In the latter case, the accuracy of the integration depends on the number of points *n* being used for evaluation. If the inclusion of large number of points is very computationally expensive, less representative sampling points can be used.

$$GPI = \frac{\int_W P \, . \, dW}{\int_W dW} \tag{1}$$

$$GPI = \frac{1}{n} \sum_{i=1}^{n} P_i \tag{2}$$

The kinematic measures indicate the kinematic behavior of the robot, whereas the dynamic measures are related to dynamic properties of the robot. The intrinsic measures indicate the inherent characteristics of the robot regardless of its task, whereas the extrinsic measures are related to the robot task. The widely used performance measures include workspace, closeness or avoidance from singularity, dexterity, manipulability, stiffness, accuracy, repeatability, and reliability. Some of them are discussed below. The performance measures should be considered during design phase of the robot. Optimal design usually considers one or several performance measures as the objective function(s) to be optimized.

2.1. Workspace

The workspace is the set of points (locations) which can be reached by the end effector. It is driven by the mobility of the robot which includes the number and types of its degrees of freedom (DOF), and constrained by the length of links, range of the joints, and interference between the components. The workspace evaluation usually includes its size (area/volume) and shape. The shape can be expressed by, for example, aspect ratio of the regular workspace. In general, larger and better-shaped workspace is required. Another way to characterize the workspace is by using workspace volume index [2] or footprint ratio [3] which is defined as the ratio between the workspace volume and the volume of the machine.

The first thing to determine in order to achieve better workspace before optimizing the geometrical parameters is selecting better topology. For example, many mechanism designs include sliders (gliders) in order to get larger workspace, such as in Hexaglide, Linapod, Pentaglide, sliding H4, and Triaglide.

The robot workspace is commonly classified into several types as follows [2]:

- Constant orientation workspace (translation workspace) which defines reachable points with constant orientation of the moving platform.
- Orientation workspace which defines reachable orientations while the center tool point is fixed.
- Reachable workspace (maximal workspace) which defines reachable points with at least one orientation of the moving platform.
- Inclusive workspace which is reachable workspace in a given orientation range.
- Dexterous workspace which defines reachable points with any orientation of the moving platform.
- Total orientation workspace which is dexterous workspace in a given orientation range.
- Useful workspace (sometimes also called used workspace) which defines part of the workspace to be used for a specified application. It is usually regular workspace such as rectangle, circle, cuboid, sphere, or cylinder.

The useful workspace is usually of the highest interest as it indicates part of the workspace which can be really utilized for application. Baek et al. [4] presented a method to find maximally inscribed rectangle in the workspace of serial kinematics manipulates (SKM) and parallel kinematics manipulators (PKM).

A complete representation of the workspace requires six-dimensional space. However, graphical representation is only possible up to three-dimensional space. For this reason, it is a common practice to represent the position workspace separately from the orientation workspace. Workspace of a planar mechanism can be plotted in a two-dimensional plot, whereas that of a spherical or spatial mechanism can be plotted in a three-dimensional plot. The workspace plot can be presented in either Cartesian or polar coordinate system in the case of 2D plot and in either Cartesian, cylindrical, or spherical coordinate system in the case of 3D plot. Plotting the graphical representation of the workspace is easier in SKMs but is not always easy in PKMs. In the latter case, the graphical illustration of the workspace can only be applied for PKMs with no more than 3-DOF. For PKMs with more than 3-DOF, n – 3-DOF should be fixed in order to be able to graphically illustrate the workspace. Depending on which DOF to be fixed, the workspace will be different [2].

In general, there are three main ways to determine and plot the workspace: geometrical approach, discretization numerical approach, and non-discretization numerical approach. Early works on the workspace determination of PKMs are conducted by geometrical approach. Bajpai and Roth [5] investigated the workspace of PKMs and the influence of legs' length to the workspace. Three years later, Gosselin [6] presented his work on the constant-orientation workspace determination of 6-DOF PKM followed by Merlet [7] who presented the orientation workspace determination, and Kim et al. [8] who proposed an algorithm to determine the reachable and dexterous workspace. As planar PKMs require different treatments, Gosselin and Jean [9] followed by Merlet et al. [10] presented the workspace determination of planar PKMs. All of the aforementioned works use geometrical approach.

In the geometrical approach, the true boundaries of a PKM workspace are obtained from the intersection of the boundaries of every open-loop chains which compose the PKM. This approach is fast and also accurate. To make it easier and much faster, CAD software might also be utilized such as the work proposed by Arrouk et al. [11]. One of the main drawbacks of this approach is its lack of general applicability since different robot topologies might need different techniques to apply this approach. In other words, this approach usually should be tailored to the considered robot. Another drawback of this approach is the difficulty to include all the constraints.

In the discretization numerical approach, a discretized bounding space which covers all possible points in the workspace is created and subsequently tested by utilizing the inverse kinematics along with the constraints whether it belongs to the workspace or not. This approach is sometimes called binary representation since it assigns binary numbers during the evaluation: "1" is assigned if it is reachable and therefore plotted and "0" is assigned if it is unreachable and therefore not plotted. The main advantage of this approach is its applicability to all types of PKMs as well as its intuitiveness. Moreover, this approach can include all the constraints. However, the accuracy of this approach depends on the size of the discretization steps. Also, small voids inside the workspace cannot be detected unless the discretization steps are small enough to capture the voids. A method similar to the discretization approach is Monte Carlo method [12, 13] in which a large number of discrete active joint points within the joint range are input to the forward kinematics and accordingly the end effector position points are plotted. Further treatment to the Monte Carlo results in order to determine the workspace boundaries or compute the workspace volume can be conducted by putting the workspace points in discretized bounding space as being used in the discretization approach.

Some recent works using the discretization numerical approach includes Bonev [14] who proposed a new approach to determine the three-dimensional orientation workspace of 6-DOF PKMs by using discretization approach. Castelli et al. [15] presented an algorithm based on the discretization approach to determine the workspace, the workspace boundaries, the workspace volume, and the workspace shape index of SKMs and PKMs. Dash et al. [16] presented a discretization method to determine the reachable workspace and detect available voids of PKM.

Beyond the aforementioned two approaches, several works proposed non-discretization numerical methods to determine the workspace of PKMs. Some of the methods are as follows [2, 17]:

- Jacobian rank deficiency method [18] but only practical for the determination of constant orientation workspace.
- Numerical continuation method [19, 20] can avoid singularity points but only practical for the determination of constant orientation workspace.
- Constrained optimization method [21] which is modified from the numerical continuation method.
- Boundary search method [17] which is based on constrained non-linear programming.
- Principle that the velocity vector of the moving platform cannot have a component along the normal of the boundary [22], but this method does not work for mechanisms with prismatic joints as well as it is difficult to include the mechanical limits and interference between links.
- Interval analysis method [23] which can deal with almost any constraints and any number of DOF.

Recently, Bohigas et al. [24] presented branch-and-prune technique which can determine all the workspace boundary points of general lower-DOF (3-DOF or lower) SKMs and PKMs. This technique overcomes the limitation of numerical continuation method. Furthermore, Gao and Zhang [25] proposed simplified boundary searching (SBS) method which integrates geometrical approach, discretization method, and inverse kinematics model of a parallel mechanism. Saputra et al. [26] proposed swarm optimization approach to determine the workspace of PKM.

2.2. Jacobian matrix

The Jacobian matrix maps the relation between the velocities at the task space (moving platform) and the velocities of the active joints. Furthermore, it also maps the relation between the active joint load and the task wrench. It is discussed here because it is related to many kinematic performance measures.

Given that the velocity kinematics of the robot is expressed by:

$$A\dot{q} + B\dot{x} = 0 \tag{3}$$

where x is the end effector twist and q is the actuator twist, then A is called forward Jacobian matrix, B is inverse Jacobian matrix, and the total Jacobian matrix J is given by:

$$J = -B^{-1}A \tag{4}$$

As long as the Jacobian matrix is unit-consistent (homogeneous), it can be directly used in the formulation of Jacobian-based performance measures such as Jacobian condition number and manipulability. The problem appears when the Jacobian matrix is unit-inconsistent (non-homogeneous) and accordingly does not have appropriate physical meanings. To address this problem, there are several ways found in the literature to normalize inconsistent (non-homogeneous) Jacobian matrix including the following:

- Using natural length or characteristic length [27–29]
- Using scaling matrix [30, 31]
- Using weighting factor [32]
- By using power transition concept [33]
- Point-based method [34–37]
- General and systematic method [38]
- Homogeneous extended Jacobian matrix [39]

2.3. Singularity

The simplest classification of singularity in PKMs is given by Gosselin and Angeles [40]. By considering only the behavior of the active joints and the end-effector, they classified the singularity in PKMs into two types which are mathematically determined by the singularity of the two Jacobian matrices in the kinematics of the robot given in Eq. (3). The three types of singularity are the following:

- Type 1 singularity (also called: direct kinematic singularity, forward kinematic singularity, serial singularity, or sub-mobility) occurs when the forward Jacobian matrix A is singular. When this kind of singularity occurs, it is not possible to generate some velocities of the end-effector. In other words, very small changes in the joint space do not affect the end-effector pose. In these configurations, the mechanism loses one or more degrees of freedom.
- Type 2 singularity (also called: inverse kinematic singularity, parallel singularity, or over-mobility) occurs when the inverse Jacobian matrix B is singular. It corresponds to the appearance of uncontrollable mobility of the end-effector because it is possible to move it while the joints are locked. At the corresponding configurations, the mechanism gains one or more uncontrollable DOF. In other words, the end-effector can move without the joints moving. Equivalently, the stiffness of the PKM is locally lost.
- Type 3 singularity (also called: combined singularity) occurs when both the forward Jacobian matrix A and the inverse Jacobian matrix B are singular. When this kind of singularity occurs, the end-effector can move when the joints are locked, and at the same time the end-effector pose does not change due to very small changes in the joints.

Furthermore, singular configurations can be obtained by observing the Jacobian matrices or by geometrical approach.

A more general discussion on the singularity was delivered by Zlatanov et al. [41] who included the passive joints in the singularity evaluation. They classified the singularity into redundant input (RI) singularity which corresponds to serial singularity, redundant output (RO) singularity which includes parallel singularity as well as so-called constraint singularity [42], and so-called actuator singularity [43] which represents non-zero passive joint velocities while the actuators are locked and the end-effector has zero velocities. Moreover, higher order singularity has also been discussed in some works, but it does not give much practical benefit [2].

2.4. Jacobian condition number

From its mathematical expression, the value of Jacobian condition number (or simply condition number) ranges from 1 to infinity, where infinity indicates singularity and 1 indicates isotropy of the Jacobian matrix. Alternatively, it can also be expressed by its inverse value, called inverse Jacobian condition number, the value of which ranges from 0 to unity where 0 indicates singularity and unity indicates isotropy of the Jacobian matrix. When the Jacobian matrix is close to singularity, it is called ill-conditioned. On the other hand, the Jacobian matrix is called well-conditioned if it is far from singularity. Furthermore, when the Jacobian matrix is isotropic, it means that the velocity and force amplification is identical in all directions.

The commonly used norms to define the Jacobian condition number are as follows:

- 2-norm, which is given by the ratio of the maximum and minimum singular values of the Jacobian matrix.
- Frobenius norm, which is very advantageous because it is an analytical function of the robot parameters and hence will not give serious concern if its gradient is evaluated [44], as well as it avoids the computation of the singular values.
- Weighted Frobenius norm, which can be rendered to specific context by adjusting its weights [45] in addition to all of the mentioned advantages of the Frobenius norm.

The Jacobian condition number is a measure of kinematic dexterity (or simply called dexterity). It indicates closeness to singularity, kinematic uniformity (isotropy), dexterity, and accuracy. The kinematic dexterity is defined as the capability of robot to move the end-effector in all directions with ease. In fact, the kinematic isotropy of the robot represents its dexterity as more isotropy indicates that the robot can move with the same ease to all directions. However, this is still not a complete information about the dexterity as it only informs how equal the ease in different directions, but not how easy. It is possible that either the motion to all directions requires small effort or the motion to all directions require large effort. Manipulability which will be reviewed soon will give more complete information about the kinematic dexterity.

Another interpretation of the Jacobian condition number is how large the error in the task space will occur due to small error in the joint space. The more ill-conditioned the Jacobian matrix, the larger the error in the task space will occur due to small error in the joint space. Based on this fact, the Jacobian condition number indicates the accuracy of the manipulator.

The Jacobian condition number is a local property. It depends on the robot configuration (posture). To evaluate globally, global condition number (or global condition index, GCI) is used. The GCI is obtained by integrating the local condition index (LCI) over the workspace. Since the Jacobian condition number is the common indicator of dexterity, GCI is also commonly called global dexterity index (GDI). A map showing the values of LCI over the workspace is commonly referred to dexterity map.

Since the manipulability is based on the Jacobian matrix, it faces unit inconsistency issue when translation and rotation are mixed. In this case, the Jacobian matrix should be homogenized.

2.5. Manipulability

Manipulability measure was first introduced by Yoshikawa [46] as a local measure (local manipulability index, LMI) which means that it is dependent on the robot configuration (posture) since it is based on Jacobian matrix. It can be evaluated globally by using global manipulability measure (GMI) which is the local manipulability measure integrated over the workspace. Another measure is uniformity of manipulability which represents how uniform the manipulability across the workspace [47]. Similar to the Jacobian matrix, the manipulability faces unit inconsistency issue when translation and rotation are mixed. In the same token with the Jacobian condition number, the Jacobian matrix should be homogenized in such a case.

The manipulability is a measure of the input-output efficiency (the ratio of output performance to input effort). In other words, it represents the quality of velocity and force transmission (amplification). The manipulability provides the information about the velocity and force amplification more than the Jacobian matrix condition number. The latter only tells how isotropic the velocity and force amplification but not the magnitude, whereas the earlier informs the magnitude in addition to the isotropy of the velocity and force amplification.

Two kinds of manipulability are well known: twist (velocity) manipulability and wrench (force) manipulability. The earlier is commonly represented by velocity manipulability ellipse/ellipsoid whereas the latter by force manipulability ellipse/ellipsoid. In the velocity manipulability ellipsoid, the velocity minimum, velocity maximum, and velocity isotropy are represented by the ellipsoid axes length, whereas the manipulability is represented by the ellipsoid volume. The major axis of the manipulability ellipsoid indicates the direction along which the mechanism can move with the least effort, while the minor axis indicates the direction along which the mechanism is stiffest, i.e., the mechanism's actuators can resist forces with minimum effort along this direction. The force manipulability uses duality relation between velocity and force (between differential kinematics and statics). Beside manipulability ellipse/ellipsoid, the manipulability can also be represented by manipulability polytope.

Tanev and Stoyanov [48] have introduced the use of normalized manipulability index which is bounded between zero and unity. Doty et al. [49] proposed weighted twist manipulability and weighted wrench manipulability. Furthermore, a new manipulability measure for parallel robot was introduced by Hong and Kim [50].

http://dx.doi.org/10.5772/intechopen.71406

2.6. Stiffness

Beside the workspace, stiffness or rigidity of a robot structure plays a very important role as it affects the accuracy and repeatability of the robot. Stiffness is defined as the ability of the robot structure to resist deformation due to wrench. A stiffness matrix relates deformation vector to wrench vector. Another term equivalent to the stiffness is compliance (flexibility). If a structure has high stiffness, it means that the structure has low compliance. A compliance matrix is simply the inverse of the stiffness matrix, and vice versa. The stiffness includes static stiffness and dynamic stiffness. For machine tool, high stiffness enables machining with high speed and feed while providing good precision, accuracy, and surface finish.

Stiffness of a mechanism depends on the robot topology, geometry, and material of the mechanism. The overall stiffness is comprised of the stiffness of the fixed base, the moving platform, the joints, and the links. The stiffness of the joints includes that of the active joints (actuators) and the passive joints. Many works discussed the influence of the passive joints on the robot stiffness. The stiffness of the links is usually defined in axial direction (axial stiffness), transversal direction (bending stiffness), or both of them. To simplify stiffness model, rigid body assumption is frequently applied to one or several components of the robot. For example, joints can be considered elastic while the links are assumed to be rigid, or vice versa. A more realistic model usually consider both of the joints and the links as elastic. In hybrid machine tools, many works have proposed the use of parallel mechanism for the spindle platform and serial mechanism for the worktable, as the most flexible part of the machine is usually the spindle platform. Furthermore, some works have suggested the use of passive legs to increase the stiffness such as in Tricept and Georg V.

Stiffness is a local property. It depends on the robot configuration (posture). To evaluate globally, global stiffness measures are used. Furthermore, stiffness varies with the direction in which it is evaluated as well as the direction of the wrench. Therefore, stiffness can be identified in different directions, either translational directions (translational stiffness) or rotational directions (rotational stiffness).

In the literature, compliance which is the inverse of stiffness is sometimes used instead of stiffness. Several different expressions of stiffness have been used in the literature, including engineering stiffness, generalized stiffness matrix, and Cartesian stiffness matrix. The engineering stiffness is a one-dimensional stiffness expression obtained by evaluating the displacement in the same direction to the applied force [51]. The generalized stiffness matrix, according to Quennouelle and Gosselin [52], includes three contributions: stiffness of the unconstrained joints, stiffness due to dependent coordinated and internal wrench, and stiffness due to external wrench. In other words, the generalized stiffness matrix is sum of the three stiffness components.

The Cartesian stiffness matrix is the most widely used expression of stiffness. Ruggiu [53] shows that Cartesian stiffness matrix of a mechanism is the Hessian of its potential energy expression. Cartesian stiffness matrix is symmetric and positive definite or positive semi-definite. However, some researchers concluded that the Cartesian stiffness matrix of the elastic structure coupling two rigid bodies is asymmetric in general and becomes symmetric if the connection is not subjected to any preloading. Different expressions of Cartesian stiffness matrix are mentioned by Klimchik [54] and Quennouelle and Gosselin [52]. The latter authors proposed a Cartesian stiffness matrix which can take into account non-zero external loads, non-constant Jacobian matrices, stiff passive joints, and additional compliances. Furthermore, the Cartesian matrix can be directly related to the generalized stiffness matrix by utilizing the Jacobian matrix.

In robots with translational and rotational DOF, the Cartesian stiffness matrix will be unit inconsistent. Consequently, evaluation of further stiffness indices such as stiffness condition number becomes nonsense. To deal with this problem, several approaches have been proposed including the following:

- Homogenizing the Jacobian matrix (such as using characteristic length) and subsequently using the homogenized Jacobian matrix to calculate the stiffness matrix [55, 56].
- Eigenscrew decomposition of the stiffness or compliance matrix [57–60].
- Principal axis decomposition through congruence transformation was proposed by making use of the eigenvectors of the translational entry in the stiffness matrix [61].
- Decomposition of the dynamic inertia matrix by transforming variables into dimensionless parameters, which can be applied to the stiffness matrix [62, 63].
- Decoupling of the stiffness matrix into translational parts and rotational parts [64–66].

Furthermore, to model the robot stiffness beyond using continuous model which works only for simple system, the following three different models are widely used in the literature:

- Jacobian matrix-based model; also called lumped parameter model or virtual joint method (VJM) model. A one-dimensional VJM was introduced by Gosselin [67], followed by Anatol et al. [68] who introduced multi-dimensional VJM. This model is widely used and preferred in robotics since it is analytical, and therefore the same expression works for all configurations of the robot and it requires lower computational cost. However, it gives lower accuracy but still acceptable. For that reason, this method is good for initial estimates of the robot stiffness as well as for design optimization purpose.
- Finite element model (FEM). As opposite of the lumped parameter model, this model discretizes the mechanism into many elements and therefore can also be called distributed model which implies more closeness to the realistic, continuous model. It is widely used in structural mechanics due to its high accuracy. However, it requires high computational cost. Furthermore, it needs new mesh at every different configuration of the robot which makes it not practical. Due to its high accuracy, this model is usually used to verify another less accurate model such as VJM model.
- Matrix structural analysis (MSA) model. This model is actually a special case of FEM model because it uses one-dimensional finite elements such as beam elements instead of two or three dimensional elements such as brick elements. As a result, the computational cost decreases. This model gives trade-off between accuracy and computational cost [69].

After that, modifications and improvements on the aforementioned methods have been conducted, such as follows:

- Online FEM by utilizing MSA using generalized springs [70].
- VJM combined with FEM-based identification technique: high accuracy with low computational cost [71].
- Virtual spring approach: spring compliance is evaluated based on FEM concept; high accuracy with low computational cost [71, 72].

Evaluation of stiffness can also be conducted by experimental method, i.e., from measurement. In this case, the stiffness is obtained from the relation between measured wrench and measured displacement. Another way to evaluate the stiffness is by estimation or identification of the stiffness model. Least squares estimation algorithm or other estimation algorithms can be utilized in the estimation based on measurement data.

As a performance measure, the robot stiffness is represented in the following different ways in the literature:

- Graphical representations including stiffness maps, by which the stiffness distribution can be plotted [67, 73], and other graphical representations such as iso-stiffness curves or surfaces (global) [2].
- Trace of the stiffness matrix.
- Weighted trace of the stiffness matrix [74].
- (Minimum, average, or maximum) eigenvalues (and eigenvectors) of the stiffness matrix [51]. For example, the evaluation of minimum and maximum eigenvalues across the workspace by Li and Xu [56].
- Mean value of the eigenvalues [70].
- Determinant of stiffness matrix, which is the product of the stiffness matrix eigenvalues), and indicates the area/volume of a stiffness ellipse/ellipsoid. It also indicates how far from singularity.
- Norm of the stiffness matrix, which can be its Euclidian norm, Frobenius norm, or Chebyshev norm [75].
- Center of stiffness or equivalently center of compliance [76].
- Global compliance index which is given by mean value and deviation of generalized compliance matrix [77].
- Virtual work stiffness index which is able to avoid the problem caused by different units of translation and orientation
- Collinear stiffness value (CSV) [78].

2.7. Stiffness condition number

Stiffness condition number is a local measure. It depends on the robot configuration. In similar token to the Jacobian condition number, the stiffness condition number can take a value ranging from 1 to infinity. Alternatively, inverse of the stiffness condition number which takes value ranging from 0 to 1 can also be used. Since the stiffness condition number represents the isotropy or uniformity of the stiffness of any point in the workspace, stiffness ellipses/ellipsoids are commonly used as the graphical representation.

Similar to Jacobian condition number, different definition of norms can be used to evaluate the stiffness condition number. The commonly used norms are 2-norm, Frobenius norm, and weighted Frobenius norm. The considerations in selecting any of them is explained earlier when the Jacobian condition number is discussed.

The global stiffness condition number is commonly expressed by global stiffness index (GSI) which is usually defined as the inverse of the condition number of the stiffness matrix integrated over the reachable workspace divided by the workspace volume. It depicts the uniformity of stiffness within the whole workspace.

3. Design optimization

In terms of the number of objectives being optimized, optimization can be either single-objective (also called single-criteria) or multi-objective (also called multi-criteria). The simplest way of design optimization is by trial and error in which we pick several values of design parameters based on intuition, knowledge, or experience, and compare the corresponding objective values. However, this approach is non-systematic as well as does not cover all possible values of the design parameters and therefore may not give optimum solutions. In the literature, performance atlas and optimization algorithms are commonly used in the design optimization of mechanisms. The performance atlas presents graphically the relation between the design parameters (length of the links) and the performance measures. Several performance atlases such as atlas of workspace, atlas of GSI, and atlas of LSI have been used in the literature. For single-objective optimization, the use of performance atlas is easy and straightforward. However, a multi-objective optimization requires inspection of several atlases which might give inconvenience, particularly when some objectives are conflicting each other.

Beyond the use of performance atlas, various algorithms for optimization of PKMs and HKMs have been utilized. Based on the search principles, those techniques fall into two main categories: gradient-based optimization techniques and population-based optimization techniques. The first category is a local search algorithm. It is deterministic and can be linear or nonlinear depending on the problem. The latter category is stochastic and does not need gradient information. One of the most popular population-based techniques is the genetic algorithm which is an evolutionary optimization technique and works based on the idea of natural selection or survival of the fittest. The genetic algorithm can be implemented for both single-objective and multi-objective optimization. For the latter implementation, several techniques have been developed such as VEGA,

NPGA, NPGA-II, NSGA, NSGA-II, PAES, PESA, PESA-II, SPEA, SPEA-II, SPEA-II+, and many others [79–84]. Beyond the genetic algorithm, several global optimization algorithms have also been proposed in the literature, such as controlled random search (CRS) [85], differential evolution (DE) [86, 87], particle swarm optimization (PSO) [88–90], quantum particle swarm optimization (QPSO) [91], and artificial neural network (ANN) [74]. The following will be showing more details on both types of optimization.

3.1. Single-objective optimization

Although single-objective optimization is straightforward, different algorithms might be used to search the optimal solution. For parallel mechanism, if only single-objective is to be optimized, then it would be usually the workspace since the main drawback of parallel mechanisms is their limited workspace.

Beyond the use of widely used gradient-based optimization algorithms, various algorithms have been proposed for single-objective optimization of PKMs. Hosseini et al. [32] used genetic algorithm to optimize the dexterous workspace of a Tricept PKM. Kordjazi et al. [92] used genetic algorithm combined with fuzzy logic algorithm to optimize the Jacobian matrix isotropy of PKM and showed that the result is better than using genetic algorithm alone. Arana [93] proposed a methodology to enlarge the workspace of parallel manipulators by means of non-singular transitions. Ghao and Zhang [25] proposed the use of particle swarm optimization (PSO) to optimize the workspace of 3-DOF spatial parallel mechanism.

3.2. Multi-objective optimization

In most cases, there are more than one objectives required to be optimized. Furthermore, some objectives quite frequently are conflicting each other. For example, most PKMs usually require not only larger workspace but also stiffer structure with lower mass. In fact, enlarging the workspace usually requires longer links which results in the reduction of the stiffness and the increase of mass. Multi-objective optimization can be bi-objective or many-objective optimization. The earlier is simpler than the latter. The inclusion of more than two objectives generally requires more computational cost. It also gives more difficulty in the visual representation. If more than three objectives are involved, graphical plots can only be done for three varying design parameters while the rest should be fixed. An alternative approach to reduce the number of objectives in the optimization is by putting performance index threshold value as optimization constraint. However, this approach is only suitable if the need is only to satisfy the threshold.

In the multi-objective optimization, different objectives (criteria) might be picked based on the priority of the objectives which depends on the application. Two main methods commonly used for multi-objective optimization are:

- Scalarization method which is commonly conducted by putting the weighted multiple objective functions into one composite objective function. While transforming the problem into one objective function gives simplicity, the determination of appropriate weights is difficult, even for those who are familiar with the problem. This approach can be conducted through gradient-based optimization methods as well as single-objective evolutionary methods (such as single-objective genetic algorithm).

- Pareto method which will give non-dominated solutions. This method can be conducted through multi-objective evolutionary methods (such as multi-objective genetic algorithm).

Different objectives, depending on the application needs, and various algorithms have been proposed for multi-objective optimization of PKMs. Hodgins [94] optimized the workspace, the stiffness, and the dexterity of a modified Delta robot by using weighted sum optimization method. Kelaiaia et al. [95] optimized the kinematic dexterity as well as the dynamic dexterity by using genetic algorithms. Wu [96] optimized the GCI and GDI of a 3RRR spherical parallel mechanism, which can be used as orienting device, by using genetic algorithm. Bounab [97] optimized the dexterous regular workspace and the stiffness of a delta mechanism by using genetic algorithm. Shijun [3] optimized GDI, GSI, and the ratio of the workspace to the work volume using genetic algorithm. Gao et al. [74] optimized the stiffness and dexterity by using genetic algorithm and artificial neural network. Abbasnejad et al. [91] implemented particle swarm optimization (PSO) and quantum particle swarm optimization (QPSO) to optimize the workspace and the dexterity as weighted sum objective, and showed that QPSO has faster convergence than PSO. Furthermore, Gao and Zhang [98] introduced a comprehensive index to integrate four different objectives.

4. Recommendations

It appears that mixed DOFs result in inconsistency of the indices while many PKMs should have mixed DOFs to do the required tasks. For this reason, the authors suggest that the introduction of any new index in the future should be able to overcome this issue in a more natural way so that the physical insights of the index will be as sound and intuitive as possible. Furthermore, the authors have not found any published work discussing the optimization of a large number of performance measures. While a good number of attempts have been done to handle up to three or four objective functions, it will be practically useful yet challenging to handle larger number of objective functions. Knowing that the determination of appropriate weights in the scalarization approach is not easy even for two to four objective functions, it definitely will be more difficult to put appropriate weights to larger number of objectives while it may be no more practical to use the Pareto approach for the larger number of objectives. Therefore, the authors suggest the introduction of new approach or technique to reliably optimize larger number of objective functions.

5. Conclusion

This chapter provided a comprehensive overview of the literature related to a good number of kinematic performance indices that designers would be interested in using during the design of parallel kinematics mechanisms. Kinematic performance indices such as workspace,

Jacobian condition number, manipulability, stiffness magnitude, and stiffness condition number are of a prime interest to designers to optimize parallel kinematic mechanism designs. However, many of these indices are conflicting such as the increase in workspace size could lead to a decrease in stiffness and/or would lead to reduced speed. Therefore, using optimization techniques is the solution to accommodate such conflicting requirements by providing appropriate weights for the different objectives to reflect the designer's interests and priorities. Nevertheless, devising the proper objective function requires extensive experience and insight into the problem. The proper combination of objectives with the suitable weights will highly impact the design parameters results and hence should be chosen carefully. In summary, this paper attempted to provide comprehensive overview of what is needed by parallel kinematic mechanism designer to optimize their design and mechanism performance.

Author details

Abdur Rosyid, Bashar El-Khasawneh* and Anas Alazzam

*Address all correspondence to: bashar.khasawneh@kustar.ac.ae

Mechanical Engineering Department, Khalifa University of Science and Technology, Abu Dhabi, United Arab Emirates

References

[1] Patel S, Sobh T. Manipulator performance measures—A comprehensive literature survey. Journal of Intelligent and Robotic Systems. 2015;**77**(3-4):547-570

[2] Merlet J-P. Parallel Robots. 2nd ed. Dordrecht: Springer; 2006

[3] Shijun Y. A Serial-Parallel Hybrid Robot for Machining of Complex Surfaces. National University of Singapore, Singapore; 2015

[4] Baek J, Iurascu C-C, Park FC. Finding the maximally inscribed rectangle in a Robot's workspace. KSME International Journal. 2001;**15**(8):1119-1131

[5] Bajpai A, Roth B. Workspace and mobility of a closed-loop manipulator. The International Journal of Robotics Research. 1986;**5**:131-142

[6] Gosselin CM. Determination of the workspace of 6-DOF parallel manipulators. ASME Journal of Mechanical Design. 1989;**112**:331-336

[7] Merlet J-P. Determination of the orientation workspace of parallel manipulators. Journal of Intelligent and Robotic Systems. 1995;**13**:143-160

[8] Kim DI, Chung WK, Youm Y. Geometrical approach for the workspace of 6-DOF parallel manipulators. In: IEEE International Conference on Robotics and Automation (ICRA); Albuquerque, IEEE, New Mexico; 1997

[9] Gosselin CM, Jean M. Determination of the workspace of planar parallel manipulators with joint limits. Robotics and Autonomous Systems. 1996;**17**:129-138

[10] Merlet J-P, Gosselin CM, Mouly N. Workspaces of planar parallel manipulators. Mechanism and Machine Theory. 1998;**33**(1-2):7-20

[11] Arrouk KA, Bouzgarrou BC, Gogu G. Workspace determination and representation of planar parallel manipulators in a CAD environment. In: New Trends in Mechanism Science. Netherlands: Springer; 2010. p. 605-612

[12] Alciatore DG, Ng C-CD. Determining manipulator workspace boundaries using the Monte Carlo method and least squares segmentation. ASME Robotics: Kinematics, Dynamics and Controls. 1994;**72**:141-146

[13] Rastegar J, Perel D. Generation of manipulator workspace boundary geometry using the Monte Carlo method and interactive computer graphics. ASME Trends and Developments in Mechanisms, Machines, and Robotics. 1990;**3**:299-305

[14] Bonev IA, Ryu J. A new approach to orientation workspace analysis of 6-DOF parallel manipulators. Mechanism and Machine Theory. 2001;**36**(1):15-28

[15] Castelli G, Ottaviano E, Ceccarelli M. A fairly general algorithm to evaluate workspace characteristics of serial and parallel manipulators#. Mechanics Based Design of Structures and Machines. 2008;**36**(1):14-33

[16] Dash AK, et al. Workspace generation and planning singularity-free path for parallel manipulators. Mechanism and Machine Theory. 2005;**40**(7):776-805

[17] Wang Z, et al. A study on workspace, boundary workspace analysis, and workpiece positioning for parallel machine tools. Mechanism and Machine Theory. 2001;**36**(5):605-622

[18] D-Y Jo, Haug EJ. Workspace analysis of closed loop mechanisms with unilateral constraints. In: ASME Design Automation Conference; Montreal; 1989. pp. 53-60.

[19] Haug EJ, et al. Numerical algorithms for mapping boundaries of manipulator workspaces. ASME Journal of Mechanical Design. 1995;**118**:228-234

[20] Jo D-Y, Haug EJ. Workspace analysis of multibody mechanical systems using continuation methods. Journal of Mechanisms, Transmissions, and Automation in Design. 1989;**111**:581-589

[21] Snyman JA, Plessis LJd, Duffy J. An optimization approach to the determination of the boundaries of manipulator workspaces. ASME Journal of Mechanical Design. 2000;**122**(4):447-456

[22] Kumar V. Characterization of workspaces of parallel manipulators. ASME Journal of Mechanical Design. 1992;**114**(3):368-375

[23] Kaloorazi MHF, Masouleh MT, Caro S. Determining the maximal singularity-free circle or sphere of parallel mechanisms using interval analysis. Robotica. 2014;**34**(01):135-149

[24] Bohigas O, Ros L, Manubens M. A complete method for workspace boundary determination on general structure manipulators. IEEE Transactions on Robotics. 2012;**28**(5): 993-1006

[25] Gao Z, Zhang D. Workspace representation and optimization of a novel parallel mechanism with three-degrees-of-freedom. Sustainability. 2011;**3**(12):2217-2228

[26] Saputra VB, Ong SK, Nee AYC. A swarm optimization approach for solving workspace determination of parallel manipulators. Robotica. 2014;**33**(03):649-668

[27] Angeles J. Fundamentals of Robotic Mechanical Systems, Mechanical Engineering Series. New York: Springer; 2002

[28] O Ma, Angeles J. Optimum architecture design of platform manipulators. In: IEEE International Conference on Advanced Robotics; Pisa; 1991. pp. 1130-1135

[29] Ranjbaran F, et al. The mechanical design of a seven-axes manipulator with kinematic isotropy. Journal of Intelligent and Robotic Systems. 1995;**14**:21-41

[30] Stocco LJ, Salcudean SE, Sassani F. Matrix normalization for optimal robot design, in IEEE International Conference on Robotics and Automation. Leuven, Belgium; 1998

[31] Stocco LJ, Saculdean SE, Sassani F. On the use of scaling matrices for task-specific robot design. IEEE Transactions on Robotics and Automation, 1999;**15**:958-965

[32] Hosseini MA, Daniali H-RM, Taghirad HD. Dexterous workspace optimization of a Tricept parallel manipulator. Advanced Robotics. 2011;**25**(13-14):1697-1712

[33] Mansouri I, Ouali M. A new homogeneous manipulability measure of robot manipulators, based on power concept. Mechatronics. 2009;**19**(6):927-944

[34] Altuzarra O, et al. Point-based Jacobian formulation for computational kinematics of manipulators. Mechanism and Machine Theory. 2006;**41**(12):1407-1423

[35] Gosselin CM. The optimum design of robotic manipulators using dexterity indices. Robotics and Autonomous Systems. 1992;**9**(4):213-226

[36] Kim S-G, Ryu J. New dimensionally homogeneous Jacobian matrix formulation by three end-effector points for optimal design of parallel manipulators. IEEE Transactions on Robotics and Automation. 2003;**19**(4):731-736

[37] Kong M, et al. A novel approach to deriving the unit-homogeneous Jacobian matrices of mechanisms. In: IEEE International Conference of Mechatronics and Automation; 2007

[38] Liu H, Huang T, Chetwynd DG. A method to formulate a dimensionally homogeneous Jacobian of parallel manipulators. IEEE Transactions on Robotics. 2011;**27**(1):150-156

[39] Nurahmi L, Caro S. Dimensionally Homogeneous Jacobian and Condition Number. Applied Mechanics and Materials. 2016;**836**:42-47

[40] Gosselin CM, Angeles J. Singularity analysis of closed-loop kinematic chains. IEEE Transactions on Robotics and Automation. 1990;**6**(3):281-290

[41] Zlatanov D, Fenton RG, Benhabib B. A unifying framework for classification and interpretation of mechanism singularities. Journal of Mechanical Design. 1995;**117**(4):566

[42] Zlatanov D, Bonev IA, Gosselin CM. Constraint singularities of parallel mechanisms. In: IEEE International Conference of Robotics and Automation (ICRA); 2002

[43] Han C, Park. Kinematic sensitivity analysis of the 3-UPU parallel manipulator. Mechanism and Machine Theory. 2002;**37**(8):787-798

[44] Altuzarra O, et al. Optimal dimensioning for parallel manipulators: Workspace, dexterity, and energy. Journal of Mechanical Design. 2011;**133**(4):041007

[45] Khan WA, Angeles J. The Kinetostatic optimization of robotic manipulators. Journal of Mechanical Design. 2006;**128**:168-178

[46] Yoshikawa T. Manipulability of robot mechanisms. The International Journal of Robotics Research. 1985;**4**(2):3-9

[47] Pham HH, Chen I-M. Optimal synthesis for workspace and manipulability of parallel flexure mechanism. In: 11th World Congress on Theory of Machines and Mechanisms; Tianjin; 2004

[48] Tanev T, Stoyanov B. On the performance indexes for robot manipulators. Problems of Engineering Cybernetics and Robotics. 2000;**49**:64-71

[49] Doty KL, et al. Robot Manipulability. IEEE Transactions on Robotics and Automation. 1995;**11**(3):462-468

[50] Hong K-S, Kim J-G. Manipulability analysis of a parallel machine tool: Application to optimal link length design. Journal of Robotic Systems. 2000;**17**(8):403-415

[51] El-Khasawneh BS, Ferreira PM. Computation of stiffness and stiffness bounds for parallel link manipulators. International Journal of Machine Tools and Manufacture. 1999;**39**(2):321-342

[52] Quennouelle C, Gosselin CM. A General Formulation for the Stiffness Matrix of Parallel Mechanisms. 2012. arXiv preprint arXiv:1212.0950

[53] Ruggiu M. On the Lagrangian and Cartesian stiffness matrices of parallel mechanisms with elastic joints. Advanced Robotics. 2012;**26**(1-2):137-153

[54] Klimchik A. Enhanced Stiffness Modeling of Serial and Parallel Manipulators for Robotic-Based Processing of High Performance Materials. Ecole Centrale de Nantes, France; 2011

[55] Li Y, Xu Q. GA-based multi-objective optimal design of a planar 3-DOF cable-driven parallel manipulator. In: IEEE International Conference on Robotics and Biomimetics; Kunming, IEEE, China; 2006

[56] Li Y, Xu Q. Stiffness analysis for a 3-PUU parallel kinematic machine. Mechanism and Machine Theory. 2008;**43**(2):186-200

[57] Ciblak N, Lipkin H. Synthesis of Cartesian stiffness for robotic applications. In: IEEE International Conference of Robotics and Automation. IEEE, Detroit, MI.1999

[58] Dai JS, Ding X. Compliance analysis of a three-legged rigidly-connected platform device. Journal of Mechanical Design. 2006;**128**(4):755

[59] Ding X, Selig JM. On the compliance of coiled springs. International Journal of Mechanical Sciences. 2004;**46**(5):703-727

[60] Huang S, Schimmels JM. The Eigenscrew decomposition of spatial stiffness matrices. IEEE Transactions on Robotics and Automation. 2000;**16**(2):146-156

[61] Chen G, et al. The principal axes decomposition of spatial stiffness matrices. IEEE Transactions on Robotics. 2015;**31**(1):191-207

[62] Taghvaeipour A, Angeles J, Lessard L. On the elastostatic analysis of mechanical systems. Mechanism and Machine Theory. 2012;**58**:202-216

[63] Wu G. Stiffness analysis and optimization of a co-axial spherical parallel manipulator. Modeling, Identification and Control: A Norwegian Research Bulletin. 2014;**35**(1):21-30

[64] Angeles J. On the nature of the Cartesian stiffness matrix. Ingenierıa Mecanica. 2010;**3**(5):163-170

[65] Wu G, Bai S, Hjornet P. On the stiffness of three or four DOF parallel pick-and-place robots with four identical limbs. In: IEEE International Conference on Robotics and Automation (ICRA); IEEE, Stockholm, Sweden; 2016

[66] Wu G, Zou P. Stiffness analysis and comparison of a Biglide parallel grinder with alternative spatial modular parallelograms. Robotica. 2016;**35**(06):1310-1326

[67] Gosselin CM. Stiffness mapping for parallel manipulators. IEEE Transactions on Robotics and Automation. 1990;**6**(3):377-382

[68] Pashkevich A, Chablat D, Wenger P. Stiffness analysis of multi-chain parallel robotic systems. Journal of Automation, Mobile Robotics and Intelligent Systems. 2009;**3**(**3**):75-82

[69] Ceccarelli M. Stiffness analysis of parallel manipulator using matrix structural analysis. In: Proceedings of EUCOMES 08. 2008. Springer, Dordrecht

[70] Taghvaeipour A, Angeles J, Lessard L. Online Computation of the Stiffness Matrix in Robotic Structures Using Finite Element Analysis. Department of Mechanical Engineering and Centre for Intelligent Machines, McGill University, Montreal; 2010

[71] Pashkevich A, Chablat D, Wenger P. Stiffness analysis of overconstrained parallel manipulators. Mechanism and Machine Theory. 2009;**44**(5):966-982

[72] Pashkevich A, Klimchik A, Chablat D. Enhanced stiffness modeling of manipulators with passive joints. Mechanism and Machine Theory. 2011;**46**(5):662-679

[73] Mekaouche A, Chapelle F, Balandraud X. FEM-based generation of stiffness maps. IEEE Transactions on Robotics. 2015;**31**(1):217-222

[74] Gao Z, Zhang D, Ge Y. Design optimization of a spatial six degree-of-freedom parallel manipulator based on artificial intelligence approaches. Robotics and Computer-Integrated Manufacturing. 2010;**26**(2):180-189

[75] Carbone G, Ceccarelli M. A comparison of indices for stiffness performance evaluation. In: 12th International federation for the promotion of mechanism and machine science (IFToMM) World Congress; Besançon, France; 2007

[76] Patterson T, Lipkin H. A classification of robot compliance. ASME Journal of Mechanical Design. 1993;**115**(3):581-584

[77] Xi F, et al. Global kinetostatic modelling of tripod-based parallel kinematic machine. Mechanism and Machine Theory. 2004;**39**(4):357-377

[78] Shneor Y, Portman VT. Stiffness of 5-axis machines with serial, parallel, and hybrid kinematics: Evaluation and comparison. CIRP Annals - Manufacturing Technology. 2010;**59**(1):409-412

[79] Amouzgar K. Multi-Objective Optimization Using Genetic Algorithms. Stockholm, Sweden: Tekniska Hogskolan; 2012

[80] Coello CA. An updated survey of GA-based multiobjective optimization techniques. ACM Computing Surveys. 2000;**32**(2):109-143

[81] Coello CA, Lamont GB, Veldhuizen DAV. Evolutionary Algorithms for Solving Multi-Objective Problems. 2nd ed. New York, NY, USA: Springer; 2007

[82] Deb K. Multi-Objective Optimization Using Evolutionary Algorithms. New York, NY, USA: John Wiley & Sons, Inc.; 2001

[83] Konak A, Coit DW, Smith AE. Multi-objective optimization using genetic algorithms: A tutorial. Reliability Engineering and System Safety. 2006;**91**:992-1007

[84] Tan KC, Khor EF, Lee TH. Multiobjective Evolutionary Algorithms and Applications. London: Springer-Verlag; 2005

[85] Lou YJ, Liu GF, Li ZX. Randomized optimal Design of Parallel Manipulators. IEEE Transactions on Automation Science and Engineering. 2008;**4**(2):625-649

[86] Bulatović RR, Dordeević SR. On the optimum synthesis of a four-bar linkage using differential evolution and method of variable controlled deviations. Mechanism and Machine Theory. 2009;**44**(1):235-246

[87] Shiakolas PS, Koladiya D, Kebrle J. On the optimum synthesis of six-bar linkages using differential evolution and the goemetric centroid of precision positions technique. Mechanism and Machine Theory. 2005;**40**(3):319-335

[88] Li Y, Xu Q. Design and analysis of a totally decoupled flexurebased XY parallel micro-manipulator. IEEE Transactions on Robotics. 2009;**25**(3):645-657

[89] Li Y, Xu Q. A totally decoupled piezo-driven XYZ flexure parallel micropositioning stage for micro/nanomanipulation. IEEE Transactions on Automation Science and Engineering. 2011;**8**(2):265-279

[90] Xu Q, Li Y. Error analysis and optimal design of a class of translational parallel kinematic machine using particle swarm optimization. Robotica. 2008;**27**(01):67

[91] Abbasnejad G, Daniali HM, Fathi A. Architecture optimization of 4PUS+1PS parallel manipulator. Robotica. 2010;**29**(05):683-690

[92] Kordjazi H, Akbarzadeh A, Rostami AS. Isotropy design and optimization of a planar parallel manipulator with combination of fuzzy logic and genetic algorithm. In: 8th World Congress on Structural and Multidisciplinary Optimization. International Society for Structural and Multidisciplinary Optimization; Lisbon, Portugal; 2009

[93] Arana MU. Methodology to Enlarge the Workspace of Parallel Manipulators by Means of Non-singular Transitions. Bilbao, Spain: University of the Basque Country (UPV/EHU); 2012

[94] Hodgins J. Design, Optimization, and Prototyping of a 3 Translational DOF Parallel Robot. University of Ontario Institute of Technology: Canada; 2012

[95] Kelaiaia R, Company O, Zaatri A. Multiobjective optimization of parallel kinematic mechanisms by the genetic algorithms. Robotica. 2011;**30**(05):783-797

[96] Wu G. Multiobjective optimum design of a 3-RRR spherical parallel manipulator with kinematic and dynamic dexterities. Modeling, Identification and Control: A Norwegian Research Bulletin. 2012;**33**(3):111-122

[97] Bounab B. Multi-objective optimal design based kineto-elastostatic performance for the delta parallel mechanism. Robotica. 2014;**34**(02):258-273

[98] Gao Z, Zhang D. Performance analysis, mapping, and multiobjective optimization of a hybrid robotic machine tool. IEEE Transactions on Industrial Electronics. 2015;**62**(1):423-433

Chapter 2

The Inertia Value Transformation in Maritime Applications

Holger Korte, Sven Stuppe, Jan-Hendrik Wesuls and
Tsutomu Takagi

Additional information is available at the end of the chapter

http://dx.doi.org/10.5772/intechopen.71445

Abstract

Due to recent developments in computer technology, computer-aided investigations of structural movements in a maritime environment have become more relevant during the last years. With regard to mechanically coupled multibody systems in fishery and offshore operations, the analysis of such systems is in the focus of research and development. To analyse multibody systems, forces and moments of all included bodies have to be defined within the same reference frame, which requires a transformation algorithm. Showing the correctness of the transformation algorithm, it must be also applicable for six degrees of freedom (6DOF) motions of a free floating single body in seaways. Therefore, the computation of irregular waves is discussed before the traditional motion description of a floating structure by using the Kirchhoff equations. With these basics, an approach to calculate the motion equations of single bodies within the earth-fixed reference frame is presented before the method of the inertia value transformation. To compare the body-fixed and earth-fixed calculation method, a free-floating crew transfer vessel in irregular waves is simulated and the results are discussed. Finally, the inertia value transformation will be proved by the energy conservation principle on the example of a pure rotating rigid body with none digital calculations.

Keywords: inertia value transformation, wave-disturbed ship motions, wave-structure interaction, six degrees of freedom (6DOF), hydromechanics, inertial kinematics, Euclidian room

1. Introduction

Modern simulation techniques enable a more profound analysis of various engineering problems in an early design stage. In maritime kinematics, the focus is on the behaviour of offshore structures under natural environment conditions like wind, waves and current. Detailed knowledge of loads and motions is required for proper dimensioning of efficient and safe systems. Furthermore, simulations are increasingly used for the design of controller-based automation and assistance systems. In addition to the accurate calculation of the structural movements, the calculation speed of a simulation is an important quality feature for these applications, e.g. real-time constraints. To determine the kinematics model of a free-floating structure in a seaway, different approaches are commonly used. The equations of motions can either be defined in the inertial reference frame or alternatively in the body-fixed reference frame. The difference between dry mechanics and maritime mechanics is based on additional hydrodynamic effects, namely, the hydrodynamic added mass, e.g. [1–3]. The hydrodynamic mass force is an inertia force. The relative acceleration between incompressible fluid and structure induces a pressure field, which results in a hydrodynamic force that is formulated as the product of the relative acceleration and a 'virtual' mass. The size of the hydrodynamic added mass depends primarily on the direction of movement and the geometry of the structure [4].

Due to the phenomenon of added or virtual masses, non-scalar and directed inertia values are necessary for describing translational and rotational motions. Probably, this results in an ordinary practice to describe floating structures within a body-fixed view, e. g. by the Kirchhoff motion equations of floating bodies [1, 3].

The investigation of mechanically coupled and multiple rigid body systems is in the focus of the authors and represents a special difficulty in this area [2, 5]. Within literature before millennium, only Paschen's algorithm was found describing dynamics of real 3D systems for fishery and mine-hunting systems with flexible, non-elastic numerical elements and rigid bodies by partially neglect of added masses [6, 7]. Later, different marine multibody systems were described with algorithms of structural mechanics [8] or as pure planar motion descriptions, e.g. [9–13]. To describe the mutual interdependencies, the equations of motions of all involved bodies have to be set up in the same coordinate system, preferably the inertial or earth-fixed system. This includes the transformation of all vectorial entities including the inertia matrices (mass matrix, moments of inertia). Therefore, the use of a transformation algorithm is required, which describes the system in the inertial reference frame. The inertia value transformation here presented, also known as Kane's method [14], was introduced by Korte and Takagi [2, 5] for fishery systems to analyse forces and motions of purse seines. Based on the principle of concentrated masses, all included nodes of the net structure are focused and connected with damped mass-spring elements. To connect the inertia values of the nodes, a transformation of the hydrodynamic added masses is required [15].

Korte et al. presented an application of the inertia value transformation for multibody systems in 2015 for a pure rotating body without elasticity [5]. The analysed scenario represents the transfer of offshore service staff from a crew transfer vessel (CTV) to a wind turbine, which is a real and environmentally affected gyro. CTV is constrained at the bow to the landing of the wind turbine in seas. The inertial reference system is located at the contact point of the structure. The ship can make an ideal rotation around all three axes of the inertial system. The motion equations of the system are defined within the inertial system to calculate the constraining forces. It was observed that temporal integration of the equations within the inertial reference frame leads to unpredictable motions of the ship. The simulation was unstable. During development of the multibody scenario, it was discovered that an additional and oppositely directed transformation of the rotational accelerations leads to a stable solution of the simulation.

To analyse a general applicability of the inertia value transformation for maritime applications, Korte et al. defined the equations of motions for a free-floating ship with six degrees of freedom (6DOF) in the inertial reference frame [17]. Motion equations were derived from momentum and angular momentum theorem. The results of the free-floating ship were compared with ordinary simulations in the body-fixed reference frame and kinematics (see [1]).

This contribution describes the applicability of the inertia value transformation in maritime applications while transferring the inertia values between a body-fixed and the inertial reference frame. The motion behaviour of the free-floating structure is in the focus of the discussion. In the following section, the modelling of the irregular seas as environmental reason of structure motions is presented. The six degrees of freedom of a floating body as well as the traditional motion equations in the body-fixed reference system are introduced afterwards. The derivation of the motion equations within the inertial reference frame by using the inertia value transformation is part of the main section. The possibility to transform the equations from inertial to body-fixed system is shown. For comparison of both methods, 6DOF simulation results of a free-floating CTV in waves are discussed. In the last section, a proof shows the energy conservation of the inertia value transformation for the case of a pure rotating rigid body.

2. Wave modelling

Depending on formation is natural irregular seas divided into wind seas and swell. Real occurred sea state phenomena are described in nautical practice as the superposition of a stochastic wind sea and two observable swells.

Wind seas are direct wind excited. The waves are rather short and steep. Due to friction, the wind transfers energy to the water surface. The resulting capillary waves with short wavelength increase the area of the free surface, whereby the effect of the energy transfer is amplified. This leads in higher and longer gravity waves, which influence the water column down to a water depth of a half wavelength. The waves propagate in the wind direction. Wind wave characteristics depend on wind speed, wind duration and fetch. The wave components of wind seas are located in the higher-frequency range of a wave spectrum. Due to containing higher frequency waves with small amplitudes, the water column oscillates inhomogeneously. Wind seas decrease by a disappearance of the wind excitation. The relation between wind classes (Beaufort wind scale) and sea-state classes (Douglas sea scale) can be seen in **Figure 1**.

Contrary to wind seas, which is described as evolving waves, is swell the full homogeneously oscillating water column. Swell is a decaying, nearly sinusoidal wave that has moved away from the formation area, e.g. a storm region. The frequencies are in the lower range of a wave spectrum. The energy density of waves is inversely proportional to the wave frequency. Long waves with a small frequency are more energetic than short waves.

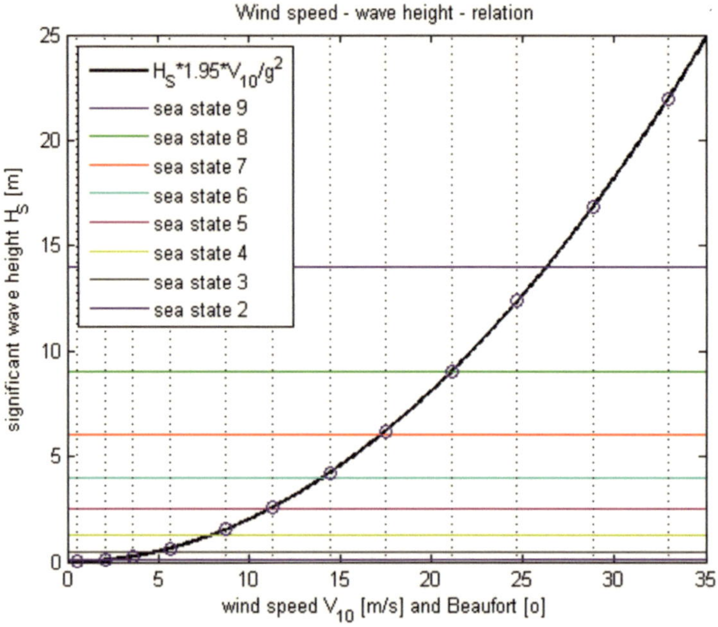

Figure 1. Relation of wind speed (Beaufort wind scale) and wave height (Douglas sea scale) for fully developed seas (modified to practice).

Irregular waves are mathematically described as superposition of a finite number of wave components with different wave frequencies, different amplitudes and different phases. The representation is carried out with energy density spectra, which have a unique distribution of the wave components for each sea area. The spectrum contains the energy $S_n(\omega)$ of all included wave frequencies. **Figure 2** shows the structure of a wave spectrum.

Relevant wave parameters can be derived from the wave spectrum. The mean wave frequency ω_0 is located at the maximum $S(\omega)$ of the spectrum. The peak period T_P is

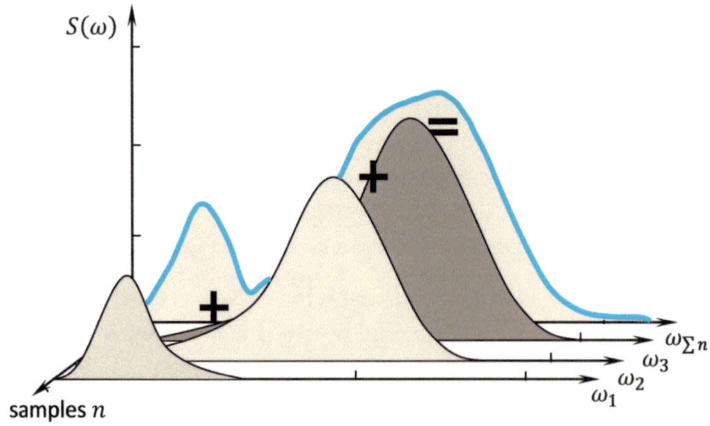

Figure 2. Structure of a wave spectrum as superposition of different wave components.

$$T_P = \frac{2\pi}{\omega_0} \tag{1}$$

Further, wave properties can be determined with the so-called spectral moments m_n, compared with [1]. Their definition is

$$m_n = \int_0^\infty \omega^n S(\omega) d\omega \tag{2}$$

The moments up to the second order, $n = 2$, are of interest to characterise the sea state. The characteristic wave height of the spectrum is

$$H_{m0} = 4\sqrt{m_0} \tag{3}$$

and the significant wave height is

$$H_S = 3.81 \cdot \sqrt{m_0} \tag{4}$$

The period of the characteristic wave height H_{m0} is defined as

$$T_1 = \frac{m_0}{m_1} \tag{5}$$

The mean zero upcrossing period is

$$T_Z = \sqrt{\frac{m_0}{m_2}} \tag{6}$$

Variance σ^2 and standard deviation σ are

$$\sigma^2 = m_0 \tag{7}$$

$$\sigma = \sqrt{m_0} = 0.263 \cdot H_S \tag{8}$$

There are several mathematical wave spectra for computer simulations of irregular waves available, namely, the Bretschneider spectrum, the Phillips spectrum, the Pierson-Moskowitz spectrum as well as the Joint North Sea Wave Project (JONSWAP) spectrum. The most important spectra are introduced in this section.

2.1. Pierson-Moskowitz spectrum

The Pierson-Moskowitz (PM) spectrum was introduced for fully developed seas and is commonly used since 1964 [18]. It is based on long-term observation data of a weather ship in the period from 1955 to 1960. It assumes the fully developed sea, including wind sea and swell, on the North Atlantic with unlimited water depth and fetch as well as steady wind for a long time. The spectrum is

$$S_{PM}(\omega) = A \cdot \omega^{-5} \cdot e^{-B\omega^{-4}} \tag{9}$$

with parameters A and B

$$A = 0.0081 \cdot g^2 \tag{10}$$

$$B = 0.74\left(\frac{g}{V_{wind}}\right)^4 = \frac{2.814}{H_S^2} \tag{11}$$

The relation between significant wave height and wind speed is:

$$H_S = \frac{1.95}{g^2} V_{wind}^2 \tag{12}$$

2.2. Modified Pierson-Moskowitz spectrum

After experiences with the Pierson-Moskowitz spectrum, the maritime community recommended a magnification of the spectrum [19, 20]. According to this recommendation, the parameters A and B were modified and can be determined from the observable weather parameters H_S and T_Z:

$$A = \frac{4\pi^3 H_S^2}{T_Z^4} \tag{13}$$

$$B = \frac{16\pi^3}{T_Z^4} \tag{14}$$

Figure 3 shows a comparison of the Pierson-Moskowitz spectrum and the Modified Pierson-Moskowitz (MPM) spectrum for several wind speeds. An increase of the energy density in the peak frequency of the MPM as well as a shift of the peak frequency to lower frequencies can be seen.

2.3. JONSWAP spectrum

The Pierson-Moskowitz spectra were developed for unlimited water depth. However, the most intensively used sea areas are often in regions of shelf seas with restrictions on the water depth and the fetch.

In 1968/1969, the characteristics of wave formations in sea areas with limited fetch and water depth were investigated exemplary for the North Sea in the international joint project 'Joint North Sea Wave Project (JONSWAP)'. During a period of approximately 10 weeks, measurements were carried out and evaluated at points reaching up to 160 km seawards in the region of the island of Sylt. The JONSWAP spectrum is based on the PM spectrum and provided with a magnification factor γ for the peak distribution.

Hasselmann's mathematical definition of the JONSWAP spectrum is [20]

$$S_{JON}(\omega) = \alpha g^2 \omega^{-5} \cdot e^{\left[-\frac{5}{4}\left(\frac{\omega_0}{\omega}\right)^4\right]} \cdot \gamma^r \tag{15}$$

with

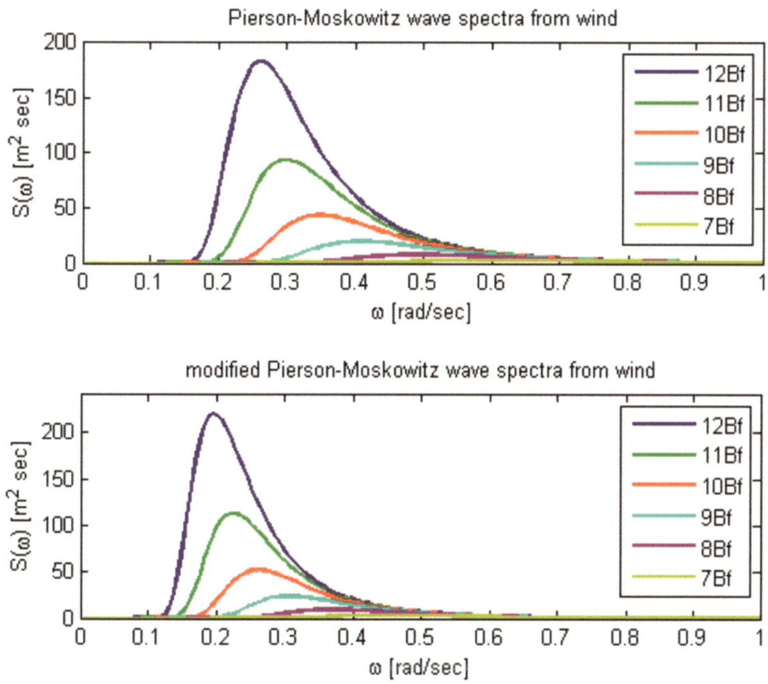

Figure 3. Comparison of Pierson-Moskowitz spectrum (top) and modified Pierson-Moskowitz spectrum (bottom) for several wind speeds.

$$r = e^{\left[-\frac{(\omega-\omega_0)^2}{2\sigma^2\omega_0^2}\right]} \tag{16}$$

$$\gamma = 3.3 \tag{17}$$

$$\alpha = 0.076\left(\frac{V_{10}^2}{F_W g}\right)^{0.22} \tag{18}$$

$$\omega_0 = 22\left(\frac{g^2}{F_W V_{10}}\right)^{\frac{1}{3}} \tag{19}$$

$$\sigma = \begin{cases} 0.07, & \omega \leq \omega_0 \\ 0.09, & \omega > \omega_0 \end{cases} \tag{20}$$

Fetch is defined in metres and increases the wave energy linearly. A direct comparison of the JONSWAP spectrum with the MPM spectrum for a large-scale wind sea on the North Sea illustrates the lower total energy in **Figure 4**. The characteristic frequencies ω_0 of the individual wind classes are higher than those of the PM spectrum. Due to limited water depth, the wave heights are reduced. Both, the maximum of the spectrum and the area under the curve are considerable smaller than those of the reference spectrum. A comparison of seaways with different fetch lengths confirms this trend (cf. **Figure 4**). A JONSWAP spectra are used for the here presented simulations.

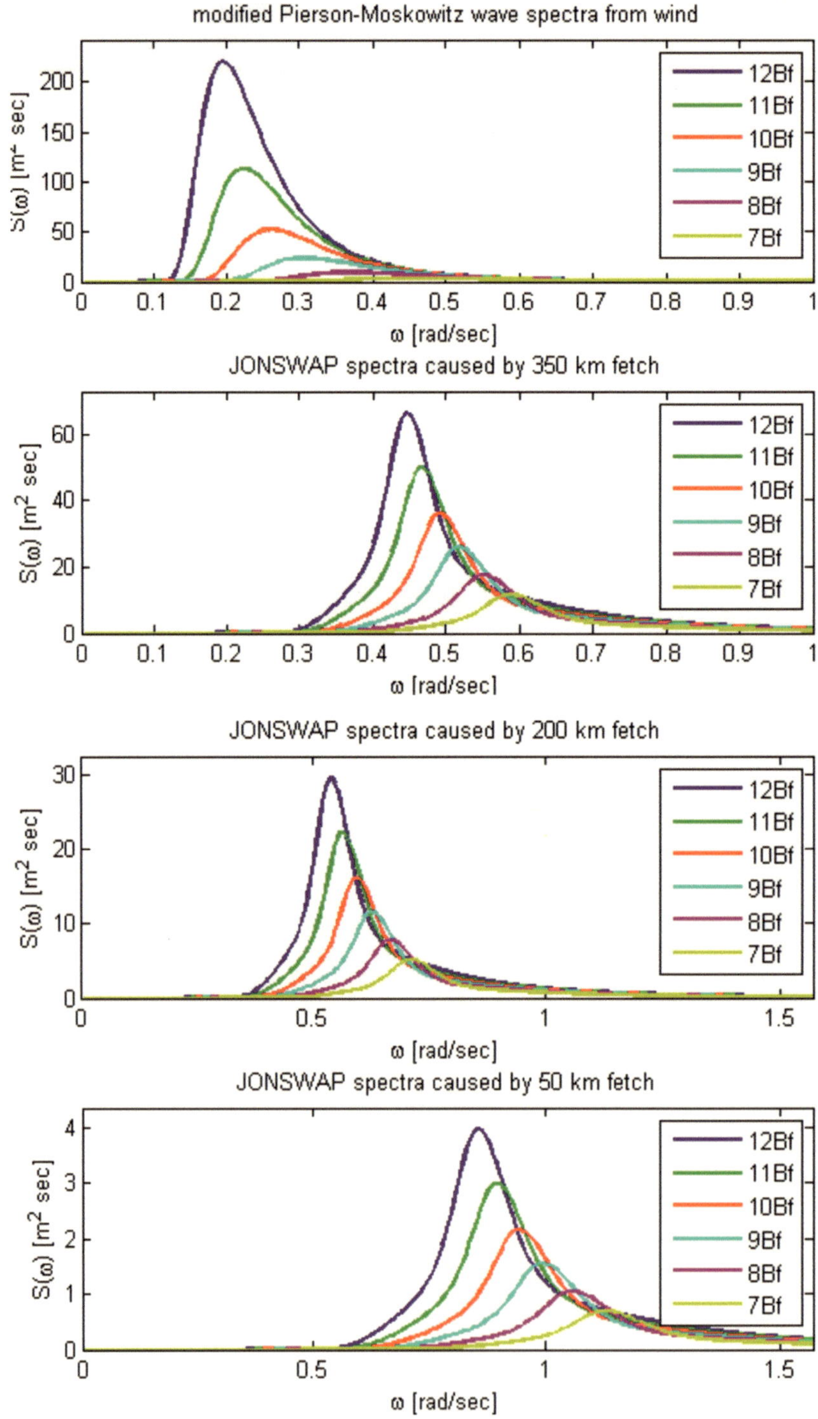

Figure 4. Comparison of modified Pierson-Moskowitz spectra and JONSWAP spectra for 350 km, 200 km and 50 km fetch and several wind speeds.

http://dx.doi.org/10.5772/intechopen.71445

3. 6DOF motions of a free-floating offshore structure

In the following section, the six degrees of freedom of a ship are introduced. **Figure 5** shows a sketch of a ship hull with a body-fixed coordinate system. Body-fixed coordinate system means that the coordinate system moves with the ship. Due to advantages in determining the moments of inertia, the origin of the system is preferably located in the ships centre of gravity. The motions of a ship are described in the body-fixed coordinate system. However, the ship's position and the orientation are referenced in an earth-fixed coordinate system.

The translational motions are as follows:

- Surge: Translation along the longitudinal x-axis
- Sway: Translation along the transversal y-axis
- Heave: Translation along the vertical z-axis

The rotational motions are as follows:

- Roll: Rotation around the longitudinal x-axis
- Pitch: Rotation around the transversal y-axis
- Yaw: Rotation around the vertical z-axis

Another distinction is the classification into horizontal (surge, sway, yaw) and vertical (heave, roll, pitch) motions. The reason for this differentiation is the restoring forces caused by gravitation. These exist only for the vertical degrees of freedom. After a perturbation of the equilibrium, a ship always tends to return to it. This does not apply for horizontal motions. The result is the drift of a free-floating structure away from the original position as well as a change of the heading in case of wind, waves or current. For a driven ship or a ship in dynamic positioning mode, the ship's actuators (propeller, rudder, thruster, etc.) control the horizontal motions.

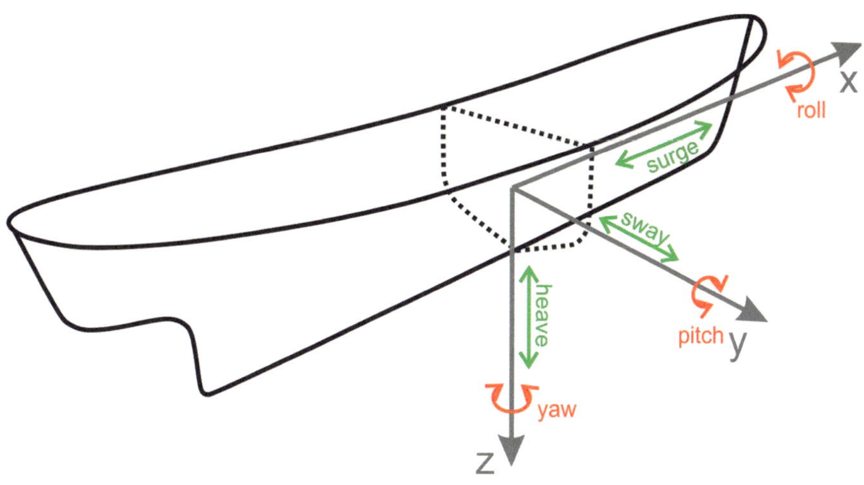

Figure 5. 6DOF motions of a free-floating ship.

4. Motion equations in the body-fixed reference system: Kirchhoff equations

As already mentioned, hydrodynamic inertial effects in the form of a hydrodynamic added mass have to be taken into account in marine applications. The mass matrix m_b (b for body) in the traditionally used body-fixed reference system is defined in Eq. (22). In the matrix, m is the physical mass and $m_{h,ij}$ is the direction-dependent content of the hydrodynamic inertia. The mass matrix is constant in the body-fixed frame. This applies analogously for the matrix of moments of inertia:

$$m_{b,ij} = m + m_{h,ij} \tag{21}$$

$$\overline{\mathbf{m}}_b = \begin{bmatrix} m + m_{h,11} & 0 & 0 \\ 0 & m + m_{h,22} & 0 \\ 0 & 0 & m + m_{h,33} \end{bmatrix} \tag{22}$$

In 1869, Kirchhoff published his work about the 'Movement of a rotating Body in a Fluid' [3]. In this work, he defined the motion equations of a floating body in a body-fixed reference system in analogy to Euler's gyro equation. The Kirchhoff equations are a system of each three equations for translation (Eq. (23)) and rotation (Eq. (24)):

$$\frac{d\vec{\mathbf{P}}}{dt} + \vec{\omega} \times \vec{\mathbf{P}} = (X, Y, Z)^{\mathsf{T}} \tag{23}$$

$$\frac{d\vec{\mathbf{L}}}{dt} + \vec{\omega} \times \vec{\mathbf{L}} + \vec{v} \times \vec{\mathbf{P}} = (K, M, N)^{\mathsf{T}} \tag{24}$$

The so-called living forces and moments depict all external forces and moments including all hydrodynamic effects, weight, buoyancy and their effect and can be found at the right-hand side of the equations inclusively. The determination of these external forces and moments is especially difficult for the horizontal degrees of freedom with a lack of restoring forces. For the simulations in this work, a simplified model is used, which considers weight, buoyancy as well as potential damping. At present, viscous effects are neglected.

The accelerations have to be integrated twice to analyse the position and orientation of the free-floating ship. The first integration is executed in the body-fixed system.

To calculate the position$(x, y, z)^{\mathsf{T}}$ and orientation$(\Phi, \Theta, \Psi)^{\mathsf{T}}$ of a floating body, the velocities have to be transformed into the inertia system. The peculiarity is the differing transformation of the rotational vector from the body-fixed system into the inertia system and vice versa, e.g. using the transformation with Euler's angles (Eq. (26), cf. [1]):

$$\vec{\omega}_e = \overline{\mathbf{C}}_{rot,be}\vec{\omega}_b \tag{25}$$

with

$$\overline{\mathbf{C}}_{rot,be} = \begin{bmatrix} 1 & s\Phi t\Theta & c\Phi t\Theta \\ 0 & c\Phi & -s\Phi \\ 0 & {}^{s\Phi}/_{c\Theta} & {}^{c\Phi}/_{c\Theta} \end{bmatrix} = \left(\overline{\mathbf{C}}_{rot,eb}\right)^{-1} \tag{26}$$

(e for earth or inertial, b for body, c for cosine, s for sine and t for tangent).

5. Motion equations in the inertial reference system

In the following section, the motion equations of a free-floating body in the earth-fixed reference system are shown. The equations are formally derived from momentum and angular momentum theorem. The transformation of the inertia values is discussed, and it is shown that a transformation of the motion equations leads to the Kirchhoff equations. It is known that the twice integration of the rotational equations results in an unstable solution. By using an additional and opposite directed transformation of the rotational accelerations, the simulations can be stabilised for longer time series.

5.1. Inertia value transformation

Kane introduced the transformation of inertia values in 1985 [14]. The used transformation matrix from the body-fixed into the earth-fixed reference system is

$$\overline{\mathbf{C}}_{be} = \begin{bmatrix} c\Theta c\Psi & s\Phi s\Theta c\Psi - c\Phi s\Psi & c\Phi s\Theta c\Psi + s\Phi s\Psi \\ c\Theta s\Psi & s\Phi s\Theta s\Psi + c\Phi c\Psi & c\Phi s\Theta s\Psi - s\Phi c\Psi \\ -s\Theta & s\Phi c\Theta & c\Phi c\Theta \end{bmatrix} \tag{27}$$

with

$$\overline{\mathbf{C}}_{be}\overline{\mathbf{C}}_{\boldsymbol{be}}^{\mathsf{T}} = \begin{bmatrix} 1 & 0 & 0 \\ 0 & 1 & 0 \\ 0 & 0 & 1 \end{bmatrix} = \overline{\mathbf{C}}_{\boldsymbol{be}}^{\mathsf{T}}\overline{\mathbf{C}}_{be} \tag{28}$$

and

$$\overline{\mathbf{C}}_{eb} = \overline{\mathbf{C}}_{\boldsymbol{be}}^{\mathsf{T}} \tag{29}$$

The transformation of the inertia matrix is [14, 21]

$$\overline{\mathbf{J}}_e = \overline{\mathbf{C}}_{be}\overline{\mathbf{J}}_{\boldsymbol{b}}\overline{\mathbf{C}}_{\boldsymbol{be}}^{\mathsf{T}} \tag{30}$$

Analogously, it follows the transformation of the mass matrix including added mass [2]:

$$\overline{\mathbf{m}}_e = \overline{\mathbf{C}}_{be}\overline{\mathbf{m}}_{\boldsymbol{b}}\overline{\mathbf{C}}_{\boldsymbol{be}}^{\mathsf{T}} \tag{31}$$

The cross product-operator relation for matrices can be found in the literature [22, 23]:

$$\dot{\overline{\mathbf{C}}}_{be} \cdot \overline{\mathbf{C}}_{be}^{\mathsf{T}} = \tilde{\omega}_e = \begin{bmatrix} 0 & -\dot{\Psi} & \dot{\Theta} \\ \dot{\Psi} & 0 & -\dot{\Phi} \\ -\dot{\Theta} & \dot{\Phi} & 0 \end{bmatrix} = \left(\vec{\omega}_e \times\right) \tag{32}$$

5.2. Rotation

The rotational equations follow from angular momentum theorem, which is defined in the inertial reference frame [21, 23]:

$$\vec{L}_e = \overline{\mathbf{J}}_e \vec{\omega}_e \tag{33}$$

Taking the temporal change of the inertia matrix $\overline{\mathbf{J}}_e$ into account follows for the moment:

$$\frac{d\vec{L}_e}{dt} = \vec{\mathbf{M}}_e = \overline{\mathbf{J}}_e \dot{\vec{\omega}}_e + \dot{\overline{\mathbf{J}}}_e \vec{\omega}_e \tag{34}$$

The temporal derivative of the inertia matrix is

$$\dot{\overline{\mathbf{J}}}_e = \dot{\overline{\mathbf{C}}}_{be} \overline{\mathbf{J}}_b \overline{\mathbf{C}}_{be}^{\mathsf{T}} + \overline{\mathbf{C}}_{be} \overline{\mathbf{J}}_b \dot{\overline{\mathbf{C}}}_{be}^{\mathsf{T}} \tag{35}$$

Substituting Eq. (35) into Eq. (34) follows:

$$\vec{\mathbf{M}}_e = \overline{\mathbf{J}}_e \dot{\vec{\omega}}_e + \left(\dot{\overline{\mathbf{C}}}_{be} \overline{\mathbf{J}}_b \overline{\mathbf{C}}_{be}^{\mathsf{T}} + \overline{\mathbf{C}}_{be} \overline{\mathbf{J}}_b \dot{\overline{\mathbf{C}}}_{be}^{\mathsf{T}} \right) \vec{\omega}_e \tag{36}$$

Using the cross product-operator relation follows the momentum equation in the inertial reference system:

$$\vec{\mathbf{M}}_e = \overline{\mathbf{J}}_e \dot{\vec{\omega}}_e + \tilde{\omega}_e \overline{\mathbf{J}}_e \vec{\omega}_e = \overline{\mathbf{J}}_e \dot{\vec{\omega}}_e + \vec{\omega}_e \times \left(\overline{\mathbf{J}}_e \vec{\omega}_e \right) \tag{37}$$

The transformation of Eq. (37) into the body-fixed reference frame is

$$\vec{\mathbf{M}}_b = \overline{\mathbf{C}}_{eb} \vec{\mathbf{M}}_e = \overline{\mathbf{C}}_{eb} \overline{\mathbf{J}}_e \overline{\mathbf{C}}_{eb}^{\mathsf{T}} \cdot \overline{\mathbf{C}}_{eb} \dot{\vec{\omega}}_e + \overline{\mathbf{C}}_{eb} \tilde{\omega}_e \overline{\mathbf{C}}_{eb}^{\mathsf{T}} \cdot \overline{\mathbf{C}}_{eb} \overline{\mathbf{J}}_e \overline{\mathbf{C}}_{eb}^{\mathsf{T}} \cdot \overline{\mathbf{C}}_{eb} \vec{\omega}_e \tag{38}$$

It follows with

$$\vec{\mathbf{M}}_b = \overline{\mathbf{J}}_b \dot{\vec{\omega}}_b + \tilde{\omega}_b \overline{\mathbf{J}}_b \vec{\omega}_b = \overline{\mathbf{J}}_b \dot{\vec{\omega}}_b + \vec{\omega}_b \times \left(\overline{\mathbf{J}}_b \vec{\omega}_b \right) \tag{39}$$

Eq. (39) is identical to Euler's gyroscope equation, which is defined in the body-fixed system and corresponds to the Kirchhoff rotation equation (Eq. (24)) if the term $\vec{v} \times \vec{P}$ is neglected.

As already mentioned, the integration of the equations in the inertial system results in unstable behaviour. The floating body makes unpredictable, chaotic movements. It is assumed that the instability is a consequence of the numerical inaccuracy as well as the twice integration of the equations. It was found within the project Mine Hunting 2000 that an additional and opposing transformation stabilises the simulation for longer time [6]. Due to the missing inertia value algorithm at that time, it could not be explained why. The authors assumed a numerical reason, which cannot be proved. That phenomenon of stabilisation was rebuilt for simulations of docked CTV at wind turbine tower, compare [24] with [16].

During the presented motion of the free-floating vessel, the changed rotational accelerations stabilise the system for longer periods (cf. Eq. (40)):

$$\vec{\omega}_e = \int \overline{\mathbf{C}}_{eb} \dot{\vec{\omega}}_b dt \equiv \int \overline{\mathbf{C}}_{eb} \dot{\vec{\omega}}_e dt \tag{40}$$

By using the additional transformation in the multibody application described in Ref. [16], the motions of the vessel on the wind turbine tower were stable at all times.

5.3. Translation

Consequently, the derivation of the translational equations follows the derivation of the rotational equations from the last section. The hydrodynamic added mass force is generally an external force, which defined the right-hand side of the motion equations. For the derivation of the motion equations, the hydrodynamic mass is considered as intrinsic property of the free-floating body, which has to be taken into account for every accelerated marine system. Analogously, as for the rotational inertia matrix, the mass matrix of a rotating body changes. This temporal change is equal to zero in case of pure translation.

The momentum theorem is defined in the inertial reference system:

$$\vec{P}_e = \overline{\mathbf{m}}_e \vec{v}_e \tag{41}$$

The force is the temporal derivative of the momentum:

$$\frac{d\vec{P}_e}{dt} = \vec{F}_e = \overline{\mathbf{m}}_e \dot{\vec{v}}_e + \dot{\overline{\mathbf{m}}}_e \vec{v}_e \tag{42}$$

With the transformation matrix $\overline{\mathbf{C}}_{be}$, the transformation of the mass matrix into the inertial system follows:

$$\overline{\mathbf{m}}_e = \overline{\mathbf{C}}_{be} \overline{\mathbf{m}}_b \overline{\mathbf{C}}_{be}^{\mathsf{T}} \tag{43}$$

The other direction is

$$\overline{\mathbf{m}}_b = \overline{\mathbf{C}}_{be}^{\mathsf{T}} \overline{\mathbf{m}}_e \overline{\mathbf{C}}_{be} \tag{44}$$

The temporal derivative of the mass matrix is

$$\dot{\overline{\mathbf{m}}}_e = \dot{\overline{\mathbf{C}}}_{be} \overline{\mathbf{m}}_b \overline{\mathbf{C}}_{be}^{\mathsf{T}} + \overline{\mathbf{C}}_{be} \overline{\mathbf{m}}_b \dot{\overline{\mathbf{C}}}_{be}^{\mathsf{T}} \tag{45}$$

By substituting Eq. (43) into Eq. (45) follows

$$\dot{\overline{\mathbf{m}}}_e = \dot{\overline{\mathbf{C}}}_{be}\overline{\mathbf{C}}_{be}{}^{\mathsf{T}}\overline{\mathbf{m}}_e + \overline{\mathbf{m}}_e\overline{\mathbf{C}}_{be}\dot{\overline{\mathbf{C}}}_{be}{}^{\mathsf{T}} \tag{46}$$

and by using the cross product-operator relation

$$\dot{\overline{\mathbf{m}}}_e = \tilde{\omega}_e\overline{\mathbf{m}}_e + \overline{\mathbf{m}}_e(-\tilde{\omega}_e) \tag{47}$$

Substituting Eq. (47) into Eq. (42) results in the translational equation for a floating body in the inertial reference frame:

$$\vec{F}_e = \overline{\mathbf{m}}_e\dot{\vec{v}}_e + \tilde{\omega}_e\left(\overline{\mathbf{m}}_e\vec{v}_e\right) - \left(\overline{\mathbf{m}}_e\tilde{\omega}_e\right)\vec{v}_e = \overline{\mathbf{m}}_e\dot{\vec{v}}_e + \vec{\omega}_e \times \left(\overline{\mathbf{m}}_e\vec{v}_e\right) - \overline{\mathbf{m}}_e\vec{\omega}_e \times \vec{v}_e \tag{48}$$

Eq. (48) contains only values of the inertial reference system. Due to change of the orientation and the non-scalar characteristic of the mass matrix, Eq. (48) contains two additional terms. These terms disappear for each application where the hydrodynamic added mass is not considered and the mass is scalar, e.g. in aerospace industries or robotics. In case of pure translation, these terms are equal to zero too, and the equation results in Newton's second law.

In the following equation, the transformation of Eq. (48) into the body-fixed reference system is shown. The relation of the velocity in both systems is

$$\vec{v}_e = \overline{\mathbf{C}}_{be}\vec{v}_b \tag{49}$$

In case of simultaneous rotation and translation, the rotation matrix changes. The acceleration in inertial frame is

$$\dot{\vec{v}}_e = \dot{\overline{\mathbf{C}}}_{be}\vec{v}_b + \overline{\mathbf{C}}_{be}\dot{\vec{v}}_b \tag{50}$$

(Remark: In case of pure translation, $\dot{\vec{v}}_e = \mathbf{C}_{be}\dot{\vec{v}}_b$).

By inserting identity matrix, Eq. (28) follows:

$$\dot{\vec{v}}_e = \dot{\overline{\mathbf{C}}}_{be}\overline{\mathbf{C}}_{be}{}^{\mathsf{T}}\overline{\mathbf{C}}_{be}\vec{v}_b + \overline{\mathbf{C}}_{be}\dot{\vec{v}}_b \tag{51}$$

This is identical to

$$\dot{\vec{v}}_e = \tilde{\omega}_e\vec{v}_e + \overline{\mathbf{C}}_{be}\dot{\vec{v}}_b = \vec{\omega}_e \times \vec{v}_e + \overline{\mathbf{C}}_{be}\dot{\vec{v}}_b \tag{52}$$

Substituting Eq. (52) into Eq. (48), the term $\left(\overline{\mathbf{m}}_e\tilde{\omega}_e\right)\vec{v}_e$ disappears, and Eq. (48) becomes

$$\vec{F}_e = \overline{\mathbf{m}}_e\overline{\mathbf{C}}_{be}\dot{\vec{v}}_b + \tilde{\omega}_e\left(\overline{\mathbf{m}}_e\vec{v}_e\right) \tag{53}$$

The transformation of Eq. (53) into the body-fixed reference frame is

http://dx.doi.org/10.5772/intechopen.71445

$$\vec{F}_b = \overline{\mathbf{C}}_{eb}\vec{F}_e = \overline{\mathbf{C}}_{eb}\overline{\mathbf{m}}_e\overline{\mathbf{C}}_{eb}^{\mathsf{T}} \cdot \overline{\mathbf{C}}_{eb}\overline{\mathbf{C}}_{be}\dot{\vec{v}}_b + \overline{\mathbf{C}}_{eb}\tilde{\omega}_e\overline{\mathbf{C}}_{eb}^{\mathsf{T}} \cdot \overline{\mathbf{C}}_{eb}\overline{\mathbf{m}}_e\overline{\mathbf{C}}_{eb}^{\mathsf{T}}\overline{\mathbf{C}}_{eb}\vec{v}_e \tag{54}$$

Replacing the values from inertial system with values from body-fixed system follows:

$$\vec{F}_b = \overline{\mathbf{m}}_b\dot{\vec{v}}_b + \tilde{\omega}_b\overline{\mathbf{m}}_b\vec{v}_b = \overline{\mathbf{m}}_b\dot{\vec{v}}_b + \vec{\omega}_b \times \left(\overline{\mathbf{m}}_b\vec{v}_b\right) \tag{55}$$

Eq. (55) corresponds to the translational Kirchhoff equation (Eq. (23)) in the body-fixed system.

6. Simulations in body-fixed and inertial reference frame

The following section shows simulation comparisons of a free-floating vessel in both reference systems. 6DOF motion simulations are performed for various wave conditions. The hydrodynamic model of the simulation is implemented in MATLAB/Simulink.

6.1. Parameterisation of the ship

The simulated ship is a crew transfer vessel, which is used for the transfer of offshore service staff in the German Bight. **Figure 6** shows a snapshot of the ship's CAD model. It is a catamaran hull with a length of $L_{oA} = 22.0\ m$, a breadth of $B = 8.3\ m$ and a mass of $m = 60\ t$. For the simulation, the hydrodynamic parameters like hydrodynamic added masses and moments of inertia, as well as the potential damping coefficients of the CTV, are required. They were determined within the project 'Safe Offshore Operations (SOOP)' using the potential radiation and diffraction programme WAMIT (cf. [24]). The ship is discretised station-wise

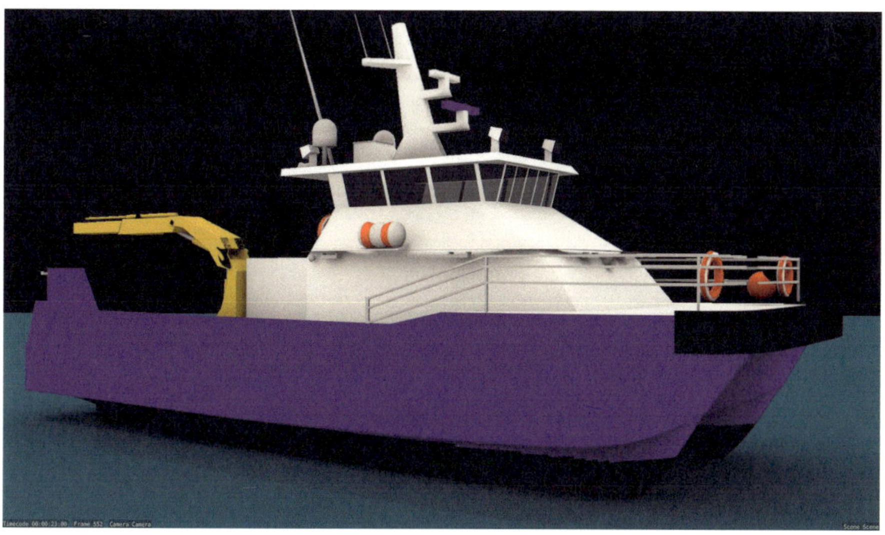

Figure 6. CAD snapshot of the simulated crew transfer vessel.

to calculate wave-induced forces and moments. The calculated forces are buoyancy and weight as well as potential damping force.

6.2. 6DOF simulations

The following figures show the comparison of the CTV 6DOF body motions in irregular seas. The ship is free floating with no initial velocity. **Figure 7** shows the simulation results for the case of head seas and **Figure 8** for beam seas. The wind wave parameters are a significant wave height of $H_S = 2.1\ m$ and a peak period of $\omega_0 = 0.7\ rad/s$. Swell is neglected.

The figures show from top to bottom the registered wave elevation at the ships centre of gravity, the x-, y- and z-position as well as the orientation angles Φ, Θ and Ψ. Blue curves are for the body-fixed reference frame and red curves for simulations in the inertial reference frame.

At the beginning of the presented simulations, the motion behaviour in both reference systems shows an identical behaviour. In further process, however, an increasing deviation of the ship's heading in the inertial reference is registered. The result is a changing encounter angle of ship and waves. For the case of head seas, the ship begins to move in transverse direction and to roll. In the beam seas, simulation leads the changed encounter angle in a pitch motion. In total, it can be seen that the comparison of vertical motions shows an identical behaviour. As already mentioned, the yaw motion is controlled in case of a driven ship. It is assumed that the motions in both reference systems are identical in this case. Analysis of yaw motion is the subject of the present work. It has to be pointed out that the simulations in the inertial reference system were

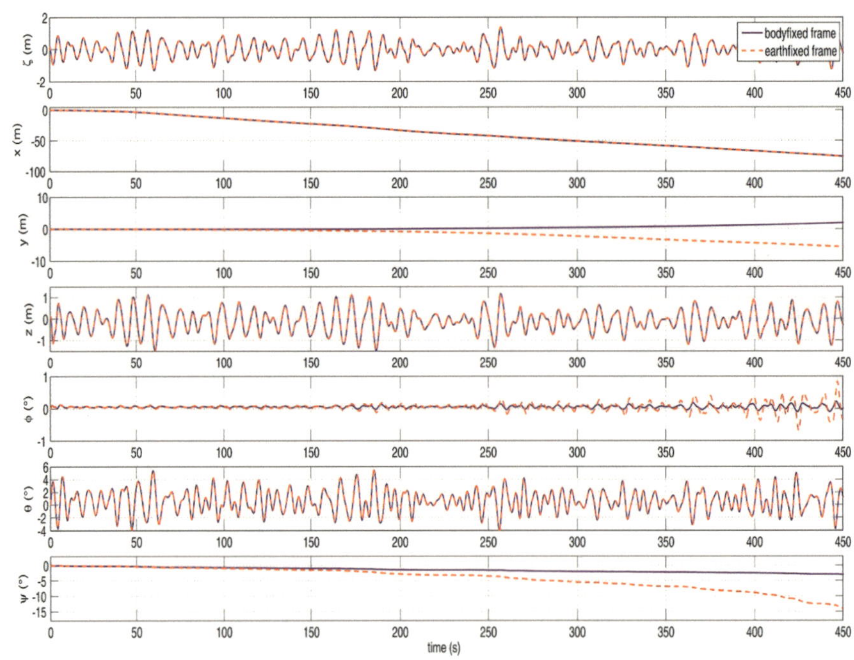

Figure 7. Comparison of the ship position and orientation for head seas. Solid line for body-fixed reference frame. Dashed line for inertial reference frame.

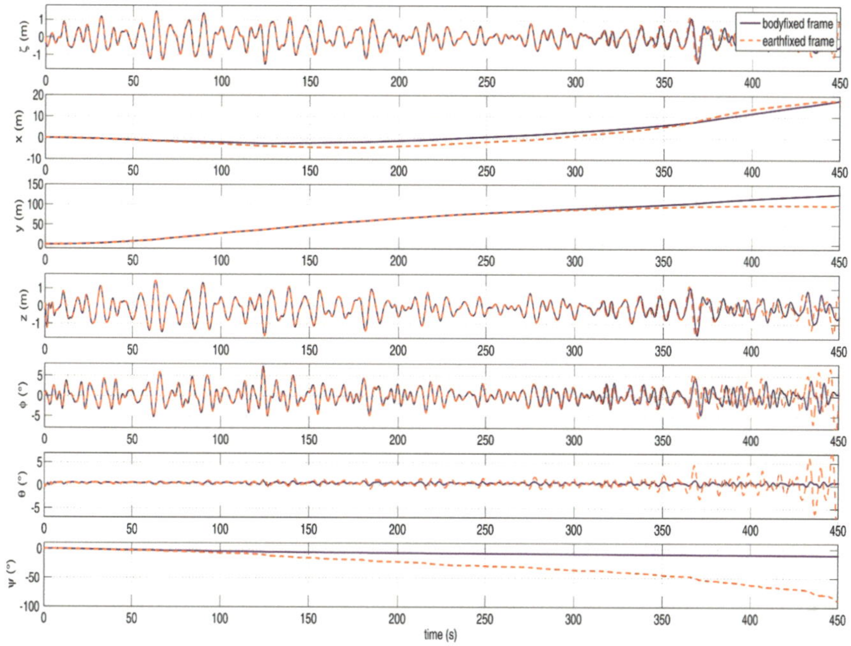

Figure 8. Comparison of the ship position and orientation for beam seas. Solid line for body-fixed reference frame. Dashed line for inertial reference frame.

unstable after longer simulation time. Instability occurs abruptly. Intensive analysis of the instability reason is also the subject of the present work.

7. Proof of energy conservation for rotatory inertia value transformation

Problem: Common practice in calculating rotational motions of rigid bodies applies the so-called *angular velocity transformation* between a first integration over time in body-fixed frame and a second within inertial frame (see [1]). This is contrary to the so-called *linear velocity transformation*, which uses the ordinary vector transformation algorithm as applied in our solution [2], e.g. Euler's angle transformation. The derivation of the rotation matrix $\overline{\mathbf{C}}_{rot,be}$ (Eq. (26)) is carried out by observations and describes the rotation around the three body-fixed coordinate axes sequentially in time.

In 1995, Blass and Gurevich postulates 'Matrix Transformation Is Complete for the Average Case' [25]:

$$\overline{A}_e = \overline{\mathbf{C}}_b^e \cdot \overline{A}_b \cdot \left(\overline{\mathbf{C}}_b^e\right)^{-1} \tag{56}$$

whereby $\overline{\mathbf{C}}_b^e$ describes the ordinary transformation matrix for vector values from coordinate system b, here body fixed, to the inertial or earth-fixed ones e. Which transformation is correct, way I in comparison to Eq. (25) or way II regarding Eqs. (30) and (49)?

I: if $\vec{\boldsymbol{\omega}}_e = \overline{\mathbf{C}}_{rot,\,be} \cdot \vec{\boldsymbol{\omega}}_b$ is true; also,

$\vec{\mathbf{M}}_e = \overline{\mathbf{C}}_{rot,\,be} \cdot \overline{\mathbf{J}}_b \cdot \left(\overline{\mathbf{C}}_{rot,\,be}\right)^{-1} \cdot \overline{\mathbf{C}}_{rot,\,be} \cdot \vec{\boldsymbol{\omega}}_b$ might be possible being true, but wrong in our opinion, because $\left|\overline{\mathbf{C}}_{rot,\,be}\right| \neq 1$ for each $\Theta \neq 0$

or II: if $\vec{\boldsymbol{\omega}}_e = \overline{\mathbf{C}}_b^e \cdot \vec{\boldsymbol{\omega}}_b$ is true, contrary to the above

$\vec{\mathbf{M}}_e = \overline{\mathbf{C}}_b^e \cdot \overline{\mathbf{J}}_b \cdot \left(\overline{\mathbf{C}}_b^e\right)^{\mathrm{T}} \cdot \overline{\mathbf{C}}_b^e \cdot \vec{\boldsymbol{\omega}}_b$ have to be true!

Matrix mathematics cannot give answer, because both assumptions are valid.

Authors expect calculus II as valid due to the correct transformation of the moment vector.

Its correctness may prove only by the general use of one physical principal of conservation. For that purpose, the energy conservation shall be applied. Therefore, the pure rotating energy from one mass point of the rigid body (see *step 1*) has to be compared with the kinetic energy of same mass point and motion applying continuum physics within FEM methods (see *step 2*). Furthermore and in accordance with authors who claim of wrong description of rotatory kinetic energy within the body-fixed frame, all used rotation speed components are time derivatives of inertial-fixed Euler's angles. Within the calculation of *step 3*, it has to be shown that all coefficients of both energy equations are equal, otherwise the proof fails.

7.1. Calculation of rotatory kinetic energy of a mass point from a rotating body (*step 1*)

Vector value transformation based on Eulerian angles $\vec{\mathbf{\Pi}}_e = (\Phi \;\; \Theta \;\; \Psi)^{\mathrm{T}}$ using the well known transformation matrix, compare with [1] (Eq. (27)). The components C_{ij} of rotation matrix are

$$\overline{\mathbf{C}}_b^e = \begin{bmatrix} C_{11} & C_{12} & C_{13} \\ C_{21} & C_{22} & C_{23} \\ C_{31} & C_{32} & C_{33} \end{bmatrix} = \begin{bmatrix} c\Theta c\Psi & s\Phi s\Theta c\Psi - c\Phi s\Psi & c\Phi s\Theta c\Psi + s\Phi s\Psi \\ c\Theta s\Psi & s\Phi s\Theta s\Psi + c\Phi c\Psi & c\Phi s\Theta s\Psi - s\Phi c\Psi \\ -s\Theta & s\Phi c\Theta & c\Phi c\Theta \end{bmatrix} \tag{57}$$

The rotatory inertia value $\overline{\mathbf{J}}_{dm}$ of the mass point calculates

$$\overline{\mathbf{J}}_{dm} \cdot dm = dm \cdot \begin{bmatrix} R_y^2 + R_z^2 & -R_x R_y & -R_x R_z \\ -R_x R_y & R_x^2 + R_z^2 & -R_y R_z \\ -R_x R_z & -R_y R_z & R_x^2 + R_y^2 \end{bmatrix} \tag{58}$$

by angular speed

$$\dot{\vec{\mathbf{\Pi}}}_e = \vec{\boldsymbol{\omega}}_e = \begin{pmatrix} \dot{\Phi} & \dot{\Theta} & \dot{\Psi} \end{pmatrix}^{\mathrm{T}} \tag{59}$$

and lever arm $\vec{\mathbf{R}}_m = \begin{pmatrix} R_x & R_y & R_z \end{pmatrix}^{\mathrm{T}}$ from the bodies' centre of gravity to the mass point in body-fixed frame b. By vectorial depiction, the rotatory kinetic energy can be formulated as follows, e.g. Ginsberg [21]:

$$E_{rot} = \frac{1}{2} dm \cdot \vec{\boldsymbol{\omega}}_e^{\mathrm{T}} \cdot \left(\overline{\mathbf{C}}_b^e \cdot \bar{\mathbf{J}}_{dm} \cdot \overline{\mathbf{C}}_e^b \right) \cdot \vec{\boldsymbol{\omega}}_e \tag{60}$$

The first calculation step can be $\overline{\mathbf{C}}_e^b \cdot \vec{\boldsymbol{\omega}}_e$ (cf. Eq. (61)):

$$\overline{\mathbf{C}}_e^b \cdot \vec{\boldsymbol{\omega}}_e = \begin{bmatrix} C_{11}\dot{\Phi} + C_{21}\dot{\Theta} + C_{31}\dot{\Psi} \\ C_{12}\dot{\Phi} + C_{22}\dot{\Theta} + C_{32}\dot{\Psi} \\ C_{13}\dot{\Phi} + C_{23}\dot{\Theta} + C_{33}\dot{\Psi} \end{bmatrix} \tag{61}$$

Now, $\bar{\mathbf{J}}_{dm} \cdot \overline{\mathbf{C}}_e^b \cdot \vec{\boldsymbol{\omega}}_e$ can be estimated:

$$\begin{bmatrix} R_y^2 + R_z^2 & -R_xR_y & -R_xR_z \\ -R_xR_y & R_x^2 + R_z^2 & -R_yR_z \\ -R_xR_z & -R_yR_z & R_x^2 + R_y^2 \end{bmatrix} \cdot \begin{bmatrix} C_{11}\dot{\Phi} + C_{21}\dot{\Theta} + C_{31}\dot{\Psi} \\ C_{12}\dot{\Phi} + C_{22}\dot{\Theta} + C_{32}\dot{\Psi} \\ C_{13}\dot{\Phi} + C_{23}\dot{\Theta} + C_{33}\dot{\Psi} \end{bmatrix} =$$
$$\begin{bmatrix} C_{11}R_y^2\dot{\Phi} + C_{11}R_z^2\dot{\Phi} + C_{21}R_y^2\dot{\Theta} + C_{21}R_z^2\dot{\Theta} + C_{31}R_y^2\dot{\Psi} + C_{31}R_z^2\dot{\Psi}\ldots \\ -C_{12}R_xR_y\dot{\Phi} - C_{22}R_xR_y\dot{\Theta} - C_{32}R_xR_y\dot{\Psi} - C_{13}R_xR_z\dot{\Phi} - C_{23}R_xR_z\dot{\Theta} - C_{33}R_xR_z\dot{\Psi} \\ -C_{11}R_xR_y\dot{\Phi} - C_{21}R_xR_y\dot{\Theta} - C_{31}R_xR_y\dot{\Psi} + C_{12}R_x^2\dot{\Phi} + C_{12}R_z^2\dot{\Phi} + C_{22}R_x^2\dot{\Theta}\ldots \\ +C_{22}R_z^2\dot{\Theta} + C_{32}R_x^2\dot{\Psi} + C_{32}R_z^2\dot{\Psi} - C_{13}R_yR_z\dot{\Phi} - C_{23}R_yR_z\dot{\Theta} - C_{33}R_yR_z\dot{\Psi} \\ -C_{11}R_xR_z\dot{\Phi} - C_{21}R_xR_z\dot{\Theta} - C_{31}R_xR_z\dot{\Psi} - C_{12}R_yR_z\dot{\Phi} - C_{22}R_yR_z\dot{\Theta} - C_{32}R_yR_z\dot{\Psi}\ldots \\ C_{13}R_x^2\dot{\Phi} + C_{13}R_y^2\dot{\Phi} + C_{23}R_x^2\dot{\Theta} + C_{23}R_y^2\dot{\Theta} + C_{33}R_x^2\dot{\Psi} + C_{33}R_y^2\dot{\Psi} \end{bmatrix} \tag{62}$$

In the following equation, the transposed vector from Eq. (61) is multiplied with the vector of Eq. (62):

$$\left(\overline{\mathbf{C}}_e^b \cdot \vec{\boldsymbol{\omega}}_e \right)^{\mathrm{T}} \cdot \bar{\mathbf{J}}_{dm} \cdot \overline{\mathbf{C}}_e^b \cdot \vec{\boldsymbol{\omega}}_e = E_1 + E_2 + E_3 \tag{63}$$

getting three components of a sum E_1 to E_3:

$$\begin{aligned} E_1 = {} & C_{11}^2R_y^2\dot{\Phi}^2 + C_{11}^2R_z^2\dot{\Phi}^2 + C_{11}C_{21}R_y^2\dot{\Phi}\dot{\Theta} + C_{11}C_{21}R_z^2\dot{\Phi}\dot{\Theta} + C_{11}C_{31}R_y^2\dot{\Phi}\dot{\Psi}\ldots \\ & + C_{11}C_{31}R_z^2\dot{\Phi}\dot{\Psi} - C_{11}C_{12}R_xR_y\dot{\Phi}^2 - C_{11}C_{22}R_xR_y\dot{\Phi}\dot{\Theta} - C_{11}C_{32}R_xR_y\dot{\Phi}\dot{\Psi}\ldots \\ & - C_{11}C_{13}R_xR_z\dot{\Phi}^2 - C_{11}C_{23}R_xR_z\dot{\Phi}\dot{\Theta} - C_{11}C_{33}R_xR_z\dot{\Phi}\dot{\Psi} + C_{11}C_{21}R_y^2\dot{\Phi}\dot{\Theta} + C_{21}^2R_z^2\dot{\Theta}^2\ldots \\ & + C_{11}C_{21}R_z^2\dot{\Phi}\dot{\Theta} + C_{21}^2R_y^2\dot{\Theta}^2 + C_{21}C_{31}R_y^2\dot{\Theta}\dot{\Psi} + C_{21}C_{31}R_z^2\dot{\Theta}\dot{\Psi} - C_{12}C_{21}R_xR_y\dot{\Phi}\dot{\Theta}\ldots \\ & - C_{21}C_{22}R_xR_y\dot{\Theta}^2 - C_{21}C_{32}R_xR_y\dot{\Theta}\dot{\Psi} - C_{13}C_{21}R_xR_z\dot{\Phi}\dot{\Theta} - C_{21}C_{23}R_xR_z\dot{\Theta}^2\ldots \\ & - C_{21}C_{33}R_xR_z\dot{\Theta}\dot{\Psi} + C_{11}C_{31}R_y^2\dot{\Phi}\dot{\Psi} + C_{11}C_{31}R_z^2\dot{\Phi}\dot{\Psi} + C_{21}C_{31}R_y^2\dot{\Theta}\dot{\Psi} + C_{21}C_{31}R_z^2\dot{\Theta}\dot{\Psi}\ldots \\ & + C_{31}^2R_y^2\dot{\Psi}^2 + C_{31}^2R_z^2\dot{\Psi}^2 - C_{12}C_{31}R_xR_y\dot{\Phi}\dot{\Psi} - C_{22}C_{31}R_xR_y\dot{\Theta}\dot{\Psi} - C_{32}C_{31}R_xR_y\dot{\Psi}^2\ldots \\ & - C_{13}C_{31}R_xR_z\dot{\Phi}\dot{\Psi} - C_{23}C_{31}R_xR_z\dot{\Theta}\dot{\Psi} - C_{31}C_{33}R_xR_z\dot{\Psi}^2 \end{aligned}$$

$$E_2 = -C_{11}C_{12}R_xR_y\dot{\Phi}^2 - C_{12}C_{21}R_xR_y\dot{\Phi}\dot{\Theta} - C_{12}C_{31}R_xR_y\dot{\Phi}\dot{\Psi} + C_{12}^2R_x^2\dot{\Phi}^2 + C_{12}^2R_z^2\dot{\Phi}^2 \ldots$$
$$+ C_{12}C_{22}R_x^2\dot{\Phi}\dot{\Theta} + C_{12}C_{22}R_z^2\dot{\Phi}\dot{\Theta} + C_{12}C_{32}R_x^2\dot{\Phi}\dot{\Psi} + C_{12}C_{32}R_z^2\dot{\Phi}\dot{\Psi} - C_{12}C_{13}R_yR_z\dot{\Phi}^2 \ldots$$
$$- C_{12}C_{23}R_yR_z\dot{\Phi}\dot{\Theta} - C_{12}C_{33}R_yR_z\dot{\Phi}\dot{\Psi} - C_{11}C_{22}R_xR_y\dot{\Phi}\dot{\Theta} - C_{21}C_{22}R_xR_y\dot{\Theta}^2 \ldots$$
$$- C_{22}C_{31}R_xR_y\dot{\Theta}\dot{\Psi} + C_{12}C_{22}R_x^2\dot{\Phi}\dot{\Theta} + C_{12}C_{22}R_z^2\dot{\Phi}\dot{\Theta} + C_{22}^2R_x^2\dot{\Theta}^2 + C_{22}^2R_z^2\dot{\Theta}^2 \ldots$$
$$+ C_{22}C_{32}R_x^2\dot{\Theta}\dot{\Psi} + C_{22}C_{32}R_z^2\dot{\Theta}\dot{\Psi} - C_{13}C_{22}R_yR_z\dot{\Phi}\dot{\Theta} - C_{22}C_{23}R_yR_z\dot{\Theta}^2 \ldots$$
$$- C_{22}C_{33}R_yR_z\dot{\Theta}\dot{\Psi} - C_{11}C_{32}R_xR_y\dot{\Phi}\dot{\Psi} - C_{21}C_{32}R_xR_y\dot{\Theta}\dot{\Psi} - C_{31}C_{32}R_xR_y\dot{\Psi}^2 \ldots$$
$$+ C_{12}C_{32}R_x^2\dot{\Phi}\dot{\Psi} + C_{12}C_{32}R_z^2\dot{\Phi}\dot{\Psi} + C_{22}C_{32}R_x^2\dot{\Theta}\dot{\Psi} + C_{22}C_{32}R_z^2\dot{\Theta}\dot{\Psi} + C_{32}^2R_x^2\dot{\Psi}^2 \ldots$$
$$+ C_{32}^2R_z^2\dot{\Psi}^2 - C_{13}C_{32}R_yR_z\dot{\Phi}\dot{\Psi} - C_{23}C_{32}R_yR_z\dot{\Theta}\dot{\Psi} - C_{32}C_{33}R_yR_z\dot{\Psi}^2$$

$$E_3 = -C_{11}C_{13}R_xR_z\dot{\Phi}^2 - C_{13}C_{21}R_xR_z\dot{\Phi}\dot{\Theta} - C_{13}C_{31}R_xR_z\dot{\Phi}\dot{\Psi} - C_{12}C_{13}R_yR_z\dot{\Phi}^2 \ldots$$
$$- C_{13}C_{22}R_yR_z\dot{\Phi}\dot{\Theta} - C_{13}C_{32}R_yR_z\dot{\Phi}\dot{\Psi} + C_{13}^2R_x^2\dot{\Phi}^2 + C_{13}^2R_y^2\dot{\Phi}^2 + C_{13}C_{23}R_x^2\dot{\Phi}\dot{\Theta} \ldots$$
$$+ C_{13}C_{23}R_y^2\dot{\Phi}\dot{\Theta} + C_{13}C_{33}R_x^2\dot{\Phi}\dot{\Psi} + C_{13}C_{33}R_y^2\dot{\Phi}\dot{\Psi} - C_{11}C_{23}R_xR_z\dot{\Phi}\dot{\Theta} \ldots$$
$$- C_{21}C_{23}R_xR_z\dot{\Theta}^2 - C_{23}C_{31}R_xR_z\dot{\Theta}\dot{\Psi} - C_{12}C_{23}R_yR_z\dot{\Phi}\dot{\Theta} - C_{22}C_{23}R_yR_z\dot{\Theta}^2 \ldots$$
$$- C_{23}C_{32}R_yR_z\dot{\Theta}\dot{\Psi} + C_{13}C_{23}R_x^2\dot{\Phi}\dot{\Theta} + C_{13}C_{23}R_y^2\dot{\Phi}\dot{\Theta} + C_{23}^2R_x^2\dot{\Theta}^2 + C_{23}^2R_y^2\dot{\Theta}^2 \ldots$$
$$+ C_{23}C_{33}R_x^2\dot{\Theta}\dot{\Psi} + C_{23}C_{33}R_y^2\dot{\Theta}\dot{\Psi} - C_{11}C_{33}R_xR_z\dot{\Phi}\dot{\Psi} - C_{21}C_{33}R_xR_z\dot{\Theta}\dot{\Psi} \ldots$$
$$- C_{31}C_{33}R_xR_z\dot{\Psi}^2 - C_{12}C_{33}R_yR_z\dot{\Phi}\dot{\Psi} - C_{22}C_{33}R_yR_z\dot{\Theta}\dot{\Psi} - C_{32}C_{33}R_yR_z\dot{\Psi}^2 \ldots$$
$$+ C_{13}C_{33}R_x^2\dot{\Phi}\dot{\Psi} + C_{13}C_{33}R_y^2\dot{\Phi}\dot{\Psi} + C_{23}C_{33}R_x^2\dot{\Theta}\dot{\Psi} + C_{23}C_{33}R_y^2\dot{\Theta}\dot{\Psi} + C_{33}^2R_x^2\dot{\Psi}^2 \ldots$$
$$+ C_{33}^2R_y^2\dot{\Psi}^2$$

Finally, the sum of these components E_1, E_2 and E_3 has to be multiplied by the factor ½ *dm* to get the rotatory kinetic energy. Because that factor is implemented also within the second energy equation (*step* 2), the energy components can be summarised and compared directly with Eq. (68). By sorting all components to the lever arm and rotating speed combinations, we get Eq. (64):

$$\frac{2E_{rot}}{dm} = (C_{32}^2 + C_{33}^2)R_x^2\dot{\Psi}^2 - 2C_{31}C_{32}R_xR_y\dot{\Psi}^2 - 2C_{31}C_{33}R_xR_z\dot{\Psi}^2 - 2C_{32}C_{33}R_yR_z\dot{\Psi}^2 \ldots$$
$$+ (C_{31}^2 + C_{33}^2)R_y^2\dot{\Psi}^2 + (C_{31}^2 + C_{32}^2)R_z^2\dot{\Psi}^2 + 2(C_{22}C_{32} + C_{23}C_{33})R_x^2\dot{\Theta}\dot{\Psi} \ldots$$
$$- 2(C_{22}C_{31} + C_{21}C_{32})R_xR_y\dot{\Theta}\dot{\Psi} - 2(C_{21}C_{33} + C_{23}C_{31})R_xR_z\dot{\Theta}\dot{\Psi} \ldots$$
$$- 2(C_{22}C_{33} + C_{23}C_{32})R_yR_z\dot{\Theta}\dot{\Psi} + 2(C_{23}C_{33} + C_{21}C_{31})R_y^2\dot{\Theta}\dot{\Psi} \ldots$$
$$+ 2(C_{22}C_{32} + C_{21}C_{31})R_z^2\dot{\Theta}\dot{\Psi} + 2(C_{13}C_{33} + C_{12}C_{32})R_x^2\dot{\Phi}\dot{\Psi} \ldots$$
$$- 2(C_{12}C_{31} + C_{11}C_{32})R_xR_y\dot{\Phi}\dot{\Psi} - 2(C_{11}C_{33} + C_{13}C_{31})R_xR_z\dot{\Phi}\dot{\Psi} \ldots$$
$$- 2(C_{12}C_{33} + C_{13}C_{32})R_yR_z\dot{\Phi}\dot{\Psi} + 2(C_{13}C_{33} + C_{11}C_{31})R_y^2\dot{\Phi}\dot{\Psi} \ldots \tag{64}$$
$$+ 2(C_{12}C_{32} + C_{11}C_{31})R_z^2\dot{\Phi}\dot{\Psi} + (C_{22}^2 + C_{23}^2)R_x^2\dot{\Theta}^2 - 2C_{21}C_{22}R_xR_y\dot{\Theta}^2 \ldots$$
$$- 2C_{21}C_{23}R_xR_z\dot{\Theta}^2 - 2C_{22}C_{23}R_yR_z\dot{\Theta}^2 + (C_{21}^2 + C_{23}^2)R_y^2\dot{\Theta}^2 + (C_{21}^2 + C_{22}^2)R_z^2\dot{\Theta}^2 \ldots$$
$$+ 2(C_{13}C_{23} + C_{12}C_{22})R_x^2\dot{\Phi}\dot{\Theta} - 2(C_{11}C_{22} + C_{12}C_{21})R_xR_y\dot{\Phi}\dot{\Theta} \ldots$$
$$- 2(C_{11}C_{23} + C_{13}C_{21})R_xR_z\dot{\Phi}\dot{\Theta} - 2(C_{12}C_{23} + C_{13}C_{22})R_yR_z\dot{\Phi}\dot{\Theta} \ldots$$
$$+ 2(C_{11}C_{21} + C_{13}C_{23})R_y^2\dot{\Phi}\dot{\Theta} + 2(C_{11}C_{21} + C_{12}C_{22})R_z^2\dot{\Phi}\dot{\Theta} + (C_{12}^2 + C_{13}^2)R_x^2\dot{\Phi}^2 \ldots$$
$$- 2C_{11}C_{12}R_xR_y\dot{\Phi}^2. - 2C_{11}C_{13}R_xR_z\dot{\Phi}^2 - 2C_{12}C_{13}R_yR_z\dot{\Phi}^2 + (C_{11}^2 + C_{13}^2)R_y^2\dot{\Phi}^2 \ldots$$
$$+ (C_{11}^2 + C_{12}^2)R_z^2\dot{\Phi}^2$$

7.2. Calculation of kinetic energy from a mass point of a pure rotating body (*step* 2)

The energy can also be calculated by the tangential speed vector, preferably within the inertial frame (Eq. (65)):

$$E_{kin} = \frac{1}{2} dm \left[\vec{\boldsymbol{\omega}}_e \times \left(\overline{\mathbf{C}}_b^e \cdot \vec{\mathbf{R}}_m \right) \right]^{\mathrm{T}} \cdot \left[\vec{\boldsymbol{\omega}}_e \times \left(\overline{\mathbf{C}}_b^e \cdot \vec{\mathbf{R}}_m \right) \right] \tag{65}$$

The cross product can be calculated by a so-called skew operator $\tilde{\omega}_e$ (Eq. (32)). So, it is possible to calculate the tangential speed vector of the mass point within the inertial frame:

$$\vec{\mathbf{v}}_{dm} = \tilde{\boldsymbol{\omega}}_e \cdot \overline{\mathbf{C}}_b^e \cdot \vec{\mathbf{R}}_m = \begin{bmatrix} -C_{21}R_x\dot{\Psi} - C_{22}R_y\dot{\Psi} - C_{23}R_z\dot{\Psi} + C_{31}R_x\dot{\Theta} + C_{32}R_y\dot{\Theta} + C_{33}R_z\dot{\Theta} \\ C_{11}R_x\dot{\Psi} + C_{12}R_y\dot{\Psi} + C_{13}R_z\dot{\Psi} - C_{31}R_x\dot{\Phi} - C_{32}R_y\dot{\Phi} - C_{33}R_z\dot{\Phi} \\ -C_{11}R_x\dot{\Theta} - C_{12}R_y\dot{\Theta} - C_{13}R_z\dot{\Theta} + C_{21}R_x\dot{\Phi} + C_{22}R_y\dot{\Phi} + C_{23}R_z\dot{\Phi} \end{bmatrix} \tag{66}$$

From Eq. (66) the square of speed can be calculated (see Eq. (67)):

$$\frac{2E_{kin}}{dm} = v_{dm}^2 = \vec{\mathbf{v}}_{dm}^{\mathrm{T}} \cdot \vec{\mathbf{v}}_{dm} = E_4 + E_5 + E_6 \tag{67}$$

with.

$$\begin{aligned} E_4 = {} & C_{21}^2 R_x^2 \dot{\Psi}^2 + 2C_{21}C_{22}R_xR_y\dot{\Psi}^2 + 2C_{21}C_{23}R_xR_z\dot{\Psi}^2 - 2C_{21}C_{31}R_x^2\dot{\Theta}\dot{\Psi} \ldots \\ & - 2C_{21}C_{32}R_xR_y\dot{\Theta}\dot{\Psi} - 2C_{21}C_{33}R_xR_z\dot{\Theta}\dot{\Psi} + C_{22}^2R_y^2\dot{\Psi}^2 + 2C_{22}C_{23}R_yR_z\dot{\Psi}^2 \ldots \\ & - 2C_{22}C_{31}R_xR_y\dot{\Theta}\dot{\Psi} - 2C_{22}C_{32}R_y^2\dot{\Theta}\dot{\Psi} - 2C_{22}C_{33}R_yR_z\dot{\Theta}\dot{\Psi} + C_{23}^2R_z^2\dot{\Psi}^2 \ldots \\ & - 2C_{23}C_{31}R_xR_z\dot{\Theta}\dot{\Psi} - 2C_{23}C_{32}R_yR_z\dot{\Theta}\dot{\Psi} - 2C_{23}C_{33}R_z^2\dot{\Theta}\dot{\Psi} + C_{31}^2R_x^2\dot{\Theta}^2 \ldots \\ & + 2C_{31}C_{32}R_xR_y\dot{\Theta}^2 + 2C_{31}C_{33}R_xR_z\dot{\Theta}^2 + C_{32}^2R_y^2\dot{\Theta}^2 + 2C_{32}C_{33}R_yR_z\dot{\Theta}^2 + C_{33}^2R_z^2\dot{\Theta}^2 \end{aligned}$$

$$\begin{aligned} E_5 = {} & C_{11}^2 R_x^2 \dot{\Psi}^2 + 2C_{11}C_{12}R_xR_y\dot{\Psi}^2 + 2C_{11}C_{13}R_xR_z\dot{\Psi}^2 - 2C_{11}C_{31}R_x^2\dot{\Phi}\dot{\Psi} \ldots \\ & - 2C_{11}C_{32}R_xR_y\dot{\Phi}\dot{\Psi} - 2C_{11}C_{33}R_xR_z\dot{\Phi}\dot{\Psi} + C_{12}^2R_y^2\dot{\Psi}^2 + 2C_{12}C_{13}R_yR_z\dot{\Psi}^2 \ldots \\ & - 2C_{12}C_{31}R_xR_y\dot{\Phi}\dot{\Psi} - 2C_{12}C_{32}R_y^2\dot{\Phi}\dot{\Psi} - 2C_{12}C_{33}R_yR_z\dot{\Phi}\dot{\Psi} + C_{13}^2R_z^2\dot{\Psi}^2 \ldots \\ & - 2C_{13}C_{31}R_xR_z\dot{\Phi}\dot{\Psi} - 2C_{13}C_{32}R_yR_z\dot{\Phi}\dot{\Psi} - 2C_{13}C_{33}R_z^2\dot{\Phi}\dot{\Psi} + C_{31}^2R_x^2\dot{\Phi}^2 \ldots \\ & + 2C_{31}C_{32}R_xR_y\dot{\Phi}^2 + 2C_{31}C_{33}R_xR_z\dot{\Phi}^2 + C_{32}^2R_y^2\dot{\Phi}^2 + 2C_{32}C_{33}R_yR_z\dot{\Phi}^2 + C_{33}^2R_z^2\dot{\Phi}^2 \end{aligned}$$

$$\begin{aligned} E_6 = {} & C_{11}^2 R_x^2 \dot{\Theta}^2 + 2C_{11}C_{12}R_xR_y\dot{\Theta}^2 + 2C_{11}C_{13}R_xR_z\dot{\Theta}^2 - 2C_{11}C_{21}R_x^2\dot{\Phi}\dot{\Theta} \ldots \\ & - 2C_{11}C_{22}R_xR_y\dot{\Phi}\dot{\Theta} - 2C_{11}C_{23}R_xR_z\dot{\Phi}\dot{\Theta} + C_{12}^2R_y^2\dot{\Theta}^2 + 2C_{12}C_{13}R_yR_z\dot{\Theta}^2 \ldots \\ & - 2C_{12}C_{21}R_xR_y\dot{\Phi}\dot{\Theta} - 2C_{12}C_{22}R_y^2\dot{\Phi}\dot{\Theta} - 2C_{12}C_{23}R_yR_z\dot{\Phi}\dot{\Theta} + C_{13}^2R_z^2\dot{\Theta}^2 \ldots \\ & - 2C_{13}C_{21}R_xR_z\dot{\Phi}\dot{\Theta} - 2C_{13}C_{22}R_yR_z\dot{\Phi}\dot{\Theta} - 2C_{13}C_{23}R_z^2\dot{\Phi}\dot{\Theta} + C_{21}^2R_x^2\dot{\Phi}^2 \ldots \\ & + 2C_{21}C_{22}R_xR_y\dot{\Phi}^2 + 2C_{21}C_{23}R_xR_z\dot{\Phi}^2 + C_{22}^2R_y^2\dot{\Phi}^2 + 2C_{22}C_{23}R_yR_z\dot{\Phi}^2 + C_{23}^2R_z^2\dot{\Phi}^2 \end{aligned}$$

Summarising E_4, E_5 and E_6, follows Eq. (68).

$$\begin{aligned}
\frac{2E_{kin}}{dm} &= (C_{11}^2 + C_{21}^2)R_x^2\dot{\Psi}^2 + 2(C_{11}C_{12} + C_{21}C_{22})R_xR_y\dot{\Psi}^2 + 2(C_{11}C_{13} + C_{21}C_{23})R_xR_z\dot{\Psi}^2 \ldots \\
&+ 2(C_{12}C_{13} + C_{22}C_{23})R_yR_z\dot{\Psi}^2 + (C_{12}^2 + C_{22}^2)R_y^2\dot{\Psi}^2 + (C_{13}^2 + C_{23}^2)R_z^2\dot{\Psi}^2 \ldots \\
&- 2C_{21}C_{31}R_x^2\dot{\Theta}\dot{\Psi} - 2(C_{21}C_{32} + C_{22}C_{31})R_xR_y\dot{\Theta}\dot{\Psi} - 2(C_{21}C_{33} + C_{23}C_{31})R_xR_z\dot{\Theta}\dot{\Psi} \ldots \\
&- 2(C_{22}C_{33} + C_{23}C_{32})R_yR_z\dot{\Theta}\dot{\Psi} - 2C_{22}C_{32}R_y^2\dot{\Theta}\dot{\Psi} - 2C_{23}C_{33}R_z^2\dot{\Theta}\dot{\Psi} - 2C_{11}C_{31}R_x^2\dot{\Phi}\dot{\Psi} \ldots \\
&- 2(C_{11}C_{32} + C_{12}C_{31})R_xR_y\dot{\Phi}\dot{\Psi} - 2(C_{11}C_{33} + C_{13}C_{31})R_xR_z\dot{\Phi}\dot{\Psi} \ldots \\
&- 2(C_{12}C_{33} + C_{13}C_{32})R_yR_z\dot{\Phi}\dot{\Psi} - 2C_{12}C_{32}R_y^2\dot{\Phi}\dot{\Psi} - 2C_{13}C_{33}R_z^2\dot{\Phi}\dot{\Psi} + (C_{11}^2 + C_{31}^2)R_x^2\dot{\Theta}^2 \ldots \\
&+ 2(C_{11}C_{12} + C_{31}C_{32})R_xR_y\dot{\Theta}^2 + 2(C_{11}C_{13} + C_{31}C_{33})R_xR_z\dot{\Theta}^2 \ldots \\
&+ 2(C_{12}C_{13} + C_{32}C_{33})R_yR_z\dot{\Theta}^2 + (C_{12}^2 + C_{32}^2)R_y^2\dot{\Theta}^2 + (C_{13}^2 + C_{33}^2)R_z^2\dot{\Theta}^2 \ldots \\
&- 2C_{11}C_{21}R_x^2\dot{\Phi}\dot{\Theta} - 2(C_{11}C_{22} + C_{12}C_{21})R_xR_y\dot{\Phi}\dot{\Theta} - 2(C_{11}C_{23} + C_{13}C_{21})R_xR_z\dot{\Phi}\dot{\Theta} \ldots \\
&- 2(C_{12}C_{23} + C_{13}C_{22})R_yR_z\dot{\Phi}\dot{\Theta} - 2C_{12}C_{22}R_y^2\dot{\Phi}\dot{\Theta} - 2C_{13}C_{23}R_z^2\dot{\Phi}\dot{\Theta} + (C_{21}^2 + C_{31}^2)R_x^2\dot{\Phi}^2 \ldots \\
&+ 2(C_{21}C_{22} + C_{31}C_{32})R_xR_y\dot{\Phi}^2 + 2(C_{21}C_{23} + C_{31}C_{33})R_xR_z\dot{\Phi}^2 \ldots \\
&+ 2(C_{22}C_{23} + C_{32}C_{33})R_yR_z\dot{\Phi}^2 + (C_{22}^2 + C_{32}^2)R_y^2\dot{\Phi}^2 + (C_{23}^2 + C_{33}^2)R_z^2\dot{\Phi}^2
\end{aligned} \tag{68}$$

7.3. Comparison of the coefficients from both energy calculations for the mass point of pure rotating body (*step 3*)

To show that energy is identical in both approaches (*step 1* and *step 2*), the coefficients of all 36 combinations of lever arms and angular velocities are compared individually with each other in *step 3*. The coefficients of the rotation energy (cf. Eq. (64)) of *step 1* can be found on the left-hand side and the coefficients of kinetic energy (cf. Eq. (68)) of *step2* on the right-hand side. As described previously, the condition of identical energy is only proved, if all coefficients ar equal.

$$\begin{aligned}
& C_{32}^2 + C_{33}^2 = C_{11}^2 + C_{21}^2 \\
R_x^2\dot{\Psi}^2 : \; & s^2\Phi c^2\Theta + c^2\Phi c^2\Theta = c^2\Theta c^2\Psi + c^2\Theta s^2\Psi \\
& c^2\Theta = c^2\Theta
\end{aligned} \tag{69}$$

$$\begin{aligned}
& -2C_{31}C_{32} = 2(C_{11}C_{12} + C_{21}C_{22}) \\
R_xR_y\dot{\Psi}^2 : \; & s\Theta c\Theta s\Phi = s\Phi s\Theta c\Theta c^2\Psi - c\Phi c\Theta s\Psi c\Psi + s\Phi s\Theta c\Theta s^2\Psi \ldots \\
& \qquad + c\Phi c\Theta s\Psi c\Psi \\
& s\Theta c\Theta s\Phi = s\Phi s\Theta c\Theta
\end{aligned} \tag{70}$$

$$\begin{aligned}
& -2C_{31}C_{33} = 2(C_{11}C_{13} + C_{21}C_{23}) \\
R_xR_z\dot{\Psi}^2 : \; & s\Theta c\Theta c\Phi = c\Psi s\Psi c\Theta s\Phi + c^2\Psi c\Theta s\Theta c\Phi - s\Psi c\Psi c\Theta s\Phi \ldots \\
& \qquad + s^2\Psi c\Theta s\Theta c\Phi \\
& s\Theta c\Theta c\Phi = c\Theta s\Theta c\Phi
\end{aligned} \tag{71}$$

$$R_yR_z\dot{\Psi}^2: \begin{aligned} -2C_{32}C_{33} &= 2(C_{12}C_{13} + C_{22}C_{23}) \\ -c^2\Theta s\Phi c\Phi &= -s^2\Psi c\Phi s\Phi - s\Psi c\Psi c^2\Phi s\Theta + s\Psi c\Psi s^2\Phi s\Theta \ldots \\ &\quad + c^2\Psi s^2\Theta s\Phi c\Phi - c^2\Psi c\Phi s\Phi + s\Psi c\Psi c^2\Phi s\Theta \ldots \\ &\quad - s\Psi c\Psi s^2\Phi s\Theta + s^2\Psi s^2\Theta s\Phi c\Phi \\ &= -1 \cdot c\Phi s\Phi + s^2\Theta s\Phi c\Phi \\ -c^2\Theta s\Phi c\Phi &= -c^2\Theta s\Phi c\Phi \end{aligned} \tag{72}$$

$$R_y^2\dot{\Psi}^2: \begin{aligned} C_{31}^2 + C_{33}^2 &= C_{12}^2 + C_{22}^2 \\ s^2\Theta + c^2\Theta c^2\Phi &= s^2\Psi c^2\Phi - 2s\Psi c\Psi s\Theta c\Phi s\Phi + c^2\Psi s^2\Theta s^2\Phi \ldots \\ &\quad + c^2\Psi c^2\Phi + 2s\Psi c\Psi s\Theta s\Phi c\Phi + s^2\Phi s^2\Theta s^2\Psi \\ &= c^2\Phi + s^2\Theta s^2\Phi = c^2\Phi \cdot 1 + s^2\Theta s^2\Phi \\ &= c^2\Phi c^2\Theta + c^2\Phi s^2\Theta + s^2\Theta \cdot 1 - s^2\Theta c^2\Phi \\ s^2\Theta + c^2\Theta c^2\Phi &= c^2\Phi c^2\Theta + s^2\Theta \end{aligned} \tag{73}$$

$$R_z^2\dot{\Psi}^2: \begin{aligned} C_{31}^2 + C_{32}^2 &= C_{13}^2 + C_{23}^2 \\ s^2\Theta + c^2\Theta s^2\Phi &= s^2\Psi s^2\Phi + 2s\Psi c\Psi s\Theta s\Phi c\Phi + c^2\Psi s^2\Theta c^2\Phi \ldots \\ &\quad + c^2\Psi s^2\Phi - 2s\Psi c\Psi s\Theta s\Phi c\Phi + s^2\Theta s^2\Psi c^2\Phi \\ &= s^2\Phi + s^2\Theta c^2\Phi = s^2\Phi \cdot 1 + s^2\Theta c^2\Phi \\ &= s^2\Phi c^2\Theta + s^2\Phi s^2\Theta + s^2\Theta \cdot 1 - s^2\Theta s^2\Phi \\ s^2\Theta + c^2\Theta s^2\Phi &= s^2\Phi c^2\Theta + s^2\Theta \end{aligned} \tag{74}$$

$$R_x^2\dot{\Theta}\dot{\Psi}: \begin{aligned} 2(C_{22}C_{32} + C_{23}C_{33}) &= -2C_{21}C_{31} \\ c\Psi c\Theta s\Phi c\Phi + s\Psi s\Theta c\Theta s^2\Phi - c\Psi c\Theta s\Phi c\Phi. + s\Psi s\Theta c\Theta c^2\Phi &= s\Psi c\Theta s\Theta \\ s\Psi s\Theta c\Theta &= s\Psi c\Theta s\Theta \end{aligned} \tag{75}$$

$$R_xR_y\dot{\Theta}\dot{\Psi}: -2(C_{22}C_{31} + C_{21}C_{32}) = -2(C_{21}C_{32} + C_{22}C_{31}) \tag{76}$$

$$R_xR_z\dot{\Theta}\dot{\Psi}: -2(C_{21}C_{33} + C_{23}C_{31}) = -2(C_{21}C_{33} + C_{23}C_{31}) \tag{77}$$

$$R_yR_z\dot{\Theta}\dot{\Psi}: -2(C_{22}C_{33} + C_{23}C_{32}) = -2(C_{22}C_{33} + C_{23}C_{32}) \tag{78}$$

$$R_y^2\dot{\Theta}\dot{\Psi}: \begin{aligned} 2(C_{23}C_{33} + C_{21}C_{31}) &= -2C_{22}C_{32} \\ -c\Psi c\Theta s\Phi c\Phi + s\Psi s\Theta c\Theta c^2\Phi - s\Psi s\Theta c\Theta &= -c\Psi c\Theta c\Phi s\Phi - s\Psi c\Theta s\Theta s^2\Phi \\ s\Psi s\Theta c\Theta\left(1 - s^2\Phi - 1\right) &= -s\Psi c\Theta s\Theta s^2\Phi \\ -s\Psi c\Theta s\Theta s^2\Phi &= -s\Psi c\Theta s\Theta s^2\Phi \end{aligned} \tag{79}$$

$$R_z^2\dot{\Theta}\dot{\Psi}: \begin{aligned} 2(C_{22}C_{32}+C_{21}C_{31}) &= -2C_{23}C_{33} \\ c\Psi c\Theta s\Phi c\Phi + s\Psi s\Theta c\Theta s^2\Phi - s\Psi s\Theta c\Theta &= c\Psi c\Theta s\Phi c\Phi - s\Psi s\Theta c\Theta c^2\Phi \\ s\Psi s\Theta c\Theta\left(1-c^2\Phi-1\right) &= -s\Psi c\Theta s\Theta c^2\Phi \\ -s\Psi c\Theta s\Theta c^2\Phi &= -s\Psi c\Theta s\Theta c^2\Phi \end{aligned} \tag{80}$$

$$R_x^2\dot{\Phi}\dot{\Psi}: \begin{aligned} 2(C_{13}C_{33}+C_{12}C_{32}) &= -2C_{11}C_{31} \\ \begin{pmatrix} s\Psi c\Theta s\Phi c\Phi + c\Psi s\Theta c\Theta c^2\Phi - s\Psi c\Theta s\Phi c\Phi\ldots \\ +c\Psi s\Theta c\Theta s^2\Phi \end{pmatrix} &= c\Psi c\Theta s\Theta \\ c\Psi s\Theta c\Theta &= c\Psi c\Theta s\Theta \end{aligned} \tag{81}$$

$$R_xR_y\dot{\Phi}\dot{\Psi}: -2(C_{12}C_{31}+C_{11}C_{32}) = -2(C_{11}C_{32}+C_{12}C_{31}) \tag{82}$$

$$R_xR_z\dot{\Phi}\dot{\Psi}: -2(C_{11}C_{33}+C_{13}C_{31}) = -2(C_{11}C_{33}+C_{13}C_{31}) \tag{83}$$

$$R_yR_z\dot{\Phi}\dot{\Psi}: -2(C_{12}C_{33}+C_{13}C_{32}) = -2(C_{12}C_{33}+C_{13}C_{32}) \tag{84}$$

$$R_y^2\dot{\Phi}\dot{\Psi}: \begin{aligned} 2(C_{13}C_{33}+C_{11}C_{31}) &= -2C_{12}C_{32} \\ s\Psi c\Theta s\Phi c\Phi + c\Psi s\Theta c\Theta c^2\Phi - c\Psi c\Theta s\Theta &= s\Psi c\Theta s\Phi c\Phi - c\Psi s\Theta c\Theta s^2\Phi \\ c\Psi s\Theta c\Theta\left(1-s^2\Phi-1\right) &= -c\Psi s\Theta c\Theta s^2\Phi \\ -c\Psi s\Theta c\Theta s^2\Phi &= -c\Psi s\Theta c\Theta s^2\Phi \end{aligned} \tag{85}$$

$$R_z^2\dot{\Phi}\dot{\Psi}: \begin{aligned} 2(C_{12}C_{32}+C_{11}C_{31}) &= -2C_{13}C_{33} \\ -s\Psi c\Theta s\Phi c\Phi + c\Psi s\Theta c\Theta s^2\Phi - c\Psi c\Theta s\Theta &= -s\Psi c\Theta s\Phi c\Phi - c\Psi s\Theta c\Theta c^2\Phi \\ c\Psi s\Theta c\Theta\left(1-c^2\Phi-1\right) &= -c\Psi s\Theta c\Theta c^2\Phi \\ -c\Psi s\Theta c\Theta c^2\Phi &= -c\Psi s\Theta c\Theta c^2\Phi \end{aligned} \tag{86}$$

$$R_x^2\dot{\Theta}^2: \begin{aligned} C_{22}^2+C_{23}^2 &= C_{11}^2+C_{31}^2 \\ \begin{pmatrix} c^2\Psi c^2\Phi + 2s\Psi c\Psi s\Theta s\Phi c\Phi + s^2\Phi s^2\Theta s^2\Psi\ldots \\ +c^2\Psi s^2\Phi - 2s\Psi c\Psi s\Theta s\Phi c\Phi + s^2\Theta s^2\Psi c^2\Phi \end{pmatrix} &= c^2\Psi c^2\Theta + s^2\Theta \\ c^2\Psi + s^2\Theta s^2\Psi &= \\ c^2\Psi\cdot 1 + s^2\Theta s^2\Psi &= \\ c^2\Psi s^2\Theta + c^2\Psi c^2\Theta + s^2\Theta\cdot 1 - s^2\Theta c^2\Psi &= \\ c^2\Psi c^2\Theta + s^2\Theta &= c^2\Psi c^2\Theta + s^2\Theta \end{aligned} \tag{87}$$

$$R_xR_y\dot{\Theta}^2: \begin{aligned} -2C_{21}C_{22} &= 2(C_{11}C_{12}+C_{31}C_{32}) \\ -s\Psi c\Psi c\Theta c\Phi - s^2\Psi c\Theta s\Theta s\Phi &= -c\Psi s\Psi c\Theta c\Phi + c^2\Psi s\Theta c\Theta s\Phi - s\Theta c\Theta s\Phi \\ -s^2\Psi c\Theta s\Theta s\Phi &= \left(1-s^2\Psi-1\right)s\Theta c\Theta s\Phi \\ -s^2\Psi c\Theta s\Theta s\Phi &= -s^2\Psi s\Theta c\Theta s\Phi \end{aligned} \tag{88}$$

http://dx.doi.org/10.5772/intechopen.71445

$$R_xR_z\dot{\Theta}^2: \begin{aligned} -2C_{21}C_{23} &= 2(C_{11}C_{13}+C_{31}C_{33}) \\ s\Psi c\Psi c\Theta s\Phi - s^2\Psi s\Theta c\Theta c\Phi &= s\Psi c\Psi c\Theta s\Phi + c^2\Psi c\Theta s\Theta c\Phi - s\Theta c\Theta c\Phi \\ -s^2\Psi s\Theta c\Theta c\Phi &= \left(1-s^2\Psi-1\right)s\Theta c\Theta c\Phi \\ -s^2\Psi s\Theta c\Theta c\Phi &= -s^2\Psi s\Theta c\Theta c\Phi \end{aligned} \tag{89}$$

$$R_yR_z\dot{\Theta}^2: \begin{aligned} -2C_{22}C_{23} &= 2(C_{12}C_{13}+C_{32}C_{33}) \\ \begin{pmatrix} c^2\Psi s\Phi c\Phi - s\Psi c\Psi s\Theta c^2\Phi\ldots \\ +s\Psi c\Psi s\Theta s^2\Phi - s^2\Psi s^2\Theta s\Phi c\Phi \end{pmatrix} &= \begin{pmatrix} -s^2\Psi s\Phi c\Phi - s\Psi c\Psi s\Theta c^2\Phi\ldots \\ +s\Psi c\Psi s\Theta s^2\Phi + c^2\Psi s^2\Theta s\Phi c\Phi\ldots \\ +c^2\Theta s\Phi c\Phi \end{pmatrix} \\ c^2\Psi s\Phi c\Phi - s^2\Psi s^2\Theta s\Phi c\Phi &= \begin{pmatrix} -s^2\Psi s\Phi c\Phi + c^2\Psi s^2\Theta s\Phi c\Phi\ldots \\ +c^2\Theta s\Phi c\Phi \end{pmatrix} \\ s\Phi c\Phi\left(c^2\Psi - s^2\Psi s^2\Theta\right) &= s\Phi c\Phi\left(-s^2\Psi + c^2\Psi s^2\Theta + c^2\Theta\right) \\ c^2\Psi - s^2\Psi s^2\Theta &= -s^2\Psi + c^2\Psi s^2\Theta + c^2\Theta \\ 1 - s^2\Psi - s^2\Theta + c^2\Psi s^2\Theta &= \\ -s^2\Psi + c^2\Theta + c^2\Psi s^2\Theta &= -s^2\Psi + c^2\Psi s^2\Theta + c^2\Theta \end{aligned} \tag{90}$$

$$R_y^2\dot{\Theta}^2: \begin{aligned} C_{21}^2 + C_{23}^2 &= C_{12}^2 + C_{32}^2 \\ \begin{pmatrix} s^2\Psi c^2\Theta + c^2\Psi s^2\Phi\ldots \\ -2s\Psi c\Psi s\Theta s\Phi c\Phi\ldots \\ +s^2\Psi s^2\Theta c^2\Phi \end{pmatrix} &= \begin{pmatrix} s^2\Psi c^2\Phi - 2s\Psi c\Psi s\Theta s\Phi c\Phi\ldots \\ +c^2\Psi s^2\Theta s^2\Phi + c^2\Theta s^2\Phi \end{pmatrix} \\ \begin{pmatrix} s^2\Psi c^2\Theta + c^2\Psi s^2\Phi\ldots \\ +s^2\Psi s^2\Theta c^2\Phi \end{pmatrix} &= \begin{pmatrix} s^2\Psi c^2\Phi + c^2\Psi s^2\Theta s^2\Phi\ldots \\ +c^2\Theta s^2\Phi \end{pmatrix} \\ \begin{pmatrix} s^2\Psi c^2\Theta\cdot 1 + c^2\Psi\cdot 1\cdot s^2\Phi\ldots \\ +s^2\Psi s^2\Theta c^2\Phi \end{pmatrix} &= \begin{pmatrix} s^2\Psi\cdot 1\cdot c^2\Phi + c^2\Psi s^2\Theta s^2\Phi\ldots \\ +1\cdot c^2\Theta s^2\Phi \end{pmatrix} \\ \begin{pmatrix} s^2\Psi c^2\Theta s^2\Phi_1 + s^2\Psi c^2\Theta c^2\Phi_2\ldots \\ +c^2\Psi s^2\Theta s^2\Phi_3 + c^2\Psi c^2\Theta s^2\Phi_4\ldots \\ +s^2\Psi s^2\Theta c^2\Phi_5 \end{pmatrix} &= \begin{pmatrix} s^2\Psi s^2\Theta c^2\Phi_5 + s^2\Psi c^2\Theta c^2\Phi_2\ldots \\ +c^2\Psi s^2\Theta s^2\Phi_3 + s^2\Psi c^2\Theta s^2\Phi_1\ldots \\ +c^2\Psi c^2\Theta s^2\Phi_4 \end{pmatrix} \end{aligned} \tag{91}$$

$$R_z^2\dot{\Theta}^2: \begin{aligned} C_{21}^2 + C_{22}^2 &= C_{13}^2 + C_{33}^2 \\ \begin{pmatrix} s^2\Psi c^2\Theta + c^2\Psi c^2\Phi\ldots \\ +2s\Psi c\Psi s\Theta s\Phi c\Phi + s^2\Psi s^2\Theta s^2\Phi \end{pmatrix} &= \begin{pmatrix} s^2\Psi s^2\Phi + 2s\Psi c\Psi s\Theta s\Phi c\Phi\ldots \\ +c^2\Psi s^2\Theta c^2\Phi + c^2\Theta c^2\Phi \end{pmatrix} \\ s^2\Psi c^2\Theta + c^2\Psi c^2\Phi + s^2\Psi s^2\Theta s^2\Phi &= s^2\Psi s^2\Phi + c^2\Psi s^2\Theta c^2\Phi + c^2\Theta c^2\Phi \\ \begin{pmatrix} s^2\Psi c^2\Theta\cdot 1 + c^2\Psi\cdot 1\cdot c^2\Phi\ldots \\ +s^2\Psi s^2\Theta s^2\Phi \end{pmatrix} &= \begin{pmatrix} s^2\Psi\cdot 1\cdot s^2\Phi + c^2\Psi s^2\Theta c^2\Phi\ldots \\ +1\cdot c^2\Theta c^2\Phi \end{pmatrix} \\ \begin{pmatrix} s^2\Psi c^2\Theta s^2\Phi_1 + s^2\Psi c^2\Theta c^2\Phi_2\ldots \\ +c^2\Psi s^2\Theta c^2\Phi_3 + c^2\Psi c^2\Theta c^2\Phi_4\ldots \\ +s^2\Psi s^2\Theta s^2\Phi_5 \end{pmatrix} &= \begin{pmatrix} s^2\Psi s^2\Theta s^2\Phi_5 + s^2\Psi c^2\Theta s^2\Phi_1\ldots \\ +c^2\Psi s^2\Theta c^2\Phi_3 + s^2\Psi c^2\Theta c^2\Phi_2\ldots \\ +c^2\Psi c^2\Theta c^2\Phi_4 \end{pmatrix} \end{aligned} \tag{92}$$

$$2(C_{13}C_{23}+C_{12}C_{22})=-2C_{11}C_{21}$$

$$R_x^2\dot{\Phi}\dot{\Theta}:\ \begin{pmatrix} -s\Psi c\Psi s^2\Phi + s^2\Psi s\Theta s\Phi c\Phi - c^2\Psi s\Theta s\Phi c\Phi + s\Psi c\Psi s^2\Theta c^2\Phi\ldots \\ -s\Psi c\Psi c^2\Phi - s^2\Psi s\Theta s\Phi c\Phi + c^2\Psi s\Theta s\Phi c\Phi + s\Psi c\Psi s^2\Theta s^2\Phi \end{pmatrix} = -c\Psi s\Psi c^2\Theta$$

$$\begin{aligned} -s\Psi c\Psi s^2\Phi + s\Psi c\Psi s^2\Theta c^2\Phi - s\Psi c\Psi c^2\Phi + s\Psi c\Psi s^2\Theta s^2\Phi &= \\ -s\Psi c\Psi + s\Psi c\Psi s^2\Theta &= \\ s\Psi c\Psi\left(-1+1-c^2\Theta\right) &= \\ -s\Psi c\Psi c^2\Theta &= -c\Psi s\Psi c^2\Theta \end{aligned} \tag{93}$$

$$R_xR_y\dot{\Phi}\dot{\Theta}:\ -2(C_{11}C_{22}+C_{12}C_{21}) = -2(C_{11}C_{22}+C_{12}C_{21}) \tag{94}$$

$$R_xR_z\dot{\Phi}\dot{\Theta}:\ -2(C_{11}C_{23}+C_{13}C_{21}) = -2(C_{11}C_{23}+C_{13}C_{21}) \tag{95}$$

$$R_yR_z\dot{\Phi}\dot{\Theta}:\ -2(C_{12}C_{23}+C_{13}C_{22}) = -2(C_{12}C_{23}+C_{13}C_{22}) \tag{96}$$

$$R_y^2\dot{\Phi}\dot{\Theta}:\ \begin{aligned} 2(C_{11}C_{21}+C_{13}C_{23}) &= -2C_{12}C_{22} \\ \begin{pmatrix} c\Psi s\Psi c^2\Theta - s\Psi c\Psi s^2\Phi + s^2\Psi s\Theta s\Phi c\Phi\ldots \\ -c^2\Psi s\Theta s\Phi c\Phi + s\Psi c\Psi s^2\Theta c^2\Phi \end{pmatrix} &= \begin{pmatrix} s\Psi c\Psi c^2\Phi + s^2\Psi s\Theta s\Phi c\Phi\ldots \\ -c^2\Psi s\Theta s\Phi c\Phi - s\Psi c\Psi s^2\Theta s^2\Phi \end{pmatrix} \\ c\Psi s\Psi c^2\Theta - s\Psi c\Psi s^2\Phi + s\Psi c\Psi s^2\Theta c^2\Phi &= s\Psi c\Psi c^2\Phi - s\Psi c\Psi s^2\Theta s^2\Phi \\ s\Psi c\Psi\left(c^2\Theta - s^2\Phi + s^2\Theta c^2\Phi\right) &= s\Psi c\Psi\left(c^2\Phi - s^2\Theta s^2\Phi\right) \\ c^2\Theta - s^2\Phi + s^2\Theta c^2\Phi &= c^2\Phi - s^2\Theta s^2\Phi \\ 1 - s^2\Theta - 1 + c^2\Phi + s^2\Theta c^2\Phi &= \\ c^2\Phi + s^2\Theta\left(-1+c^2\Phi\right) &= \\ c^2\Phi + s^2\Theta\left(-1+1-s^2\Phi\right) &= \\ c^2\Phi - s^2\Theta s^2\Phi &= c^2\Phi - s^2\Theta s^2\Phi \end{aligned} \tag{97}$$

$$R_z^2\dot{\Phi}\dot{\Theta}:\ \begin{aligned} 2(C_{11}C_{21}+C_{12}C_{22}) &= -2C_{13}C_{23} \\ \begin{pmatrix} s\Psi c\Psi c^2\Theta - s\Psi c\Psi c^2\Phi\ldots \\ -s^2\Psi s\Theta s\Phi c\Phi + c^2\Psi s\Theta s\Phi c\Phi\ldots \\ +s\Psi c\Psi s^2\Theta s^2\Phi \end{pmatrix} &= \begin{pmatrix} s\Psi c\Psi s^2\Phi - s^2\Psi s\Theta s\Phi c\Phi\ldots \\ +c^2\Psi s\Theta s\Phi c\Phi - s\Psi c\Psi s^2\Theta c^2\Phi \end{pmatrix} \\ c\Psi s\Psi c^2\Theta - s\Psi c\Psi c^2\Phi + s\Psi c\Psi s^2\Theta s^2\Phi &= s\Psi c\Psi s^2\Phi - s\Psi c\Psi s^2\Theta c^2\Phi \\ s\Psi c\Psi\left(c^2\Theta - c^2\Phi + s^2\Theta s^2\Phi\right) &= s\Psi c\Psi\left(s^2\Phi - s^2\Theta c^2\Phi\right) \\ c^2\Theta - c^2\Phi + s^2\Theta s^2\Phi &= s^2\Phi - s^2\Theta c^2\Phi \\ 1 - s^2\Theta - 1 + s^2\Phi + s^2\Theta s^2\Phi &= \\ s^2\Phi + s^2\Theta\left(-1+s^2\Phi\right) &= \\ s^2\Phi + s^2\Theta\left(-1+1-c^2\Phi\right) &= \\ s^2\Phi - s^2\Theta c^2\Phi &= s^2\Phi - s^2\Theta c^2\Phi \end{aligned} \tag{98}$$

$$
R_x^2\dot{\Phi}^2: \begin{array}{c} C_{12}^2 + C_{13}^2 = C_{21}^2 + C_{31}^2 \\ \begin{pmatrix} s^2\Psi c^2\Phi - 2s\Psi c\Psi s\Theta s\Phi c\Phi \\ + c^2\Psi s^2\Theta s^2\Phi + s^2\Psi s^2\Phi \ldots \\ + 2s\Psi c\Psi s\Theta s\Phi c\Phi + c^2\Psi s^2\Theta c^2\Phi \end{pmatrix} = s^2\Psi c^2\Theta + s^2\Theta \\ \begin{pmatrix} s^2\Psi c^2\Phi + c^2\Psi s^2\Theta s^2\Phi + s^2\Psi s^2\Phi \ldots \\ + c^2\Psi s^2\Theta c^2\Phi \end{pmatrix} = \\ s^2\Psi + c^2\Psi s^2\Theta = \\ s^2\Psi \cdot 1 + \left(1 - s^2\Psi\right)s^2\Theta = \\ s^2\Psi s^2\Theta + s^2\Psi c^2\Theta + s^2\Theta - s^2\Psi s^2\Theta = \\ s^2\Psi c^2\Theta + s^2\Theta = s^2\Psi c^2\Theta + s^2\Theta \end{array} \tag{99}
$$

$$
R_x R_y \dot{\Phi}^2: \begin{array}{c} -2C_{11}C_{12} = 2(C_{21}C_{22} + C_{31}C_{32}) \\ s\Psi c\Psi c\Theta c\Phi - c^2\Psi c\Theta s\Theta s\Phi = c\Psi s\Psi c\Theta c\Phi + s^2\Psi s\Theta c\Theta s\Phi - s\Theta c\Theta s\Phi \\ -c^2\Psi c\Theta s\Theta s\Phi = \left(1 - c^2\Psi - 1\right)s\Theta c\Theta s\Phi \\ -c^2\Psi c\Theta s\Theta s\Phi = -c^2\Psi s\Theta c\Theta s\Phi \end{array} \tag{100}
$$

$$
R_x R_z \dot{\Phi}^2: \begin{array}{c} -2C_{11}C_{13} = 2(C_{21}C_{23} + C_{31}C_{33}) \\ -s\Psi c\Psi c\Theta s\Phi - c^2\Psi s\Theta c\Theta c\Phi = -s\Psi c\Psi c\Theta s\Phi + s^2\Psi s\Theta c\Theta c\Phi - s\Theta c\Theta c\Phi \\ -c^2\Psi c\Theta s\Theta c\Phi = \left(1 - c^2\Psi - 1\right)s\Theta c\Theta c\Phi \\ -c^2\Psi c\Theta s\Theta c\Phi = -c^2\Psi s\Theta c\Theta c\Phi \end{array} \tag{101}
$$

$$
R_y R_z \dot{\Phi}^2: \begin{array}{c} -2C_{12}C_{13} = 2(C_{22}C_{23} + C_{32}C_{33}) \\ \begin{pmatrix} s^2\Psi s\Phi c\Phi + s\Psi c\Psi s\Theta c^2\Phi \ldots \\ -s\Psi c\Psi s\Theta s^2\Phi - c^2\Psi s^2\Theta s\Phi c\Phi \end{pmatrix} = \begin{pmatrix} - c^2\Psi s\Phi c\Phi + s\Psi c\Psi s\Theta c^2\Phi \ldots \\ - s\Psi c\Psi s\Theta s^2\Phi + s^2\Psi s^2\Theta s\Phi c\Phi \ldots \\ + c^2\Theta s\Phi c\Phi \end{pmatrix} \\ s\Phi c\Phi\left(s^2\Psi - c^2\Psi s^2\Theta\right) = s\Phi c\Phi\left(-c^2\Psi + s^2\Psi s^2\Theta + c^2\Theta\right) \\ s^2\Psi - c^2\Psi s^2\Theta = -c^2\Psi + s^2\Psi s^2\Theta + c^2\Theta \\ = -1 + s^2\Psi + s^2\Psi s^2\Theta + 1 - s^2\Theta \\ = s^2\Psi + s^2\Theta\left(1 - c^2\Psi - 1\right) \\ s^2\Psi - c^2\Psi s^2\Theta = s^2\Psi - s^2\Theta c^2\Psi \end{array} \tag{102}
$$

$$C_{11}^2 + C_{13}^2 = C_{22}^2 + C_{32}^2$$

$$\begin{pmatrix} c^2\Psi c^2\Theta + s^2\Psi s^2\Phi \ldots \\ \\ +2s\Psi c\Psi s\Theta s\Phi c\Phi + c^2\Psi s^2\Theta c^2\Phi \end{pmatrix} = \begin{pmatrix} c^2\Psi c^2\Phi + 2s\Psi c\Psi s\Theta s\Phi c\Phi \ldots \\ \\ + s^2\Psi s^2\Theta s^2\Phi + c^2\Theta s^2\Phi \end{pmatrix}$$

$$c^2\Psi c^2\Theta + s^2\Psi s^2\Phi + c^2\Psi s^2\Theta c^2\Phi = c^2\Psi c^2\Phi + s^2\Psi s^2\Theta s^2\Phi + c^2\Theta s^2\Phi$$

$$R_y^2\dot{\Phi}^2: \quad \begin{pmatrix} c^2\Psi c^2\Theta \cdot 1 + s^2\Psi \cdot 1 \cdot s^2\Phi \ldots \\ + c^2\Psi s^2\Theta c^2\Phi \end{pmatrix} = \begin{pmatrix} c^2\Psi \cdot 1 \cdot c^2\Phi + s^2\Psi s^2\Theta s^2\Phi \ldots \\ + 1 \cdot c^2\Theta s^2\Phi \end{pmatrix} \tag{103}$$

$$\begin{pmatrix} c^2\Psi c^2\Theta s^2\Phi_1 + c^2\Psi c^2\Theta c^2\Phi_2 \ldots \\ + s^2\Psi s^2\Theta s^2\Phi_3 + s^2\Psi c^2\Theta s^2\Phi_4 \\ + c^2\Psi s^2\Theta c^2\Phi_5 \end{pmatrix} = \begin{pmatrix} c^2\Psi s^2\Theta c^2\Phi_5 + c^2\Psi c^2\Theta c^2\Phi_2 \ldots \\ + s^2\Psi s^2\Theta s^2\Phi_3 + s^2\Psi c^2\Theta s^2\Phi_4 \ldots \\ + c^2\Psi c^2\Theta s^2\Phi_1 \end{pmatrix}$$

$$C_{11}^2 + C_{12}^2 = C_{23}^2 + C_{33}^2$$

$$\begin{pmatrix} c^2\Psi c^2\Theta + s^2\Psi c^2\Phi \ldots \\ \\ -2s\Psi c\Psi s\Theta s\Phi c\Phi + c^2\Psi s^2\Theta s^2\Phi \end{pmatrix} = \begin{pmatrix} c^2\Psi s^2\Phi - 2s\Psi c\Psi s\Theta s\Phi c\Phi \ldots \\ \\ + s^2\Psi s^2\Theta c^2\Phi + c^2\Theta c^2\Phi \end{pmatrix}$$

$$R_z^2\dot{\Phi}^2: \quad c^2\Psi c^2\Theta + s^2\Psi c^2\Phi + c^2\Psi s^2\Theta s^2\Phi = c^2\Psi s^2\Phi + s^2\Psi s^2\Theta c^2\Phi + c^2\Theta c^2\Phi$$

$$\begin{pmatrix} c^2\Psi c^2\Theta \cdot 1 + s^2\Psi \cdot 1 \cdot c^2\Phi \ldots \\ \\ + c^2\Psi s^2\Theta s^2\Phi \end{pmatrix} = \begin{pmatrix} c^2\Psi \cdot 1 \cdot s^2\Phi + s^2\Psi s^2\Theta c^2\Phi \ldots \\ \\ + 1 \cdot c^2\Theta c^2\Phi \end{pmatrix}$$

$$\begin{pmatrix} c^2\Psi c^2\Theta s^2\Phi_1 + c^2\Psi c^2\Theta c^2\Phi_2 \ldots \\ \\ + s^2\Psi s^2\Theta c^2\Phi_3 + s^2\Psi c^2\Theta c^2\Phi_4 \ldots \\ \\ + c^2\Psi s^2\Theta s^2\Phi_5 \end{pmatrix} = \begin{pmatrix} c^2\Psi s^2\Theta s^2\Phi_5 + c^2\Psi c^2\Theta s^2\Phi_1 \ldots \\ \\ + s^2\Psi s^2\Theta c^2\Phi_3 + s^2\Psi c^2\Theta c^2\Phi_4 \ldots \\ \\ + c^2\Psi c^2\Theta c^2\Phi_2 \end{pmatrix} \tag{104}$$

The previous calculation demonstrated the validity of Eq. (35).

8. Conclusion

This work presents the method of inertial value transformation for maritime applications. Firstly, an introduction in irregular seas and an overview of ships degrees of freedom are given. In the following section, the traditional Kirchhoff motion equations in the body-fixed reference frame are introduced, which represent a hydrodynamic affected Eulerian gyro tied up a Newtonian body within the body-fixed view. The formal derivation of motion equations of a free-floating body in inertial coordinate system is presented in the main part. It is shown that the transformation of the equations into the body-fixed system leads to the well-known Kirchhoff motion equations. 6DOF simulations for a crew transfer vessel in head seas and beam seas illustrate the comparison of motions in both reference systems. It is mentioned that temporal integration of the motion equations in inertial system leads to unstable and chaotic motions of the ship. Rebuilding the Eulerian gyro at first within the project SOOP [6], Korte et al. [24] by introducing an additional and opposite directed transformation of the rotational accelerations, the present work shows their general applicability and necessity for free-moving bodies (6DOF) within inertial frame. By its consequent use, the motion behaviour of the ship can be stabilised over longer periods. A proof presents the energy conservation of inertia value transformation for a rotating body. Finally, the contribution has shown a failure in common motion calculation practice for vehicles.

The intention of motion equations in inertial reference system is the simulation of mechanically coupled multibody systems in seas. To analyse the interaction effects, the forces and moments of all included bodies have to be defined in the same reference system. This requires a transformation of the motion equations that can be realised with the presented method of inertia value transformation. For a multibody system, a CTV is fixed at the bow to an offshore wind turbine tower and can make ideal rotations; the method including the additional transformation leads to a stable system. The scenario describes the interaction of a fixed and a floating body. Further investigations in the field of multibody dynamics are planned for the future. A scenario of two ships, which are mechanically coupled in tandem and rotate ideally, is developed currently. The challenge in comparison to the first multibody system is the interaction of two floating bodies. Other applications of the method, which are planned in further work of the authors, are simulations of offshore crane processes or 3D simulations of a ROV, which is coupled to a mother ship via umbilical. For the parameterisation of controllers, the question of the real-time application is still in focus of research.

Acknowledgements

The authors wish to thank the Japanese Ministry of Education, Culture, Sports, Science, and Technology for supporting the 21st Century COE Program *Center of Aquaculture Science and Technology for Bluefin Tuna and Other Cultivated Fish*, the Lower Saxony Ministry for Science and Culture for supporting research programme *Safe Offshore Operations* (SOOP) and the German Federal Ministry for Economic Affairs and Energy for supporting the ZIM project *Information System for Near Real-Time Logistics* (IeK).

Author details

Holger Korte[1]*, Sven Stuppe[1], Jan-Hendrik Wesuls[1] and Tsutomu Takagi[2]

*Address all correspondence to: holger.korte@jade-hs.de

1 Department of Maritime and Logistics Studies Elsfleth, Jade University of Applied Sciences Wilhelmshaven/Oldenburg/Elsfleth, Elsfleth, Germany

2 Graduate School of Fisheries Sciences, Hokkaido University, Hakodate, Japan

References

[1] Fossen TI. Handbook of Marine Craft Hydrodynamics and Motion Control. 1st ed. Chichester UK: J. Wiley & Sons; 1994. DOI: ISBN 0 471 94113 1

[2] Korte H, Takagi T. Dynamic motion calculation of a flexible structure using inertia transformation algorithm. In: Workshop on Fishing and Marine Production Technology; 9-10. August; Trondheim. NTNU. 2004. pp. 1-11. DOI: 10.13140/RG.2.1.4006.4724

[3] Kirchhoff G. Über die Bewegung eines Rotationskörpers in einer Flüssigkeit. Crelles Journal. 1869;**71**:237-273

[4] Wendel K. Hydrodynamische Massen und hydrodynamische Massenträgheitsmomente. Jahrbuch der Schiffbautechnischen Gesellschaft. 1950;207-255

[5] Korte H, Takagi T. Transformation von Trägheitsgrößen zur Lösung der Mehrkörperdynamik in der Meerestechnik. In: 101. Jahrbuch der Schiffbautechnischen Gesellschaft. Hamburg: Schiffahrts-Verlag "Hansa"; 2007. pp. 313-325

[6] Korte H. Track control of a towed underwater sensor carrier. In: IFAC Control in Transportation Systems. Braunschweig: IFAC; 2000. pp. 89-94

[7] Paschen M. Contribution to Predict Trajectories of Pelagic Fishing Gears Affected by Ship Manoeuvres (in German) [Dissertation]. Rostock: University of Rostock; 1982

[8] Suzuki K, Takagi T, Shimizu T, Hiraishi T, Yamamoto K, Nashimoto K. Validity and visualization of a numerical model used to determine dynamic configuration of fishing net. Fisheries Science. 2003;**69**(4):1-18. DOI: 10.1046/j.1444-2906.2003.00676.x · Source: OAI

[9] Aamo O, Fossen TI. Finite element modelling of moored vessels. Mathematical and Computer Modelling of Dynamical Systems. 2001;**7**(1):47-75

[10] Abreu P, Morishita H, Pascoal A, Ribero J, Silva H. Marine vehicles with streamers for geotechnical surveys: Modeling, positioning and control. In: 10th IFAC CAMS; September; Trondheim. IFAC-PapersOnLine 49-23; 2016. p. 458-464

[11] Berntsen P, Aamo O, Leira B, Sørensen A. Structural reliability - based control of moored interconnected structures. Control Engineering Practice. 2008;**16**(4):495-504

[12] Korte H. Motion modelling of a manoeuvring ship at anchor in flowing waters. In: Yamane T, editor. Contributions on the Theory of Fishing Gears and Related Marine Systems; 5.-7. November; University of Kinki, Nara (Jp). Nara:2009. pp. 207-216. DOI: ISBN 4-946-421-13-6

[13] Yang K, Wang X, Wu C. Simulation platform for the underwater snakelike robot swimming based on Kane's dynamic model and central pattern generator. Journal Shanghai Jiaotong University. 2014;**19**(3):294-301

[14] Kane TR, Levison DA. Dynamics: Theory and Applications. 1st ed. New York: McGraw-Hill; 1985

[15] Takagi T, Miyata S, Fusejima I, Oshima T, Uehara T, Suzuki K, Nomura Y, Kanechiku M, Torisawa S. Potential of computer simulation for buoy-line type of purse seine fishing. In: Paschen M, editors. Contributions on the Theory of Fishing Gears and Related Marine Systems; 9-12. Ocktober; Rostock. Aachen: Shaker Verlag; 2013. p. 55-62. DOI: ISBN 978-3-8440-2251-3

[16] Korte H, Stuppe S, Wesuls J-H, Takagi T. Application of the inertia value transformation to simulate the transfer scenario of service staff from a crew transfer vessel to offshore wind turbines. In: Paschen M, O'Neil FG, Ivanovic A, editors. Contributions on the theory of fishing gears and related marine systems; 27-29. October; Aberdeen. Aachen: Shaker Verlag; 2015. p. 227-237. DOI: ISBN 978-3-8440-3955-9

[17] Korte H, Wesuls J-H, Stuppe S, Takagi T. Towards an approach to 6 DOF inertial kinematics. In: 10th IFAC CAMS; September; Trondheim. IFAC-PapersOnLine 49; 2016. p. 440-445

[18] Piersson WJ, Moskowitz L. A proposed spectral form for fully developed wind seas based on the similarity of S.A. Kitagorodskii. Journal of Geophysical Research. 1964;**96**(2)

[19] ITTC. Report of the Committee on Waves. Final Report and Recommendations to the 23rd International Towing Tank Conference. 2002;505-551

[20] Hasselmann K, Barnet TP, Bouws E, Carlson H, Cartwright DE, Enke K, Ewing JA, Gienapp H, Hasselmann DE, Krusemann P, Meerburg A, Müller P, Olbers DJ, Richter K, Seil W, Walden H. Measurements of wind-wave growth and swell decay during the joint North Sea wave project (JONSWAP). Ergänzungsheft zur Deutschen Hydrographischen Zeitschrift. 1973;**Reihe A(8)**(12):95

[21] Ginsberg JH. Advanced Engineering Dynamics. 2nd ed. New York: Cambridge University Press; 2004

[22] Woernle C. Dynamic of Multi-Body Systems. Rostock: University of Rostock, Lesson script; 2001

[23] Schiehlen W, Eberhard P. Technische Dynamik–Rechnergestützte Modellierung mechanischer Systeme im Maschinen- und Fahrzeugbau. 4th ed. Springer Vieweg; 2014. pp. 24-28. DOI: ISBN 978-3-658-06184-5

[24] Korte H, Ihmels I, Richter J, Zerhusen B, Hahn A. Offshore training simulations. In: 9th IFAC Manoeuvring and Control of Marine Crafts; 19–21 September 2012; Arenzano (IT). IFAC; 2012. pp. 37-42. DOI: ISSN 1474-6670

[25] Blass A, Gurevich Y. Matrix transformation is complete for the average case. SIAM Journal on Computing. 1995;**24**(1):3-29

Chapter 3

Path Planning in the Local-Level Frame for Small Unmanned Aircraft Systems

Laith R. Sahawneh and Randal W. Beard

Additional information is available at the end of the chapter

http://dx.doi.org/10.5772/intechopen.71895

Abstract

In this chapter, we propose a 3D path planning algorithm for small unmanned aircraft systems (UASs). We develop the path planning logic using a body fixed relative coordinate system which is the unrolled, unpitched body frame. In this relative coordinate system, the ownship is fixed at the center of the coordinate system, and the detected intruder is located at a relative position and moves with a relative velocity with respect to the ownship. This technique eliminates the need to translate the sensor's measurements from local coordinates to global coordinates, which saves computation cost and removes the error introduced by the transformation. We demonstrate and validate this approach using predesigned encounter scenarios in the Matlab/Simulink environment.

Keywords: small unmanned aircraft systems, path planning, collision avoidance, cell decomposition, Dijkstra's search algorithm

1. Introduction

The rapid growth of the unmanned aircraft systems (UASs) industry motivates the increasing demand to integrate UAS into the U.S. national airspace system (NAS). Most of the efforts have focused on integrating medium or larger UAS into the controlled airspace. However, small UASs weighing less than 55 pounds are particularly attractive, and their use is likely to grow more quickly in civil and commercial operations because of their versatility and relatively low initial cost and operating expense.

Currently, UASs face limitations on their access to the NAS because they do not have the ability to sense-and-avoid collisions with other air traffic [1]. Therefore, the Federal Aviation Administration (FAA) has mandated that UASs were capable of an equivalent level of safety to the see-and-avoid (SAA) required for manned aircraft [2, 3]. This sense-and-avoid (SAA) mandate is similar to a pilot's ability to visually scan the surrounding airspace for possible intruding aircraft and take action to avoid a potential collision.

Typically, a complete functional sense-and-avoid system is comprised of sensors and associated trackers, collision detection, and collision avoidance algorithms. In this chapter, our main focus is on collision avoidance and path planning. Collision avoidance is an essential part of path planning that involves the computation of a collision-free path from a start point to a goal point while optimizing an objective function or performance metric. A robust collision avoidance logic considers the kinematic constraints of the host vehicle, the dynamics of the intruder's motion, and the uncertainty in the states estimate of the intruder. The subject of path planning is very broad, and in particular collision, avoidance has been the focus of a significant body of research especially in the field of robotics and autonomous systems. Kuchar and Yang [4] provided a detailed survey of conflict detection and resolution approaches. Albaker and Rahim [5] conducted a thorough survey of collision avoidance methods for UAS. The most common collision avoidance methods are geometric-based guidance methods [6–13], potential field methods [14, 15], sampling-based methods [16, 17], cell decomposition techniques, and graph-search algorithms [18–20].

Geometric approaches to collision avoidance are straightforward and intuitive. They lend themselves to fast analytical solutions based on the kinematics of the aircraft and the geometry of the encounter scenario. The approach utilizes the geometric relationship between the encountering aircraft along with intuitive reasoning [8, 21]. Generally, geometric approach assumes a straight-line projection to determine whether the intruder will penetrate a virtual zone surrounding an ownship. Then, the collision avoidance can be achieved by changing the velocity vector, assuming a constant velocity model. Typically, geometric approaches do not account for uncertainty in intruder flight plans and noisy sensor information.

The potential field method is another widely used approach for collision avoidance in robotics. A typical potential field works by exerting virtual forces on the aircraft, usually an attractive force from the goal and repelling forces from obstacles or nearby air traffic. Generally, the approach is very simple to describe and easy to implement. However, the potential field method has some fundamental issues [22]. One of these issues is that it is a greedy strategy that is subject to local minima. However, heuristic developments to escape the local minima are also proposed in the literature [23]. Another problem is that typical potential field approaches do not account for obstacle dynamics or uncertainly in observation or control. In the context of airborne path planning and collision avoidance, Bortoff presents a method for modeling a UAS path using a series of point masses connected by springs and dampers [24]. This algorithm generates a stealthy path through a set of enemy radar sites of known locations. McLain and Beard present a trajectory planning strategy suitable for coordinated timing for multiple UAS [25]. The paths to the target are modeled using a physical analogy of a chain. Similarly, Argyle et al. present a path planner based on a simulated chain of unit masses placed in a force field [26]. This planner tries to find paths that go through maxima of an underlying bounded differentiable reward function.

Sampling-based methods like probability road maps (PRM) [16] and rapidly exploring random trees (RRTs) [17] have shown considerable success for path planning and obstacle

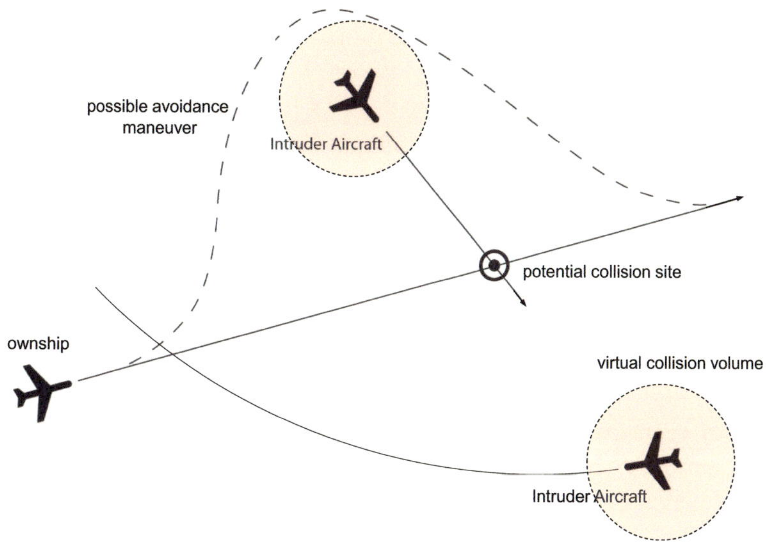

Figure 1. The geometry of an encounter scenario.

avoidance, especially for ground robots. They often require significant computation time for replanning paths, making them unsuitable for reactive avoidance. However, recent extensions to the basic RRT algorithm, such as chance-constrained RRT* [27] and close-loop RRT [28], show promising results for uncertain environments and nontrivial dynamics [28–30]. Cell decomposition is another widely used path planning approach that partitions the free area of the configuration space into cells, which are then connected to generate a graph [20]. Generally, cell decomposition techniques are considered to be global path planners that require a priori knowledge of the environment. A feasible path is found from the start node to the goal node by searching the connectivity graph using search algorithms like A* or Dijkstra's algorithm [18].

The proposed approach in this work will consider encounter scenarios such as the one depicted in **Figure 1**, where the ownship encounters one or more intruders. The primary focus of this work is to develop a collision avoidance framework for unmanned aircraft. The design, however, will be specifically tailored for small UAS. We assume that there exists a sensor(s) and tacking system that provide states estimate of the intruder's track.

2. Local-level path planning

A collision event occurs when two aircraft or more come within the minimum allowed distance between each other. The current manned aviation regulations do not provide an explicit value for the minimum allowed distance. However, it is generally understood that the minimum allowed or safe distance is required to be at least 500 ft. to 0.5 nautical miles (nmi) [21, 31]. For example, the near midair collision (NMAC) is defined as the proximity of less than 500 ft. between two or more aircraft [32]. Similarly and since the potential UAS and intruder aircraft cover a wide range of vehicle sizes, designs, airframes, weights, etc., the choice of a virtual fixed volume boundary around the aircraft is a substitute for the actual dimensions of the intruder.

As shown in **Figure 2**, the choice for this volume is a *hockey-puck* of radius d_s and height h_s that commonly includes a horizontal distance of 500 ft. and a vertical range of 200 ft. [1, 33, 34]. Accordingly, a collision event is defined as an incident that occurs when two aircraft pass less than 500 ft. horizontally and 100 ft. vertically.

In this work, we develop the path planning logic using a body-centered relative coordinate system. In this body-centered coordinate system, the ownship is fixed at the center of the coordinate system, and the intruder is located at a relative position $\mathbf{p}_r$ and moves with a relative velocity $\mathbf{v}_r$ with respect to the ownship [35].

We call this body-centered coordinate frame the local-level frame because the environment is mapped to the unrolled, unpitched local coordinates, where the ownship is stationary at the center. As depicted in **Figure 3**, the origin of the local-level reference is the current position of the ownship. In this configuration, the x-axis points out the nose of the unpitched airframe, the y-axis points points out the right wing of the unrolled airframe, and the z-axis points down forming a right-handed coordinate system. In the following discussion, we assume that the collision volume is centered at the current location of the intruder. A collision occurs when the origin of the local-level frame penetrates the collision volume around the intruder.

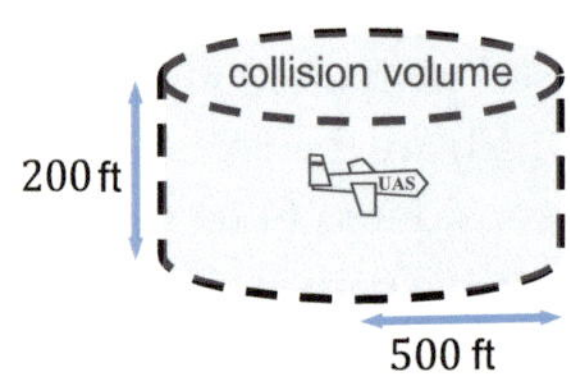

Figure 2. A typical collision volume or protection zone is a virtual fixed volume boundary around the aircraft.

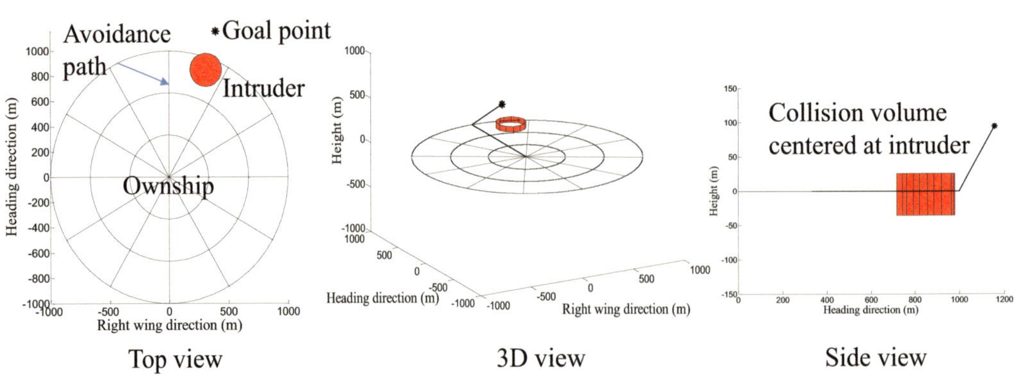

Figure 3. Local-level reference frame.

The detection region is divided into concentric circles that represent maneuvers points at increasing range from the ownship as shown in **Figure 4**, where the radius of the outmost circle can be thought of as the sensor detection range. Let the region in the space covered by the sensor be called the workspace. Then, this workspace is discretized using a cylindrical grid in which the ownship is commanded to move along the edges of the grid. The result is a directed weighted graph, where the edges represent potential maneuvers, and the associated weights represent the maneuver cost and collision risk. The graph can be described by the tuple $\mathcal{G}(\mathcal{N}, \mathcal{E}, \mathcal{C})$, where $\mathcal{N}$ is a finite nonempty set of nodes, and $\mathcal{E}$ is a collection of ordered pairs of distinct nodes from $\mathcal{N}$ such that each pair of nodes in $\mathcal{E}$ is called a directed edge or link, and $\mathcal{C}$ is the cost associated with traversing each edge.

The path is then constructed from a sequence of nonrepeated nodes $(\mathbf{n}_1, \mathbf{n}_2, \cdots, \mathbf{n}_N)$ such that each consecutive pair $(\mathbf{n}_i, \mathbf{n}_{i+1})$ is an edges in $\mathcal{G}$. Let the detection range d_r be the radius of the outermost circle, and r be the radius of the innermost circle so that $d_r = mr$. As shown in **Figure 6**, let $\mathcal{L}_l$, $l = 1, 2, \cdots, m$ be the lth level curve of the concentric circles. Assume that the level curves are equally partitioned by a number of points or nodes such that any node on the lth level curve, $\mathcal{L}_l$ connects to a predefined number of nodes k in the next level, that is, in the forward direction along the heading axis as depicted in **Figure 4**. The nodes on the graph can be thought of as predicted locations of the ownship over a look-ahead time window. Additionally, we assume that only nodes along the forward direction of the heading axis, that is, $x = 0$ connect to nodes in the vertical plane. This assumption allows to command the aircraft to climb or descend by connecting to nodes in the vertical plane as shown in **Figure 4**. Let the first level curve of the innermost circle be discretized into $|\mathcal{L}_1| = k + 2$ nodes including nodes in the vertical plane. Then, using the notation $|\mathcal{A}|$ to denote the cardinality of the discrete set $\mathcal{A}$, the number of nodes in the lth level curve is given by

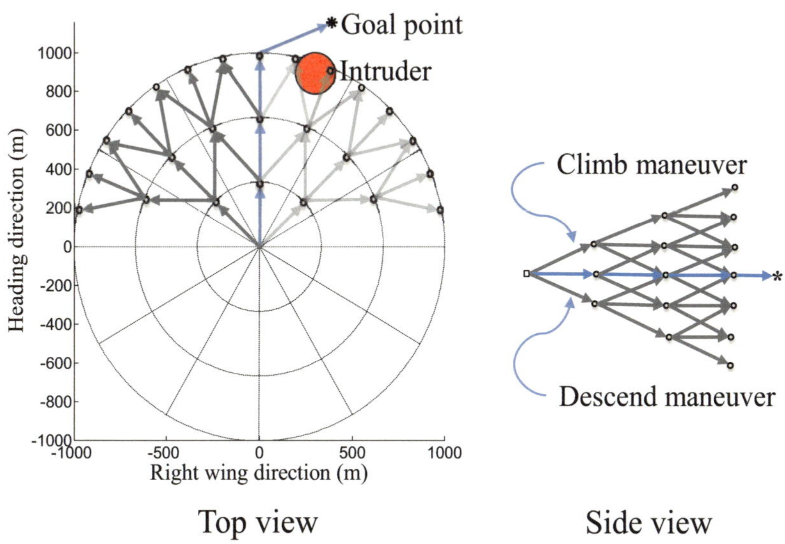

Figure 4. Discretized local-level reference workspace. The three concentric circles represent three maneuvers points.

$$|\mathcal{L}_l| = \begin{cases} k+2 & \text{if } l=1, \\ 2|\mathcal{L}_{l-1}|+2l+1 & \text{if } l=2,3,\cdots,m, \end{cases} \tag{1}$$

where the total number of nodes is $|\mathcal{N}| = \sum_{l=1}^{m} |\mathcal{L}_l|$. For example, assuming that the start node is located at the origin of the reference map and given that $k = 3$, that is, allowing the ownship to fly straight or maneuver right or left. The total number of nodes in the graph including the start and destination node is given by

$$|\mathcal{N}| = \left(\sum_{l=1}^{m+1} 2^l + 2l - 3 \right) + 1. \tag{2}$$

Figure 5 shows an example of a discretized local-level map. In this example, $k = 3$ and $m = 3$, and the total number of nodes in the graph $|\mathcal{N}|$ is 39.

Assuming that the ownship travels between the nodes with constant velocity and climb rate, the location of the ith node at the lth level curve, and $\mathbf{n}_{i,l}$ in the horizontal plane of the graph is given by

$$\mathbf{n}_{i,l} = \left[lr \sin \psi_j^{\mathcal{L}_l}, lr \cos \psi_j^{\mathcal{L}_l}, 0 \right]^T, \tag{3}$$

where $\psi_j^l = \frac{j\psi^d}{2^{(l-1)}}$ and $j = \left\{ -\frac{|\mathcal{L}_l|-1}{2}, -\frac{|\mathcal{L}_l|-1}{2}+1, \cdots, \frac{|\mathcal{L}_l|-1}{2}-1, \frac{|\mathcal{L}_l|-1}{2} \right\}$ and ψ^d is the allowed heading. In the vertical plane, the location of nodes is $\mathbf{n}_{\bar{j},l} = \left[0, 0, \pm \bar{j} l h^d \right]^T$, where $\bar{j} = \{1, 2, \cdots, l\}$ and h^d are the altitude change at each step as shown in **Figure 6**.

For example, if $\psi^d = \pi/4$, $h^d = 50$ m, $r = 500$ m, $k = 3$, and $|\mathcal{L}_1| = 5$, then we have $j = \{-1, 0, 1\}$, $\bar{j} = \{-1, 1\}$, $\psi_j^1 = \{-\pi/4, 0, -\pi/4\}$, and the locations of nodes at $\mathcal{L}_1$ in the

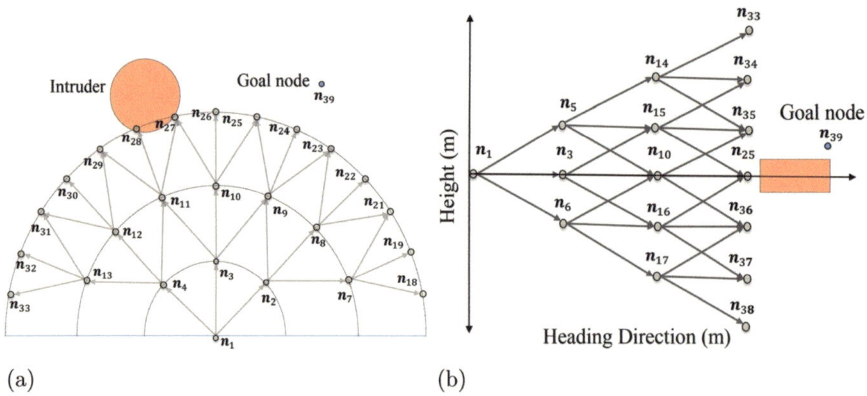

Figure 5. Example of discretized local-level map. (a) Top view: location and index of nodes and (b) side view: location and index of nodes.

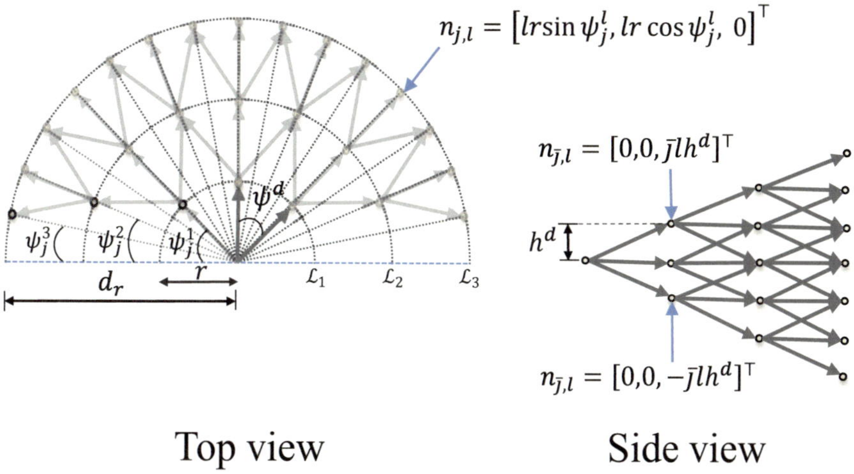

Figure 6. Nodes location in the local-level reference frame.

horizontal plane are $\left\{(-500\sin\pi/4, 500\cos\pi/4, 0)^T,\ (0, 500, 0)^T,\ (500\sin\pi/4, 500\cos\pi/4, 0)^T\right\}$, and in the vertical plane are $\left\{(0,0,50)^T, (0,0,-50)^T\right\}$.

The main priority of the ownship where it is under distress is to maneuver to avoid predicted collisions. This is an important note to consider when assigning a cost of each edge in the resulting graph. The cost associated with traveling along an edge is a function of the edge length and the collision risk. The cost associated with the length of the edge $e_{i,i+1}$ that connects between the consecutive pair nodes $(\mathbf{n}_i, \mathbf{n}_{i+1})$ is simply the Euclidean distance between the nodes $\mathbf{n}_i$ and $\mathbf{n}_{i+1}$ expressed as

$$C_L(e_{i,i+1}) = \|\mathbf{n}_{i+1} - \mathbf{n}_i\|. \tag{4}$$

The collision cost for traveling along an edge is determined if at any future time instant, the future position of the ownship along that edge is inside the collision volume of the predicted location of an intruder. An exact collision cost computation would involve the integration of collision risk along each edge over the look-ahead time window $\tau \in [t, t + mT]$.

A simpler approach involves calculating the collision risk cost at several locations along each edge, taking into account the projected locations of the intruder over the time horizon τ. Assuming a constant velocity model, a linear extrapolation of the current position and velocity of the detected intruders are computed at evenly spaced time instants over the look-ahead time window. The look-ahead time interval is then divided into several discrete time instants. At each discrete time instant, all candidate locations of the ownship along each edge are checked to determine whether it is or will be colliding with the propagated locations of the intruders. For the simulation results presented in this chapter, the collision risk cost is calculated at three points along each edge in $\mathcal{G}$. If v_o is the speed of the ownship, then the distance along an edge is given by v_oT, where $T = r/v_o$. The three points are computed as

$$\mathbf{p}_1 = \mathbf{n}_i + \mathbf{v}_o T_s \frac{\mathbf{n}_{i+1} - \mathbf{n}_i}{\|\mathbf{n}_{i+1} - \mathbf{n}_i\|}, \tag{5}$$

$$\mathbf{p}_2 = \mathbf{p}_1 + v_o T_s \frac{\mathbf{n}_{i+1} - \mathbf{n}_i}{\|\mathbf{n}_{i+1} - \mathbf{n}_i\|}, \tag{6}$$

$$\mathbf{p}_3 = \mathbf{p}_2 + v_o T_s \frac{\mathbf{n}_{i+1} - \mathbf{n}_i}{\|\mathbf{n}_{i+1} - \mathbf{n}_i\|}, \tag{7}$$

where $T_s = T/3$. Let the relative horizontal and vertical position of the intruder with respect to the ownship at the current time t be $\mathbf{p}_r(t)$ and $p_{r_z}(t)$, respectively. Define the collision volume as

$$\mathcal{C}(\mathbf{p}_r(t)) = \left\{ d \in \mathbb{R}^2 : \|\mathbf{p}_r(t)\| - d \leq d_s \text{ and } h \in \mathbb{R} : \left| p_{r_z} - h \right| \leq h_s/2 \right\}. \tag{8}$$

The predicted locations of each detected intruder over time horizon T at three discrete time samples T_s are

$$\mathbf{p}_{r3D}(t + (1 + 3(l-1))T_s) = \mathbf{p}_{r3D}(t) + \mathbf{v}_{r3D}(t)(1 + 3(l-1))T_s, \tag{9}$$

$$\mathbf{p}_{r3D}(t + (2 + 3(l-1))T_s) = \mathbf{p}_{r3D}(t) + \mathbf{v}_{r3D}(t)(2 + 3(l-1))T_s, \tag{10}$$

$$\mathbf{p}_{r3D}(t + (3 + 3(l-1))T_s) = \mathbf{p}_{r3D}(t) + \mathbf{v}_{r3D}(t)(3 + 3(l-1))T_s, \tag{11}$$

where $\mathbf{p}_{r3D}(t) = \left[\mathbf{p}_r(t), p_{r_z}(t)\right]^T \in \mathbb{R}^3$ and $\mathbf{v}_{r3D}(t) = [v_r(t), \mathbf{v}_{r_z}(t)]^T \in \mathbb{R}^3$ be the 3D relative position and velocity of the intruder with respect to the ownship in the relative coordinate system, where $\mathbf{v}_r(t)$ and $v_{r_z}(t)$ are the relative horizontal velocity and vertical speed at the current time t.

In Eqs. (9)–(11), if $e_{i,i+1}$ is the current edge being evaluated, then the node $\mathbf{n}_{i+1}$ determines the value of l. In other words, if $\mathbf{n}_{i+1} \in \mathcal{L}_1$, then $l = 1$. For example, if we are to compute the three points along the edge $e_{1,2}$ in Eqs (5)–(7), then $\mathbf{n}_2 \in \mathcal{L}_1$ and $l = 1$. Using the definition of the binary cost function, the collision risk cost associated with the $e_{i,i+1}$ edge with respect to each detected intruder is given by the expression

$$C_{col}(\text{int}, e_{i,i+1}) = \begin{cases} \infty & \text{if any of } \mathbf{p}_1, \mathbf{p}_2, \text{ or } \mathbf{p}_3 \in \mathcal{C}\left(\mathbf{p}_{r3D}(t + (\ell + 3(l-1))T_s)\right), \\ 0 & \text{otherwise,} \end{cases} \tag{12}$$

where $\ell = \{1, 2, 3\}$. In Eq. (12), the ∞ or the maximum allowable cost is assigned to any edge that leads to a collision, basically eliminating that edge and the path passing through it. The total collision risk associated with the ith edge is given by

$$C_{col}(e_{i,i+1}) = \sum_{\text{int}=1}^{M} C_{col}(\text{int}, e_{i,i+1}), \tag{13}$$

where M is the number of detected intruders.

A visual illustration of the collision risk computation is shown in **Figure 7**. The propagated collision volume of a detected intruder and the candidate locations of the ownship over the first-time interval $[t+T_s, t+3T_s]$ both in the horizontal and vertical plane is depicted in **Figure 7a** and **b**. Clearly, there is no intersection between these candidate points the ownship may occupy and the propagated locations of the collision volume over the same interval. Then, according to Eq. (13), the cost assigned to these edges is zero. Next, all candidate locations of the ownship along each edge over the second time interval $[t+4T_s, t+6T_s]$ are investigated. As shown in **Figure 7c**, edges $e_{2,7}$, $e_{2,8}$, and $e_{2,9}$ intersect with the predicted intruder location at time $t+4T_S$ and $t+5T_S$, respectively. Similarly, edges $e_{3,15}$ and $e_{3,16}$ in the horizontal plane intersect with the predicted intruder location at time $t+4T_S$ as shown in **Figure 7d**. Accordingly, the maximum allowable costs will be assigned to these edges, which eliminate these edges and the path passing through them. All the candidate locations of the ownship over the time interval $[t+7T_s, t+9T_s]$ do not intersect with the predicted locations of the intruder as shown in **Figure 7e** and **f**. Therefore, by the time, the ownship will reach these edges the detected intruder will be leaving the map, and consequently, a cost of zero is assigned to edges belonging to the third level curve $\mathcal{L}_3$.

To provide an increased level of robustness, an additional threat cost is added to penalize edges close to the propagated locations of the intruder even if they are not within the collision volume. At each discrete time instant, we compute the distances from the candidate locations of the ownship to all the propagated locations of the intruders at that time instant. The cost of collision threat along each edge is then given by the sum of the reciprocal of the associated distances to each intruder

$$C_{th}(\text{int}, e_{i,i+1}) = \frac{1}{d_1} + \frac{1}{d_2} + \frac{1}{d_3}. \tag{14}$$

where d_1, d_2, and d_3 are given by

$$d_1 = \left\| p_1 - \mathbf{p}_{r_{3D}}(t + (1 + 3(l-1))T_s) \right\|,$$

$$d_2 = \left\| p_2 - \mathbf{p}_{r_{3D}}(t + (2 + 3(l-1))T_s) \right\|,$$

$$d_3 = \left\| p_3 - \mathbf{p}_{r_{3D}}(t + (3 + 3(l-1))T_s) \right\|,$$

and the total collision risk cost associated with the ith edge with regard to all intruders is given by

$$C_{th}(e_{i,i+1}) = \sum_{\text{int}=1}^{M} C_{th}(\text{int}, e_{i,i+1}). \tag{15}$$

For example, the edges $e_{1,2}$, $e_{1,3}$, $e_{1,4}$, $e_{1,5}$, and $e_{1,6}$ shown in **Figure 7a** are not intersecting with the propagated collision volume locations over the first-time interval, yet they will be penalized based on their distances to the predicated locations of the intruder according to Eq. (15). Note that edge $e_{1,2}$ will have greater cost as it is the closest to the intruder among other candidate edges.

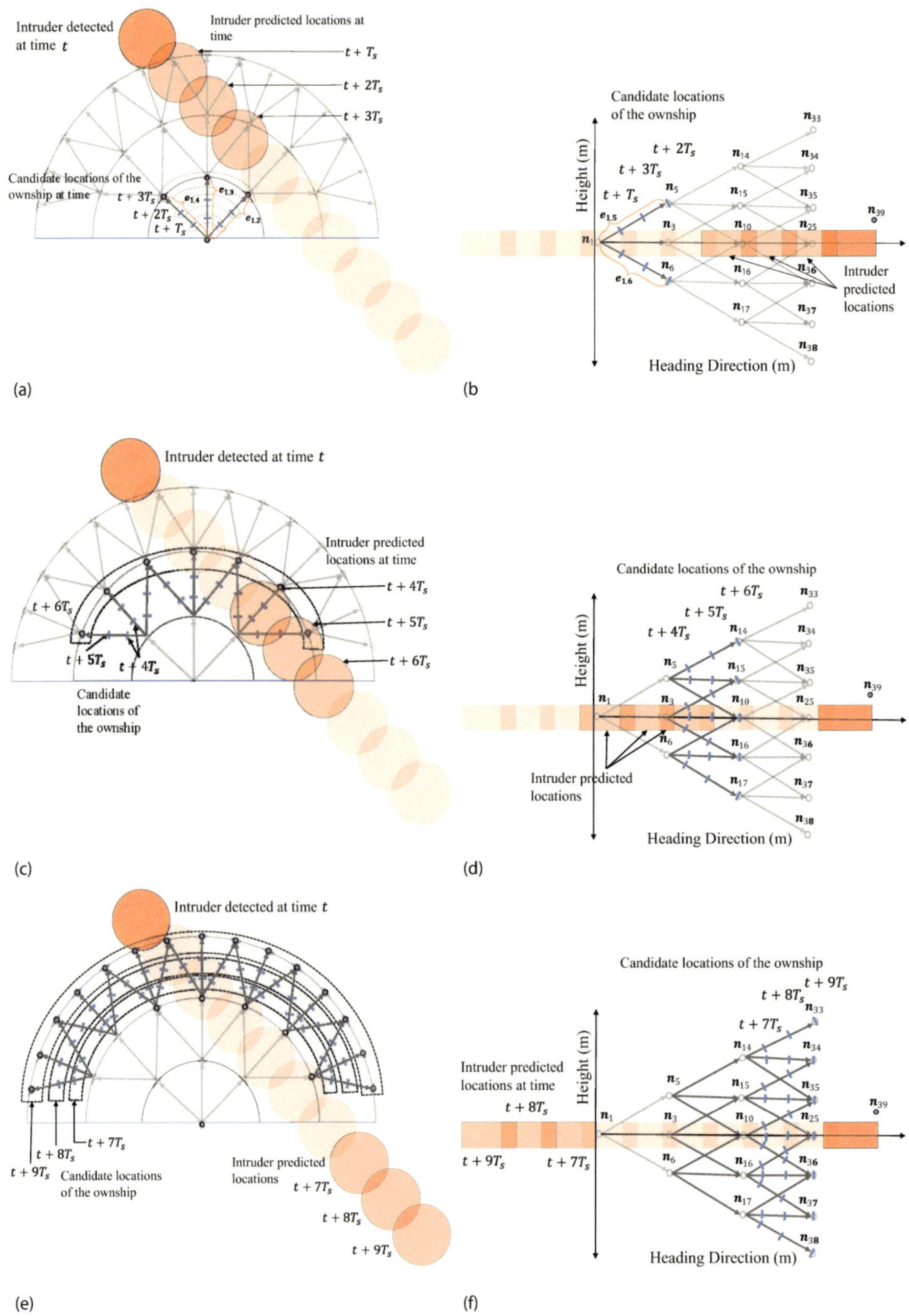

Figure 7. Example illustrating the steps to compute the collision risk. In this example, we have $k = 3$ and $m = 3$. (a) Top view: predicted locations of intruder (less transparent circles), and candidate locations of ownship; (b) side view: predicted locations of intruder (less transparent rectangles), and candidate locations of ownship; (c) predicted locations of intruder and candidate locations of ownship over time window ($t + 4Ts$, $t + 6Ts$); (d) time window ($t + 4Ts$, $t + 6Ts$); (e) time window ($t + 7Ts$, $t + 9Ts$); (f) time window ($t + 7Ts$, $t + 9Ts$).

Another objective of a path planning algorithm is to minimize the deviation from the original path, that is, the path the ownship was following before it detected a collision. Generally, the path is defined as an ordered sequence of waypoints $\mathcal{W} = \mathbf{w}_1, \mathbf{w}_2, \cdots . \mathbf{w}_f$, where $\mathbf{w}_i = (w_{n,i}, w_{e,i}, w_{d,i})^T \in \mathbb{R}^3$ is the north-east-down location of the ith waypoint in a globally known NED reference frame. The transformation from the global frame to the local-level frame is given by

$$\mathbf{w}_i^b = \mathcal{R}_g^b(\psi_o)\mathbf{w}_i, \tag{16}$$

where

$$\mathbf{R}_g^b(\psi_o) = \begin{pmatrix} \cos\psi_o & \sin\psi_o & 0 \\ -\sin\psi_o & \cos\psi_o & 0 \\ 0 & 0 & 1 \end{pmatrix}$$

where ψ_o is the heading angle of the ownship. Let $\mathbf{w}_s$ be the location waypoint of the ownship at the current time instant t and $\mathbf{w}_f \in \mathcal{W}$ be the next waypoint the ownship is required to follow. Assuming a straight-line segment between the waypoints $\mathbf{w}_s$ and $\mathbf{w}_f$, then any point on this segment can be described as $\mathcal{L}(\varrho) = (1-\varrho)\mathbf{w}_s + \boldsymbol{\rho} w_f$ where $\varrho \in [0, 1]$, and the minimum distance between an arbitrary node $\mathbf{n}_i$ in $\mathcal{G}$ can be expressed by [36]

$$D(\mathbf{w}_s, \mathbf{w}_f, n_i) \triangleq \begin{cases} D(\varrho^*), & \text{if } \varrho^* \in [0, 1], \\ \|\mathbf{n}_i - \mathbf{w}_s\|, & \text{if } \varrho^* < 0, \\ \|\mathbf{n}_i - \mathbf{w}_f\|, & \text{if } \varrho^* > 1, \end{cases} \tag{17}$$

where

$$D(\varrho^*) = \sqrt{\|\mathbf{n}_i - \mathbf{w}_s\|^2 - \frac{\left((\mathbf{w}_s - \mathbf{n}_i)^T(\mathbf{w}_s - \mathbf{w}_f)\right)^2}{\|\mathbf{w}_s - \mathbf{w}_f\|^2}},$$

and

$$\varrho^* = \frac{(\mathbf{w}_s - \mathbf{n}_i)^T(\mathbf{w}_s - \mathbf{w}_f)}{\|\mathbf{w}_s - \mathbf{w}_f\|^2}.$$

Then, the cost that penalizes the deviation of an edge in $\mathcal{G}$ from the nominal path is given by

$$C_{dev}(e_{i,i+1}) = D(\mathbf{w}_s, \mathbf{w}_f, n_i). \tag{18}$$

If small UASs are to be integrated seamlessly alongside manned aircraft, they may require to follow right-of-way rules. Therefore, an additional cost can be also added to penalize edges that violate right-of-way rules. In addition, this cost can be used to favor edges in the horizontal plane over those in the vertical plane. Since the positive direction of the y-axis in the local-level frame is the right-wing direction, it is convenient to define right and left maneuvers as the positive and the negative directions along the right-wing axis, respectively. Let $\vec{\mathbf{e}}_i \triangleq \mathbf{n}_{i+1} - \mathbf{n}_i$ be the direction vector associated with the edge $e_{i,i+1}$ in $\mathcal{G}$, where $\mathbf{n}_i \triangleq (x_i, y_i, z_i)^T \in \mathbb{R}^3$ is the location of ith node in the local-level reference frame. Let the direction vector $\vec{\mathbf{e}}_i$ be expressed as $\vec{\mathbf{e}}_i = \left(e_{i_x}, e_{i_y} e_{i_z}\right)^T \in \mathbb{R}^3$. We define $\mathbf{E} \triangleq (e_{i_x}, L, R, e_{i_z})^T \in \mathbb{R}^4$, where e_{i_x} and e_{i_z} are the x and the z components of $\vec{\mathbf{e}}_i$. The y-component of $\vec{\mathbf{e}}_i$ is decomposed into two components: left L and right R, that are defined by

$$L, R \triangleq \begin{cases} L = e_{i_y}, R = 0 & \text{if } e_{i_y} \le 0, \\ L = 0, R = e_{i_y} & \text{if } e_{i_y} > 0. \end{cases} \tag{19}$$

If we define the maneuvering design matrix to be $\mathbf{J} = \text{diag}([0, c_L, c_R, c_z])$, then the maneuvering cost associated with each edge is given by

$$C_m(e_{i,i+1}) = \sqrt{\mathbf{E}^T \mathbf{J} \mathbf{E}}, \tag{20}$$

The costs c_L and c_R allow the designer to place more or less cost on the left or right edges. Similarly, c_z allows the designer to penalize vertical maneuvers. Multiple values of these cost parameters may be saved in a look-up table, and the collision avoidance algorithm choses the appropriate value based on the geometry of the encounter.

The overall cost for traveling along an edge comes from the weighted sum of all costs given as [35]

$$C(e_{i,i+1}) = C_L(e_{i,i+1}) + C_{col}(e_{i,i+1}) + k_1 C_{th}(e_{i,i+1}) + k_2 C_{dev}(e_{i,i+1}) + k_3 C_m(e_{i,i+1}), \tag{21}$$

where k_1, k_2, and k_3 are positive design parameters that allow the designer to place weight on collision risk or deviation from path or maneuvering preferences depending on the encounter scenario. Once the cost is assigned to each edge in $\mathcal{G}$, then a graph-search method can be used to find the least cost path from a predefined start point to the destination point. In this work, we have used Dijkstra's algorithm.

Dijkstra's algorithm solves the problem of shortest path in a directed graph in polynomial time given that there are not any negative weights assigned to the edges. The main idea in Dijkstra's algorithm is to generate the nodes in the order of increasing value of the cost to reach them. It starts by assigning some initial values for the distances from the start node and to every other node in the graph. It operates in steps, where at each step, the algorithm updates the cost values of the edges. At each step, the least cost from one node to another node is determined and saved such that all nodes that can be reached from the start node are labeled with cost from the start node. The algorithm stops either when the node set is empty or when every node is examined exactly once. A naive implementation of Dijkstra's algorithm runs in a total time complexity of $O\left(|\mathcal{N}|^2\right)$. However, with suitable data structure implementation, the overall time complexity can be reduced to $O(|\mathcal{E}| + |\mathcal{N}| \log_2 |\mathcal{N}|)$ [23, 35].

The local-level path planning algorithm generates an ordered sequence of waypoints $\mathcal{W}_c = \mathbf{w}_{c1}, \mathbf{w}_{c2}, \cdots, \mathbf{w}_{ci}$. Then, these waypoints are transformed from the relative reference frame to the global coordinate frame and added to the original waypoints path $\mathcal{W}$. When the ownship is avoiding a potential collision, the avoidance waypoints overwrite some or all of the original waypoints. Next, a path manager is required to follow the waypoints path and a smoother to make the generated path flyable by the ownship. One possible approach to follow waypoints path is to transit when the ownship enters a ball around the waypoint $\mathbf{W}_i$ or a better strategy is to use the half-plane switching criteria that is not sensitive to tracking error [36]. Flyable or smoothed transition between the waypoints can be achieved by implementing the fillet maneuver or using Dubins paths. For further analysis on these topics, we refer the interested reader to Ref. [36].

3. Simulation results

To demonstrate the performance of the proposed path planning algorithm, we simulate an encounter scenario similar to the planner geometry shown in **Figure 8**. The aircraft dynamics are simulated using a simplified model that captures the flight characteristics of an autopilot-controlled UAS. The kinematic guidance model that we considered assumes that the autopilot controls airspeed, v_a, altitude, h, and heading angle, ψ. Under zero-wind conditions, the corresponding equations of motion are given by

$$\dot{p}_n = v_a \cos\psi, \tag{22}$$

$$\dot{p}_e = v_a \sin\psi, \tag{23}$$

$$\dot{\psi} = \frac{g}{v_a}\phi, \tag{24}$$

$$\dot{v}_a = b_v\left(v_a^c - v_a\right) \tag{25}$$

$$\dot{\phi} = b_\phi\left(\phi^c - \phi\right) \tag{26}$$

$$\ddot{h} = b_{\dot{h}}\left(\dot{h}^c - \dot{h}\right) + b_h\left(h^c - h\right), \tag{27}$$

where p_n, p_e are the north-east position of the aircraft. The inputs are the commanded altitude, h^c, the commanded airspeed, v_a^c, and the commanded roll angel, ϕ^c. The parameters b_v, b_ϕ, b_h, and $b\dot{h}$ are positive constants that depend on the implementation of the autopilot and the state estimation scheme. For further analysis on the kinematic and dynamic guidance models for UAS, we refer the interested reader to [36]. In the following simulation, the ownship starts at $(0, 0, -200)^T$ in the NED coordinate system, with an initial heading of 0 deg. measured from north and follows a straight-line path at a constant speed of 22 m/s to reach the next waypoint located at $(1500, 0, -200)^T$. The encounter geometry includes three intruders flying at different altitudes: the first is approaching head-on, the second is converging from the right, and the third is overtaking from the left. We chose the intruders's speed similar to the known cruise speed of ScanEagle UAS, Cessna SkyHawk 172R, and Raven RQ-11B UAS. The speed of the

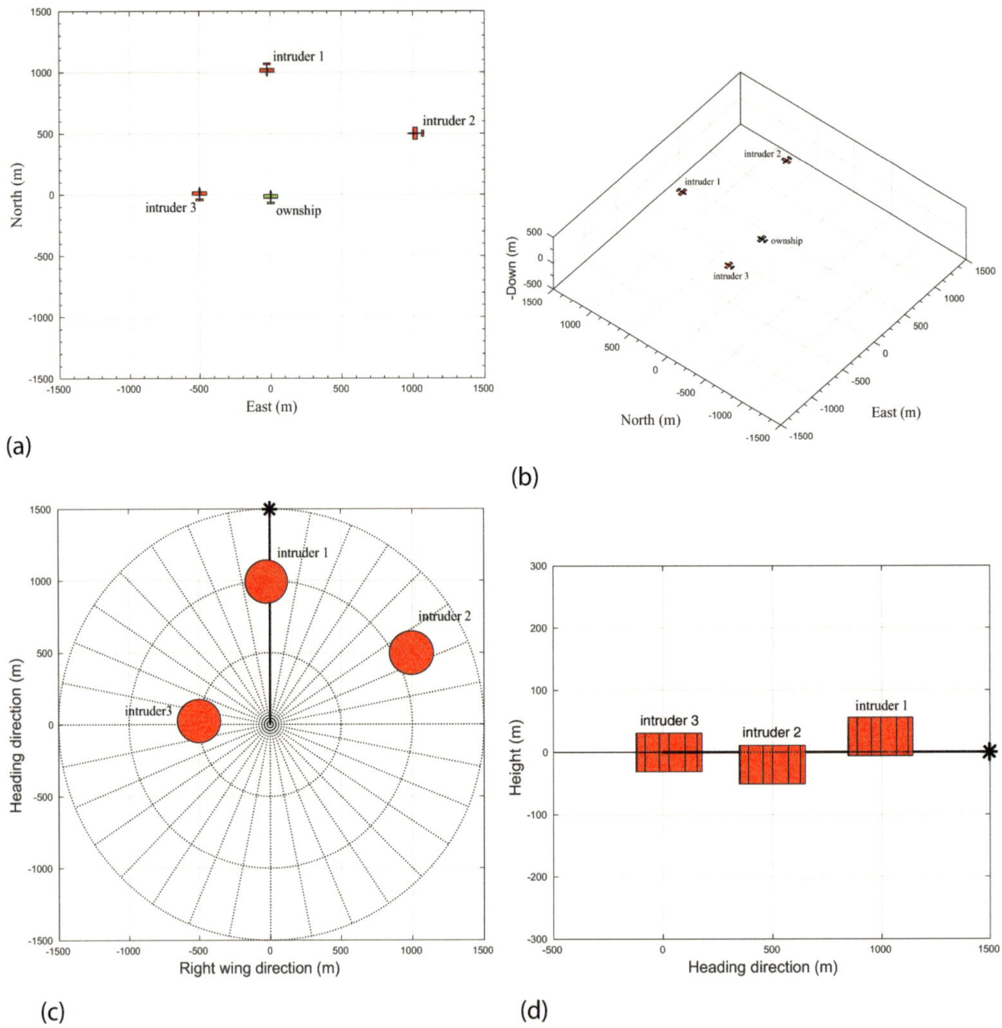

Figure 8. Encounter geometry for the ownship and three intruders at t = 0.1 s. (a) Overhead view of initial locations of aircraft; (b) 3D view of initial locations of aircraft; (c) overhead view of reference frame; (d) side view of relative reference frame.

intruders is 41, 65, and 22 m/s, respectively. In addition, the intruders are assumed to fly at a constant speed the entire simulation period. As shown in **Figure 8**, the initial locations of intruders in the NED coordinate system are $(-25, 1000, -225)^T$, $(500, 1000, -180)^T$, and $(25, -500, -200)^T$, respectively, with initial heading of 180, −90, and 0°, respectively.

In the following simulation, our choice of the collision volume is a cylinder of radius d_s = 152.4 m (500 ft) and height h_s = 61 m (200 ft) centered on each of the intruders. A collision incident occurs when the horizontal relative range and altitude to the ownship are simultaneously below horizontal and vertical minimum safe distances d_s and $h_s/2$. We assume that there exists a sensor and tracking system that provides the states of the detected intruders.

However, not every aircraft that is observed by the sensing system presents a collision threat. Therefore, we implemented a geometric-based collision detection algorithm to determine whether an approaching intruder aircraft is on a collision course. The collision detection approach is beyond the scope of this work, and we refer the interested reader to [37].

At the beginning of simulation, the predicted relative range and altitude at the closest point of approach (CPA) are shown in **Table 1**. Imminent collisions are expected to occur with the first and second intruders as their relative range and altitude with respect to the ownship are below the defined horizontal and vertical safe distances. The time remaining to the closest point of approach t_{CPA} with respect to the first and second intruders is 15.77 and 16.56 s, respectively. The scenario requires that the ownship plans and executes an avoidance maneuver well before the t_{CPA}. This example demonstrates the need for an efficient and computationally fast avoidance planning algorithm. **Table 2** shows the total time required to run the avoidance algorithm, and the maximum and average time required to execute one cycle. The results show that the proposed algorithm takes a significantly reduced time in computation with an average and maximum time to execute one cycle of the code of 20 ms and 0.1326 s, respectively, and a total time of 0.3703 s to resolve the collision conflict.

Figure 9 shows the planned avoidance path by the ownship. These results show that the avoidance path safely maneuvers the ownship without any collisions with the intruders. In addition, the ownship should plan an avoidance maneuver that does not lead to a collision with intruders that were not on a collision course initially such as the case with the third intruder. Initially, the third intruder and the ownship are flying on near parallel courses. The relative range and altitude at CPA with respect to the third intruder are 437.14 and 4361.07 m, respectively, and the time remaining to the CPA is 1982.25 s. Obviously, both aircrafts are not on a collision course. However, the third intruder is descending and changing its heading toward the ownship. The path planner, however, accounts for predicted locations of the detected intruder over the look-ahead time window, allowing the ownship to maintain a safe distance from the third intruder. This example demonstrates that the proposed path planner can handle unanticipated maneuvering intruders. Once collisions are resolved the path planner returns the ownship to the next waypoint of its initial path.

Intruder	$\lVert \mathbf{p}_r(t_{\mathrm{CPA}}) \rVert$ (m)	$\lvert p_{r_z}(t_{\mathrm{CPA}}) \rvert$ (m)	t_{CPA} (s)
1	24.90	25	15.77
2	141.33	20	16.56
3	437.14	4361.07	1982.25

Table 1. Relative range and altitude, and the time remaining to the closest point of approach.

Total run time (s)	Max. run time (one cycle) (s)	Average run time (one cycle) (s)
0.3703	0.1326	0.0206

Table 2. Collision avoidance algorithm run time.

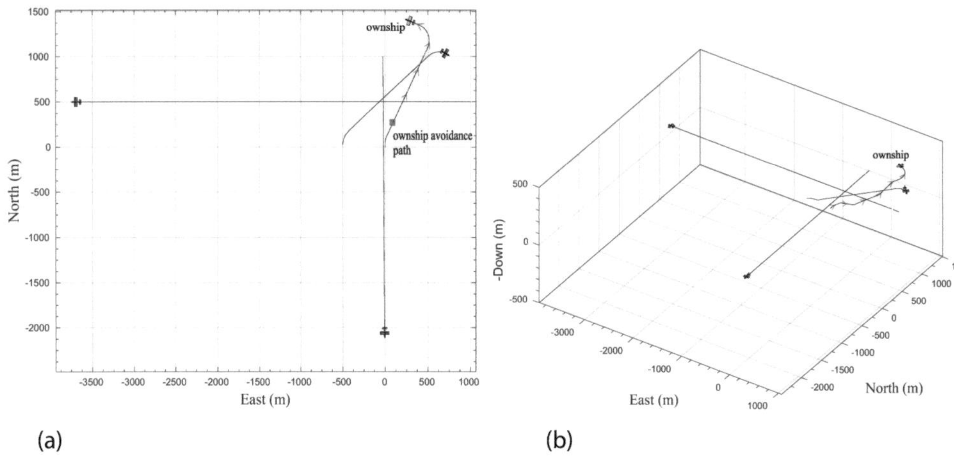

Figure 9. Avoidance path followed by the ownship and path tracks of the intruders at t = 75 s. (a) Overhead view of avoidance path and (b) 3D view of of avoidance path.

The relative range between the ownship and the intruders is shown in **Figure 10**. The results show that no collisions have occurred, and that the ownship successfully planned an avoidance maneuver. The avoidance planner ensures that when the relative horizontal range is less than d_s, the relative altitude is greater than $h_s/2$. For example, as shown in **Figure 10b**, the relative range to the first intruder over time interval [16.2, 18] s is below d_s. However, over the same time interval, the relative altitude is above $h_s/2$.

Another important aspect to evaluate the performance of the proposed algorithm is its ability to reduce the length of the avoidance path while avoiding the intruders. This is important because it reduces the amount of deviation from the original path and ultimately the flight time, which is of critical importance for the small UAS with limited power resources. **Table 3** shows that the length of the avoidance paths is fairly acceptable compared to the initial path length.

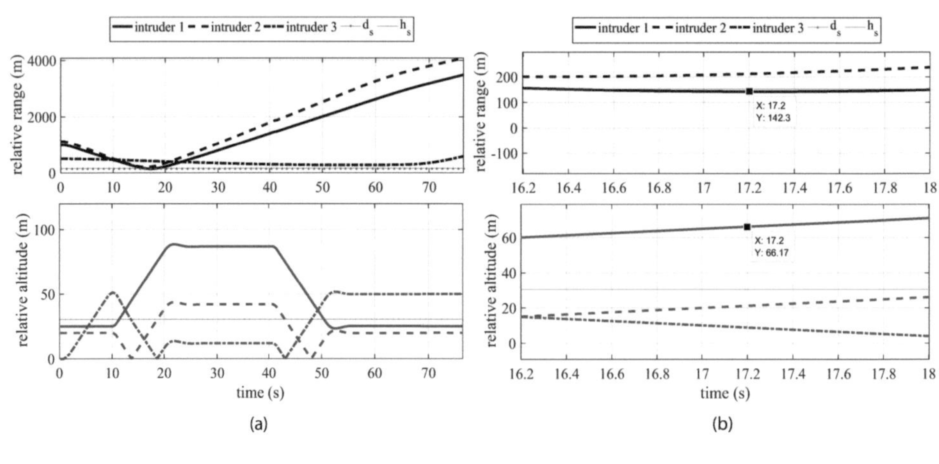

Figure 10. Relative horizontal range and altitude between the ownship and intruders. (a) Horizontal range and relative altitude to intruders and (b) a close up view of Figure 10a.

Scenario number	Initial path length (m)	Avoidance path length (m)
1	1500	1955

Table 3. Length of the avoidance path.

4. Conclusions

In this chapter, we have presented a path planning approach suitable for small UAS. We have developed a collision avoidance logic using an ownship-centered coordinate system. The technique builds a maneuver graph in the local-level frame and use Dijkstra's algorithm to find the path with the least cost.

A key feature of the proposed approach is that the future motion of the ownship is constrained to follow nodes on the map that are spaced by a constant time. Since the path is represented using waypoints that are at fixed time instants, it is easy to determine roughly where the ownship will be at any given time. This timing information is used when assigning cost to edges to better plan paths and prevent collisions.

An advantage of this approach is that collision avoidance is inherently a local phenomenon and can be more naturally represented in local coordinates than global coordinates. In addition, the algorithm accounts for multiple intruders and unanticipated maneuvering in various encounter scenarios. The proposed algorithm runs in near real time in Matlab. Considering the small runtime shown in the simulation results, we expect that implementing these algorithms in a compiled language, such as C or C++, will show that real-time execution is feasible using hardware. That makes the proposed approach a tractable solution in particular for small UAS.

An important step forward to move toward a deployable UAS is to test and evaluate the performance of the close-loop of sensor, tracker, collision detection, path planning, and collision avoidance. Practically, the deployment of any UAS requires a lengthy and comprehensive development process followed by a rigorous certification process and further analysis including using higher fidelity models of encounter airspace, representative number of simulations, and hardware-in-the-loop simulation. Unlike existing collision manned aviation collision detection and avoidance systems, an encounter model cannot be constructed solely from observed data, as UASs are not yet integrated in the airspace system and good data do not exist. An interesting research problem would be to design encounter models similar to those developed to support the evaluation and certification of manned aviation traffic alert and collision avoidance system (TCAS).

Acknowledgements

This research was supported by the Center for Unmanned Aircraft Systems (C-UAS), a National Science Foundation-sponsored industry/university cooperative research center (I/UCRC) under NSF Award No. IIP-1161036 along with significant contributions from C-UAS industry members.

Author details

Laith R. Sahawneh[1]* and Randal W. Beard[2]

*Address all correspondence to: lsahawneh@ufl.edu

1 Department of Mechanical and Aerospace Engineering, University of Florida, Florida, USA

2 Department of Electrical and Computer Engineering, Brigham Young University, Utah, USA

References

[1] George S. FAA Workshop on Sense and Avoid (SAA) for Unmanned Aircraft Systems (UAS). 2009

[2] Hottman SB, Hansen KR, Berry M. Literature review on detect, sense, and avoid technology for Unmanned Aircraft Systems. In: Technical Report. 2009

[3] Federal Aviation Administration. Subchapter F-Air Traffic and General Operating Rules. 2015

[4] Kuchar JK, Yang LC. A review of conflict detection and resolution modeling methods. IEEE Transactions on Intelligent Transportation Systems. Dec. 2000;**1**(4):179-189

[5] Albaker BM, Rahim NA. A survey of collision avoidance approaches for unmanned aerial vehicles. International Conference for Technical Postgraduates (TECHPOS). 2009:1-7

[6] Hyunjin YK. Reactive collision avoidance of unmanned aerial vehicles using a single vision sensor. AIAA Guidance, Control, and Dynamics. 2013;**36**(4):1234-1240

[7] Rajnikant S, Saunders JB, Randal Beard W. Reactive path planning for micro air vehicles using bearing-only measurements. International Robotic Systems. 2012;**65**(1–4):409-416

[8] White BA, Antonios HS. UAV obstacle avoidance using differential geometry concepts. In: 18th IFAC World Congress; Milano, Italy; 2011. Vol. 3. pp. 6325-6330

[9] Saunders J, Beard RW. Vision-based reactive multiple obstacle avoidance for micro air vehicles. In: IEEE American Control Conference ACC'09; St. Louis, MO, June 10–12; 2009, pp. 5253-5258

[10] George J, Ghose D. A reactive inverse PN algorithm for collision avoidance among multiple unmanned aerial vehicles. In: American Control Conference; St. Louis, MO; June 10–12; IEEE. pp. 3890-3895

[11] Bilimoria KD. A geometric optimization approach to aircraft conflict resolution. In: Proceedings of the AIAA Guidance, Navigation and Control Conference and Exhibit; 2010

[12] Fiorini P, Shiller Z. Motion planning in dynamic environments using velocity obstacles. The International Journal of Robotics Research. 1998;**17**(7):760-772

[13] Chakravarthy A, Ghose D. Obstical avoidance in a dynamic environment: A collision cone approach. IEEE Transactions on System, Man and Cybernitics, Part A: Systems and Humans. 1998;**28**(5):562-572

[14] Lam TM, Mulder M, Van Paassen M, Mulder JA, Van Der FC. Force-stiffness feedback in uninhabited aerial vehicle teleoperation with time delay. AIAA Guidance, Control, and Dynamics. 2009;**32**(3):821-835

[15] Sahawneh LR, Beard RW, Avadhanamz S, He B. Chain-based collision avoidance for UAS sense and avoid systems. In: AIAA Guidance, Navigation, and Control (GNC) Conference; Boston, MA; 2013

[16] Kavraki LE, Svestka P, Latombe JC, Overmars MH. Probabilistic roadmaps for path planning in high-dimensional configuration spaces. IEEE Transactions on Robotics and Automation. 1996;**12**(4):566-580

[17] LaValle SM. Rapidly-exploring random trees: A new tool for path planning. Technical Report TR 98–11. Computer Science Department, Iowa State University; October 1998

[18] Dijkstra EW. A note on two problems in connection with graphs. Numerische Mathematik. 1959;**1**:269-271

[19] Dechter R, Pearl J. Generalized best-first search strategies and the optimality of a*. Journal of the ACM (JACM). 1985;**32**(3):505-536

[20] Mirolo C, Pagello E. A cell decomposition approach to motion planning based on collision detection. In: Proceedings of the 1995 International Conference on Advanced Robotics. 1995. pp. 481-488

[21] Angelov P. Sense and Avoid in UAS: Research and Applications. Chichester, West Sussex, United Kingdom: John Wiley & Sons, Ltd; 2012

[22] Koren Y, Borenstein J. Potential field methods and their inherent limitations for mobile robot navigation. In IEEE International Conference On Robotics And Automation; IEEE. 1991;**2**:1398-1404

[23] LaValle SM. Planning Algorithms. Cambridge University Press; 2006

[24] Bortoff SA. Path planning for UAVs. In: Proceedings of the American Control Conference. Chicago, Illinois; June 2000. pp. 364-368

[25] McLain TW, Beard RW. Trajectory planning for coordinated rendezvous of unmanned air vehicles. In: Proceedings of the AIAA Guidance, Navigation, and Control Conference. AIAA Reston, VA; 2000. Vol. 4369. pp. 1-8

[26] Argyle ME, Chamberlain C, Beard RW. Chain-based path planning for multiple UAVs. In: 50th IEEE Conference on Decision and Control and European Control Conference, Orlando, FL, USA. Dec. 2011

[27] Luders BD, Karaman S, How JP. Robust sampling-based motion planning with asymptotic optimality guarantees. In: AIAA Guidance, Navigation, and Control Conference (GNC), Boston, MA. 2013

[28] Luders BD, Karaman S, Frazzoli E, How JP. Bounds on tracking error using closed-loop rapidly-exploring random trees. In: American Control Conference (ACC). IEEE; 2010. pp. 5406-5412

[29] Luders B, Karaman S, How JP. Robust sampling-based motion planning with asymptotic optimality guarantees. In Guidance, Navigation, and Control (GNC) Conference, Boston, MA. 2013. AIAA

[30] Kothari M, Postlethwaite I. A probabilistically robust path planning algorithm for UAVs using rapidly-exploring random trees. International Robotic Systems. 2013;**71**:231-253

[31] Standard Specification for Design and Performance of an Airborne Sense-and-Avoid System. Tech. Rep. TR F2411-07. West Conshohocken, PA: ASTM International; 2007

[32] US Department of Transportation and Federal Aviation Adminstration. Aeronautical Information Manual Official Guide to Basic Flight Information and ATC Procedures

[33] Lee SM, Park C, Johnson MA, Mueller ER. Investigating effects of well clear definitions on UAS sense-and-avoid operations. In: Aviation Technology, Integration, and Operations Conference, Los Angeles, CA. AIAA. 2013

[34] Consiglio M, Chamberlain J, Munoz C, and Hoffler K. Concept of integration for UAS operations in the NAS. In: 28th International Congress of the Aeronautical Sciences (ICAS); Brisbane, Australia; 2012

[35] Sahawneh LR, Airborne Collision Detection and Avoidance for Small UAS Sense and Avoid Systems [PhD Thesis] Brigham Young University; 2016

[36] Beard RW, McLain TW. Small Unmanned Aircraft: Theory and Practice. New Jersey, USA: Princeton University Press; 2012

[37] Sahawneh LR, Argyle ME, Beard RW. 3D path planning for small UAS operating in low-altitude airspace. In International Conference on Unmanned Aircraft Systems (ICUAS). IEEE, 2016. pp. 413-419

Forward and Inverse Kinematics Using Pseudoinverse and Transposition Method for Robotic Arm DOBOT

Ondrej Hock and Jozef Šedo

Additional information is available at the end of the chapter

http://dx.doi.org/10.5772/intechopen.71417

Abstract

Kinematic structure of the DOBOT manipulator is presented in this chapter. Joint coordinates and end-effector coordinates of the manipulator are functions of independent coordinates, i.e., joint parameters. This chapter explained forward kinematics task and issue of inverse kinematics task on the structure of the DOBOT manipulator. Linearization of forward kinematic equations is made with usage of Taylor Series for multiple variables. The inversion of Jacobian matrix was used for numerical solution of the inverse kinematics task. The chapter contains analytical equations, which are solution of inverse kinematics task. It should be noted that the analytical solution exists only for simple kinematic structures, for example DOBOT manipulator structure. Subsequently, simulation of the inverse kinematics of the above-mentioned kinematic structure was performed in the Matlab Simulink environment using the SimMechanics toolbox.

Keywords: forward kinematics, inverse kinematics, Matlab Simulink simulation, robotic arm, Jacobian matrix, pseudoinverse method, SimMechanics

1. Introduction

Robots and manipulators are very important and powerful instruments of today's industry. They are making lot of different tasks and operations and they do not require comfort, time for rest, or wage. However, it takes many time and capable workers for right robot function [6].

The movement of robot can be divided into forward and inverse kinematics. Forward kinematics described how robot's move according to entered angles. There is always a solution for forward kinematics of manipulator. Solution for inverse kinematics is a more difficult problem than forward kinematics. The relationship between forward kinematics and inverse kinematics is illustrated in **Figure 1**. Inverse kinematics must be solving in reverse than forward kinematics. But we know to always find some solution for inverse kinematics of manipulator. There are

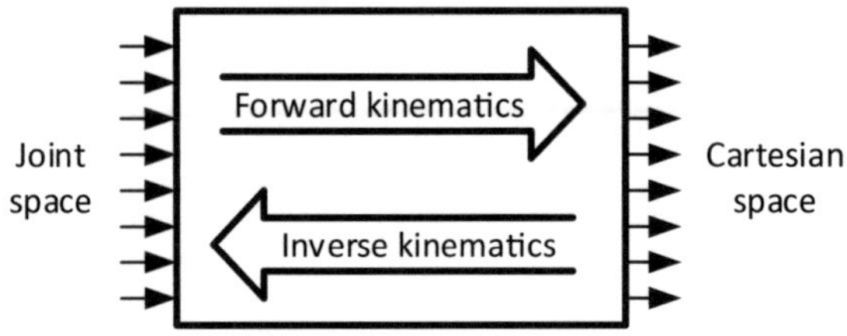

Figure 1. The schematic representation of forward and inverse kinematics.

only few groups of manipulators (manipulators with Euler wrist) with simple solution of inverse kinematics [8, 9].

Two main techniques for solving the inverse kinematics are analytical and numerical methods. In the first method, the joint variables are solved analytical, when we use classic sinus and cosine description. In the second method, the joint variables are described by the numerical techniques [9].

The whole chapter will be dedicated to the robot arm DOBOT Magician (hereafter DOBOT) shown in **Figure 2**. The basic parameters of the robotic manipulator are shown in **Figure 3** and its motion parameters are shown in **Table 1**.

This chapter is organized in the following manner. In the first section, we made the forward and inverse kinematics transformations for DOBOT manipulator. Secondly, we made the

Figure 2. DOBOT Magician [10].

http://dx.doi.org/10.5772/intechopen.71417

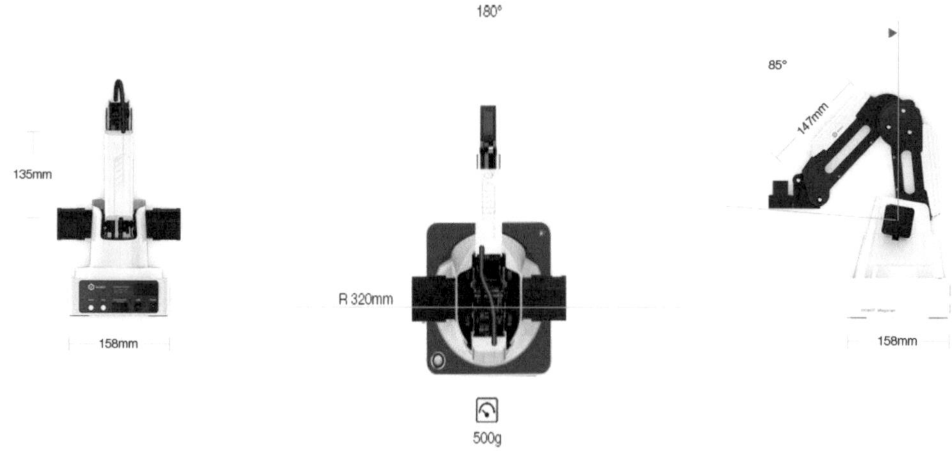

Figure 3. Simple specification of DOBOT [10].

Axis	Range	Max speed (250 g workload)
Joint 1 base	−90° to +90°	320°/s
Joint 2 rear arm	0° to +85°	320°/s
Joint 3 fore arm	−10° to +95°	320°/s
Joint 4 rotation servo	+90° to −90°	480°/s

Table 1. Axis movement of DOBOT Magician [10].

DOBOT Magician simulation in Matlab environment. Thirdly, we describe the explanation of Denavit-Hartenberg parameters. Finally, we made the pseudoinverse and transposition methods of Jacobian matrix in the inverse kinematics.

2. Kinematics structure RRR in 3D

Kinematic structure of the DOBOT manipulator is shown in **Figure 4**. It is created from three rotation joints and three links. Joint A rotates about the axis z and joints B and C rotate about the axis x_1.

Figure 5 shows a view from the direction of axis z and **Figure 6** shows a perpendicular view of the plane defined by z axis and line c.

Kinematic equations of the points B, C, and D, respectively:

$$x_0^B = 0 \tag{1}$$

$$y_0^B = 0 \tag{2}$$

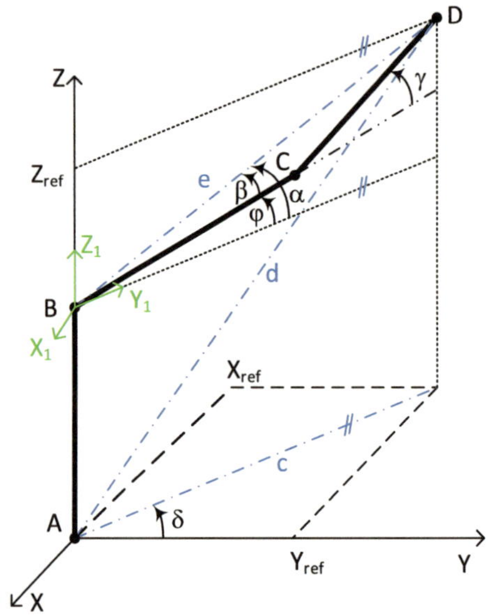

Figure 4. Representation of DOBOT manipulator in 3D view.

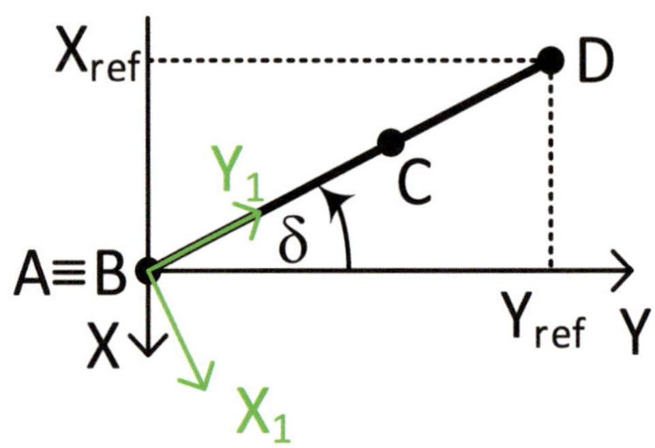

Figure 5. Representation of DOBOT footprint.

$$z_0^B = l_1 \tag{3}$$

$$x_0^C = -l_2.\cos\varphi_1.\sin\delta \tag{4}$$

$$y_0^C = l_2.\cos\varphi_1.\cos\delta \tag{5}$$

$$z_0^C = l_1 + l_2.\sin\varphi_1 \tag{6}$$

$$x_0^D = -\left(l_2.\cos\varphi_1 + l_3.\cos\varphi_2\right).\sin\delta \tag{7}$$

$$y_0^D = \left(l_2.\cos\varphi_1 + l_3.\cos\varphi_2\right).\cos\delta \tag{8}$$

$$z_0^D = l_1 + l_2.\sin\varphi_1 + l_3.\sin\varphi_2 \tag{9}$$

Where $\varphi_2 = \varphi + \gamma$.

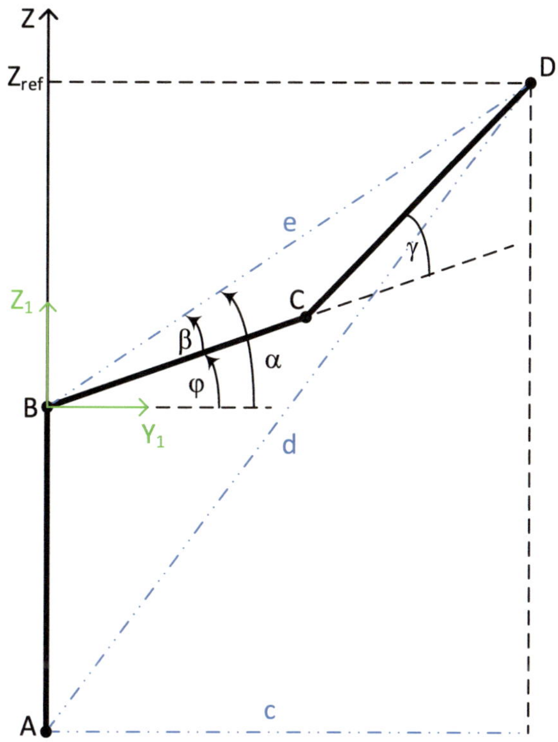

Figure 6. View of the plane defined by z axis and line c.

2.1. Forward kinematics

Forward kinematics task is defined by Eq. (10)

$$X = f(Q) \tag{10}$$

where

$$X = \begin{bmatrix} x_0^D \\ y_0^D \\ z_0^D \end{bmatrix} \tag{11}$$

X is position vector of manipulator endpoint coordinates.

$$Q = \begin{bmatrix} \varphi \\ \gamma \\ \delta \end{bmatrix} \tag{12}$$

Q is vector of independent coordinates: $\varphi = \varphi_1, \gamma, \delta$.

Because the function $X = f(Q)$ is nonlinear, it is difficult to solve the inverse task $Q = f(X)$ when looking for a vector of independent coordinates (rotation of individual manipulator joints) as a function of the desired manipulator endpoint coordinates. An analytical solution to the inverse task is possible only in the case of a relatively simple kinematic structure of the manipulator (see next chapter).

Therefore, the function $X = f(Q)$ linearized using Taylor series, taking into account only the first four (linear) members of the development:

$$x_0^D = x_0^D(\varphi,\gamma,\delta) \approx x_0^D(\varphi_0,\gamma_0,\delta_0) + \left.\frac{\partial x_0^D(\varphi,\gamma,\delta)}{\partial\varphi}\right|_{\varphi_0,\gamma_0,\delta_0}(\varphi-\varphi_0) + \left.\frac{\partial x_0^D(\varphi,\gamma,\delta)}{\partial\gamma}\right|_{\varphi_0,\gamma_0,\delta_0}(\gamma-\gamma_0) + \left.\frac{\partial x_0^D(\varphi,\gamma,\delta)}{\partial\delta}\right|_{\varphi_0,\gamma_0,\delta_0}(\delta-\delta_0) \tag{13}$$

$$y_0^D = y_0^D(\varphi,\gamma,\delta) \approx y_0^D(\varphi_0,\gamma_0,\delta_0) + \left.\frac{\partial y_0^D(\varphi,\gamma,\delta)}{\partial\varphi}\right|_{\varphi_0,\gamma_0,\delta_0}(\varphi-\varphi_0) + \left.\frac{\partial y_0^D(\varphi,\gamma,\delta)}{\partial\gamma}\right|_{\varphi_0,\gamma_0,\delta_0}(\gamma-\gamma_0) + \left.\frac{\partial y_0^D(\varphi,\gamma,\delta)}{\partial\delta}\right|_{\varphi_0,\gamma_0,\delta_0}(\delta-\delta_0) \tag{14}$$

$$z_0^D = z_0^D(\varphi,\gamma,\delta) \approx z_0^D(\varphi_0,\gamma_0,\delta_0) + \left.\frac{\partial z_0^D(\varphi,\gamma,\delta)}{\partial\varphi}\right|_{\varphi_0,\gamma_0,\delta_0}(\varphi-\varphi_0) + \left.\frac{\partial z_0^D(\varphi,\gamma,\delta)}{\partial\gamma}\right|_{\varphi_0,\gamma_0,\delta_0}(\gamma-\gamma_0) + \left.\frac{\partial z_0^D(\varphi,\gamma,\delta)}{\partial\delta}\right|_{\varphi_0,\gamma_0,\delta_0}(\delta-\delta_0) \tag{15}$$

After editing:

$$x_0^D(\varphi,\gamma,\delta) - x_0^D(\varphi_0,\gamma_0,\delta_0) = \left.\frac{\partial x_0^D(\varphi,\gamma,\delta)}{\partial\varphi}\right|_{\varphi_0,\gamma_0,\delta_0}(\varphi-\varphi_0) + \left.\frac{\partial x_0^D(\varphi,\gamma,\delta)}{\partial\gamma}\right|_{\varphi_0,\gamma_0,\delta_0}(\gamma-\gamma_0) + \left.\frac{\partial x_0^D(\varphi,\gamma,\delta)}{\partial\delta}\right|_{\varphi_0,\gamma_0,\delta_0}(\delta-\delta_0) \tag{16}$$

$$y_0^D(\varphi,\gamma,\delta) - y_0^D(\varphi_0,\gamma_0,\delta_0) = \left.\frac{\partial y_0^D(\varphi,\gamma,\delta)}{\partial\varphi}\right|_{\varphi_0,\gamma_0,\delta_0}(\varphi-\varphi_0) + \left.\frac{\partial y_0^D(\varphi,\gamma,\delta)}{\partial\gamma}\right|_{\varphi_0,\gamma_0,\delta_0}(\gamma-\gamma_0) + \left.\frac{\partial y_0^D(\varphi,\gamma,\delta)}{\partial\delta}\right|_{\varphi_0,\gamma_0,\delta_0}(\delta-\delta_0) \tag{17}$$

$$z_0^D(\varphi,\gamma,\delta) - z_0^D(\varphi_0,\gamma_0,\delta_0) = \left.\frac{\partial z_0^D(\varphi,\gamma,\delta)}{\partial\varphi}\right|_{\varphi_0,\gamma_0,\delta_0}(\varphi-\varphi_0) + \left.\frac{\partial z_0^D(\varphi,\gamma,\delta)}{\partial\gamma}\right|_{\varphi_0,\gamma_0,\delta_0}(\gamma-\gamma_0) + \left.\frac{\partial z_0^D(\varphi,\gamma,\delta)}{\partial\delta}\right|_{\varphi_0,\gamma_0,\delta_0}(\delta-\delta_0) \tag{18}$$

We denoted:

$$\Delta x_0^D = x_0^D(\varphi, \gamma, \delta) - x_0^D(\varphi_0, \gamma_0, \delta_0) \tag{19}$$

$$\Delta y_0^D = y_0^D(\varphi, \gamma, \delta) - y_0^D(\varphi_0, \gamma_0, \delta_0) \tag{20}$$

$$\Delta z_0^D = z_0^D(\varphi, \gamma, \delta) - z_0^D(\varphi_0, \gamma_0, \delta_0) \tag{21}$$

$$\Delta\varphi = \varphi - \varphi_0 \tag{22}$$

$$\Delta\gamma = \gamma - \gamma_0 \tag{23}$$

$$\Delta\delta = \delta - \delta_0 \tag{24}$$

Then, we obtained:

$$\Delta x_0^D = \left.\frac{\partial x_0^D(\varphi,\gamma,\delta)}{\partial\varphi}\right|_{\varphi_0,\gamma_0,\delta_0}\Delta\varphi + \left.\frac{\partial x_0^D(\varphi,\gamma,\delta)}{\partial\gamma}\right|_{\varphi_0,\gamma_0,\delta_0}\Delta\gamma + \left.\frac{\partial x_0^D(\varphi,\gamma,\delta)}{\partial\delta}\right|_{\varphi_0,\gamma_0,\delta_0}\Delta\delta \tag{25}$$

$$\Delta y_0^D = \left.\frac{\partial y_0^D(\varphi,\gamma,\delta)}{\partial\varphi}\right|_{\varphi_0,\gamma_0,\delta_0}\Delta\varphi + \left.\frac{\partial y_0^D(\varphi,\gamma,\delta)}{\partial\gamma}\right|_{\varphi_0,\gamma_0,\delta_0}\Delta\gamma + \left.\frac{\partial y_0^D(\varphi,\gamma,\delta)}{\partial\delta}\right|_{\varphi_0,\gamma_0,\delta_0}\Delta\delta \tag{26}$$

$$\Delta z_0^D = \left.\frac{\partial z_0^D(\varphi,\gamma,\delta)}{\partial\varphi}\right|_{\varphi_0,\gamma_0,\delta_0}\Delta\varphi + \left.\frac{\partial z_0^D(\varphi,\gamma,\delta)}{\partial\gamma}\right|_{\varphi_0,\gamma_0,\delta_0}\Delta\gamma + \left.\frac{\partial z_0^D(\varphi,\gamma,\delta)}{\partial\delta}\right|_{\varphi_0,\gamma_0,\delta_0}\Delta\delta \tag{27}$$

In matrix form:

$$\begin{bmatrix} \Delta x_0^D \\ \Delta y_0^D \\ \Delta z_0^D \end{bmatrix} = \begin{bmatrix} \frac{\partial x_0^D(\varphi,\gamma,\delta)}{\partial\varphi} & \frac{\partial x_0^D(\varphi,\gamma,\delta)}{\partial\gamma} & \frac{\partial x_0^D(\varphi,\gamma,\delta)}{\partial\delta} \\ \frac{\partial y_0^D(\varphi,\gamma,\delta)}{\partial\varphi} & \frac{\partial y_0^D(\varphi,\gamma,\delta)}{\partial\gamma} & \frac{\partial y_0^D(\varphi,\gamma,\delta)}{\partial\delta} \\ \frac{\partial z_0^D(\varphi,\gamma,\delta)}{\partial\varphi} & \frac{\partial z_0^D(\varphi,\gamma,\delta)}{\partial\gamma} & \frac{\partial z_0^D(\varphi,\gamma,\delta)}{\partial\delta} \end{bmatrix}_{\varphi_0,\gamma_0,\delta_0} \cdot \begin{bmatrix} \Delta\varphi \\ \Delta\gamma \\ \Delta\delta \end{bmatrix} \tag{28}$$

Where matrix:

$$J = \begin{bmatrix} \frac{\partial x_0^D(\varphi,\gamma,\delta)}{\partial\varphi} & \frac{\partial x_0^D(\varphi,\gamma,\delta)}{\partial\gamma} & \frac{\partial x_0^D(\varphi,\gamma,\delta)}{\partial\delta} \\ \frac{\partial y_0^D(\varphi,\gamma,\delta)}{\partial\varphi} & \frac{\partial y_0^D(\varphi,\gamma,\delta)}{\partial\gamma} & \frac{\partial y_0^D(\varphi,\gamma,\delta)}{\partial\delta} \\ \frac{\partial z_0^D(\varphi,\gamma,\delta)}{\partial\varphi} & \frac{\partial z_0^D(\varphi,\gamma,\delta)}{\partial\gamma} & \frac{\partial z_0^D(\varphi,\gamma,\delta)}{\partial\delta} \end{bmatrix}_{\varphi_0,\gamma_0,\delta_0} \tag{29}$$

is Jacobian matrix. We denoted:

$$\Delta X_0^D = \begin{bmatrix} \Delta x_0^D \\ \Delta y_0^D \\ \Delta z_0^D \end{bmatrix} \tag{30}$$

and

$$\Delta Q = \begin{bmatrix} \Delta\varphi \\ \Delta\gamma \\ \Delta\delta \end{bmatrix} \tag{31}$$

Then, we obtained the matrix equation, which represents linearized forward kinematics in incremental form:

$$\Delta X_0^D = J.\Delta Q \tag{32}$$

After we multiplied the Eq. (32) with inverse matrix J^{-1} from the left, we obtained the equation of inverse kinematics.

$$J^{-1}.\Delta X_0^D = J^{-1}.J.\Delta Q \tag{33}$$

$$J^{-1}.\Delta X_0^D = I.\Delta Q \tag{34}$$

Where I is the identity matrix. After that:

$$\Delta Q = J^{-1}.\Delta X_0^D \tag{35}$$

Derivative of the kinematic equations with respect to the independent coordinates for kinematic structure of DOBOT manipulator:

$$\frac{\partial x_0^D}{\partial\varphi} = [l_2.\sin\varphi + l_3.\sin(\varphi+\gamma)].\sin\delta \tag{36}$$

$$\frac{\partial x_0^D}{\partial\gamma} = l_3.\sin(\varphi+\gamma).\sin\delta \tag{37}$$

$$\frac{\partial x_0^D}{\partial\delta} = -[l_2.\cos\varphi + l_3.\cos(\varphi+\gamma)].\cos\delta \tag{38}$$

$$\frac{\partial y_0^D}{\partial\varphi} = -[l_2.\sin\varphi + l_3.\sin(\varphi+\gamma)].\cos\delta \tag{39}$$

$$\frac{\partial y_0^D}{\partial\gamma} = -l_3.\sin(\varphi+\gamma).\cos\delta \tag{40}$$

$$\frac{\partial y_0^D}{\partial\delta} = -[l_2.\cos\varphi + l_3.\cos(\varphi+\gamma)].\sin\delta \tag{41}$$

http://dx.doi.org/10.5772/intechopen.71417

$$\frac{\partial z_0^D}{\partial \varphi} = l_2.\cos\varphi + l_3.\cos(\varphi+\gamma) \tag{42}$$

$$\frac{\partial z_0^D}{\partial \gamma} = l_3.\cos(\varphi+\gamma) \tag{43}$$

$$\frac{\partial z_0^D}{\partial \delta} = 0 \tag{44}$$

Jacobian matrix:

$$J = \begin{bmatrix} \frac{\partial x_0^D}{\partial \varphi} & \frac{\partial x_0^D}{\partial \gamma} & \frac{\partial x_0^D}{\partial \delta} \\ \frac{\partial y_0^D}{\partial \varphi} & \frac{\partial y_0^D}{\partial \gamma} & \frac{\partial y_0^D}{\partial \delta} \\ \frac{\partial z_0^D}{\partial \varphi} & \frac{\partial z_0^D}{\partial \gamma} & \frac{\partial z_0^D}{\partial \delta} \end{bmatrix} \tag{45}$$

2.2. Analytical Solution of the Inverse Kinematics of DOBOT manipulator

The following equations are derived from **Figure 4**.

$$c^2 = x^2 + y^2 \tag{46}$$

$$d^2 = c^2 + z^2 = x^2 + y^2 + z^2 \tag{47}$$

$$e^2 = l_2^2 + l_3^2 - 2l_2l_3\cos(\pi-\gamma) \tag{48}$$

$$\gamma = \pm\arccos\left(\frac{e^2 - l_2^2 - l_3^2}{2l_2l_3}\right) \tag{49}$$

$$e^2 = c^2 + (z - l_1)^2 \tag{50}$$

$$\varphi = \alpha - \beta \tag{51}$$

$$\alpha = arctg\frac{z-l_1}{c} = arctg\frac{z-l_1}{\sqrt{x^2+y^2}} \tag{52}$$

$$l_3^2 = l_2^2 + e^2 - 2l_2.e.\cos\beta \tag{53}$$

$$\beta = \pm\arccos\left(\frac{l_2^2 + e^2 - l_3^2}{2l_2e}\right) \tag{54}$$

$$\varphi = arctg\frac{z-l_1}{\sqrt{x^2+y^2}} \mp \arccos\left(\frac{l_2^2 + e^2 - l_3^2}{2l_2e}\right) \tag{55}$$

$$\delta = arctg\frac{-x}{y} \tag{56}$$

3. Simulation DOBOT Magician in Matlab environment

For simulation movement of the manipulator, Matlab Simulink environment and SimMechanics toolbox are suitable to use. Some blocks from SimMechanics toolbox are shown in **Figure 7**, which represents the model of DOBOT manipulator.

We used basic block from SimMechanics toolbox in simulation model:

- Joint actuator
- Revolute
- Body
- Body sensor
- Machine environment

The joint actuator block transfers the requested angles to the connected joint. The revolute block defined the rotation of body in space. The body block describes the parameters of body,

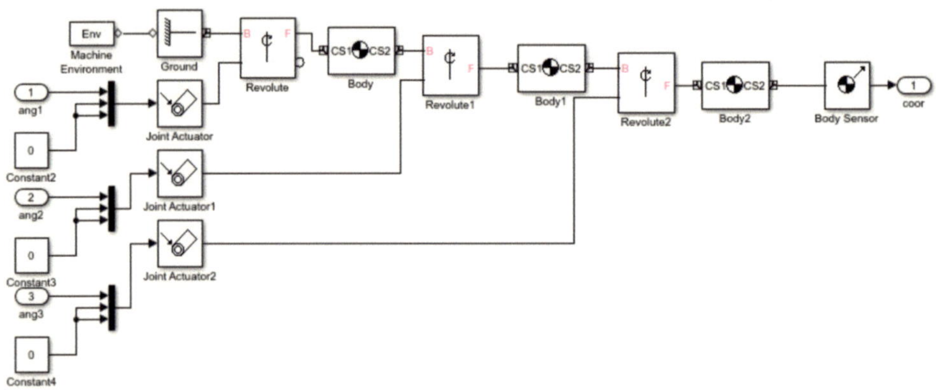

Figure 7. SimMechanics simulation model of DOBOT manipulator.

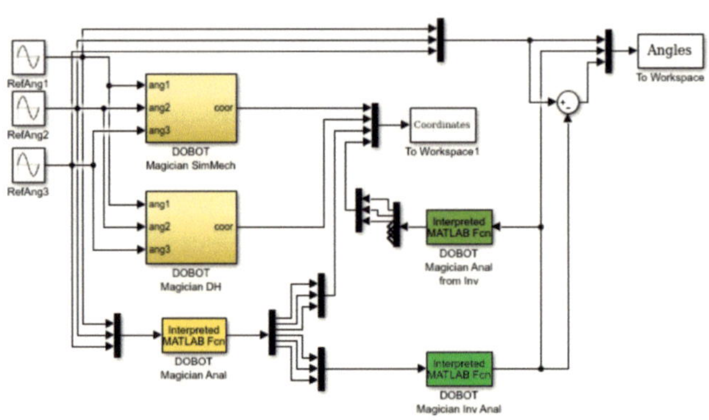

Figure 8. Simulation of DOBOT manipulator in Matlab environment.

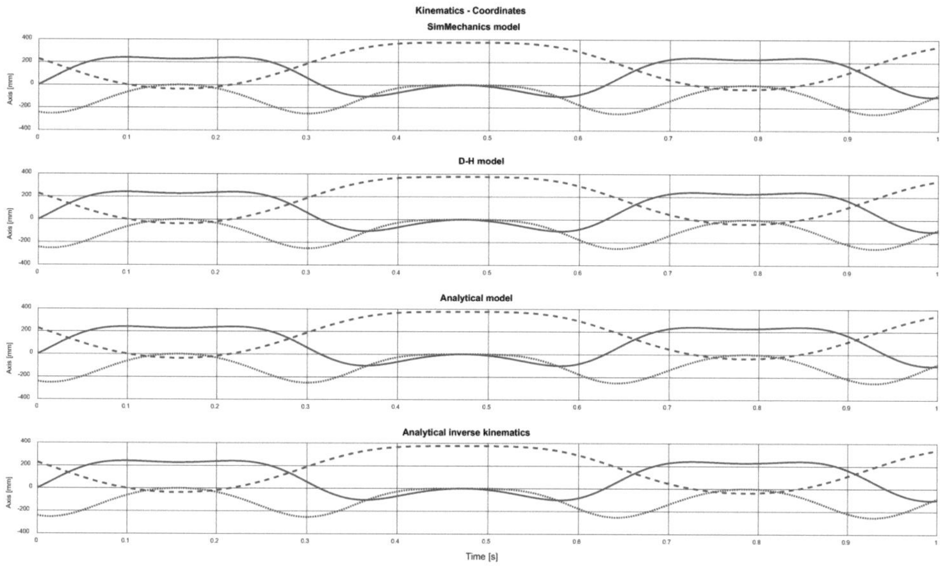

Figure 9. Endpoint coordinates of DOBOT manipulator.

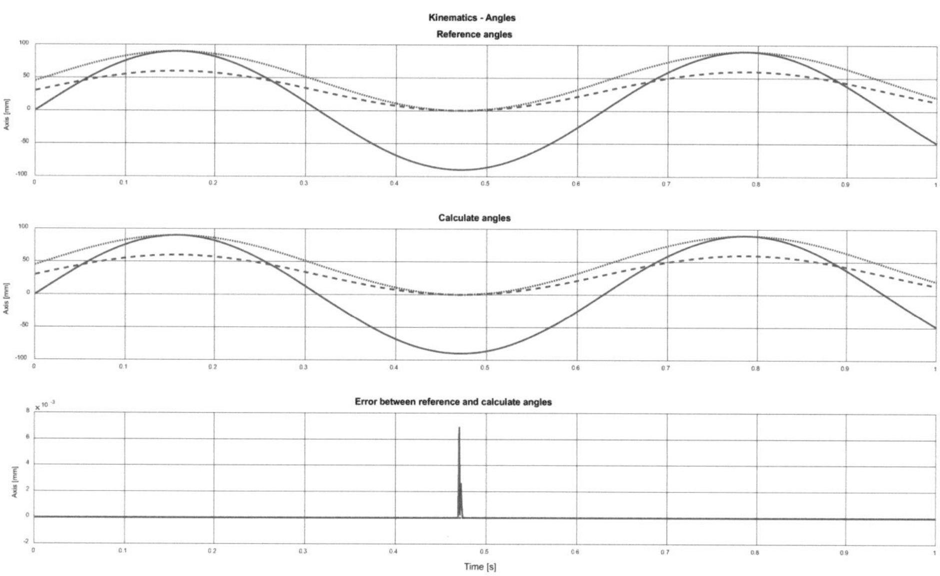

Figure 10. Reference angles, calculated angles, and error between these angles.

like dimension, inertia, etc. The body sensor block transfers the coordinates, velocity, and others to simulation, and the last block is the machine environment which defines the parameters of the environment in which the manipulator is located. You can find all necessary data about these blocks in [7].

The simulation model, shown in **Figure 8**, was designed for considerate results from SimMechanics model, model used D-H parameters and analytical model, which was described in previous chapter. Every result from these models is shown in **Figure 9**. The fourth part in **Figure 9** is the results from analytical simulation model of inverse kinematics. **Figure 10** represented the reference angles in the first part of the chart, calculated angles from analytical inverse kinematics model in the second part of chart, and finally the error between both angles. As we can see in **Figure 10**, the angles are same. This is proof that analytical model of DOBOT manipulator is usable for simulation and implementation to some DSP or microcontroller.

4. Denavit-Hartenberg parameters

The steps to get the position in using D-H convention are finding the Denavid-Hartenberg (D-H) parameters, building A matrices, and calculating T matrix with the coordinate position which is desired.

4.1. D-H parameters

D-H notation describes coordinates for different joints of a robotic manipulator in matrix entry. The method includes four parameters:

1. Twist angle α_i
2. Link length a_i
3. Link offset d_i
4. Joint angle θ_i.

Based on the manipulator geometry, twist angle and link length are constants and link offset and joint angle are variables depending on the joint, which can be prismatic or revolute. The method has provided 10 steps to denote the systematic derivation of the D-H parameters, and you can find them in [5] or [6].

4.2. A matrix

The A matrix is a homogenous 4 × 4 transformation matrix. Matrix describes the position of a point on an object and the orientation of the object in a three-dimensional space [6]. The homogenous rotation matrix along an axis is described by the Eq. (57) (**Figures 11–13**).

$$Rot_{z_{i-1}} = \begin{bmatrix} \cos\theta_i & -\cos\alpha_i\sin\theta_i & \sin\alpha_i\sin\theta_i & 0 \\ \sin\theta_i & \cos\alpha_i\cos\theta_i & -\sin\alpha_i\sin\theta_i & 0 \\ 0 & \sin\alpha_i & \cos\alpha_i & 0 \\ 0 & 0 & 0 & 1 \end{bmatrix} \quad (57)$$

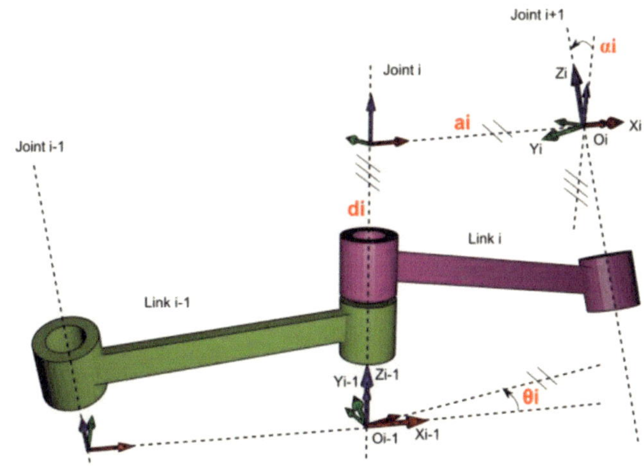

Figure 11. The four parameters of classic DH convention are θ_i, d_i, a_i, α_i [4].

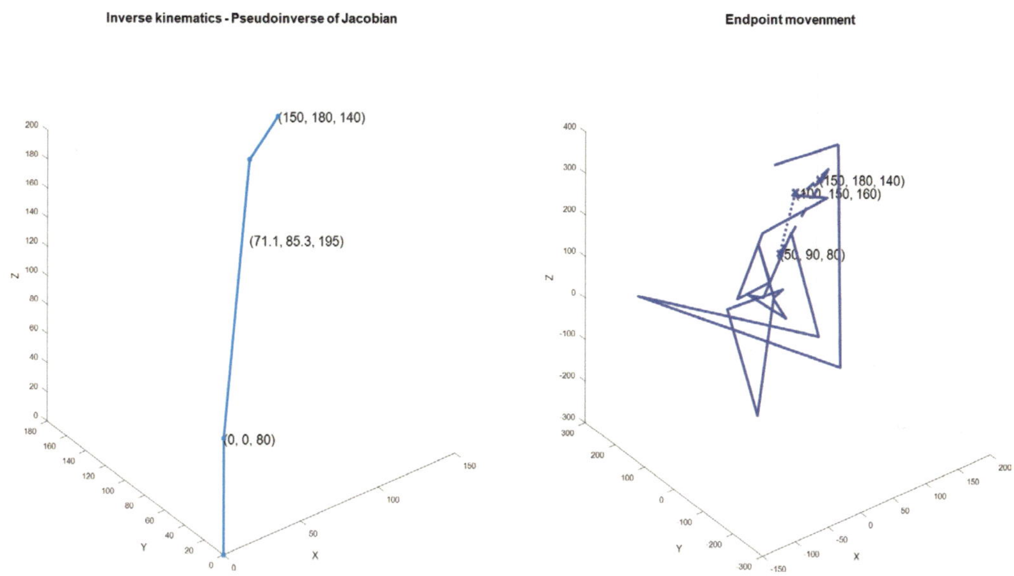

Figure 12. Simulation result of Jacobian matrix pseudoinverse in inverse kinematics model of the DOBOT manipulator.

The homogeneous translation matrix is described by Eq. (58).

$$Trans_{z_{i-1}} = \begin{bmatrix} 1 & 0 & 0 & a_i \\ 0 & 1 & 0 & 0 \\ 0 & 0 & 1 & d_i \\ 0 & 0 & 0 & 1 \end{bmatrix} \tag{58}$$

In rotation matrix and translation matrix, we can find the four parameters θ_i, d_i, a_i, and α_i. These parameters derive from specific aspects of the geometric relationship between two coordinate

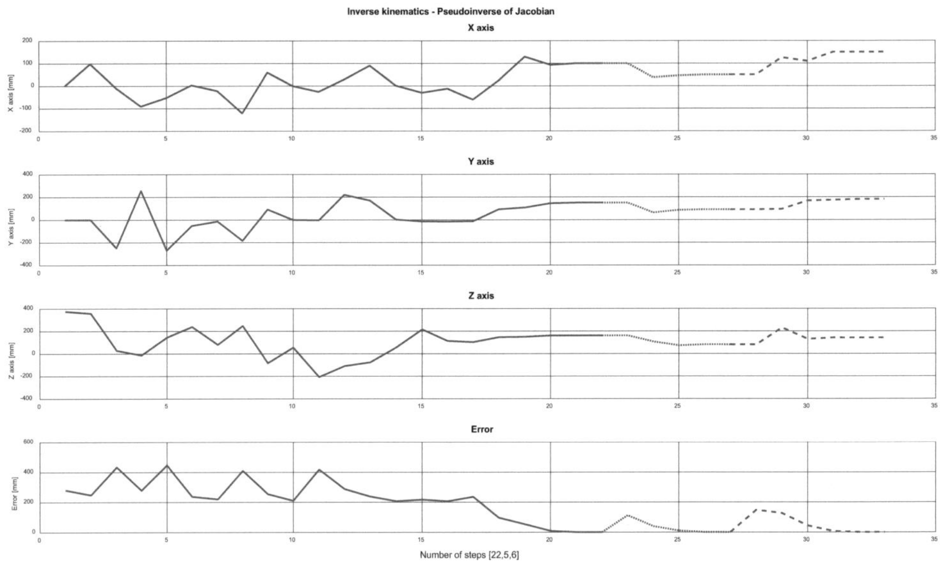

Figure 13. Simulation results of DOBOT axis and total error of coordinates for Pseudoinverse method.

frames. The four parameters are associated with link i and joint i. In Denavit-Hartenberg convention, each homogeneous transformation matrix A_i is represented as a product of four basic transformations as follows [6]:

$$T_i^{i-1} = Trans_{z_{i-1}}(d_i).Rot_{z_{i-1}}(\theta_i).Trans_{x_i}(r_i).Rot_{x_i}(\alpha_i) \tag{59}$$

D-H convention matrix is given in Eq. (60).

$$T_i^{i-1} = \begin{bmatrix} \cos\theta_i & -\sin\theta_i\cos\alpha_i & \sin\theta_i\sin\alpha_i & r_i\cos\theta_i \\ \sin\theta_i & \cos\theta_i\cos\alpha_i & -\cos\theta_i\sin\alpha_i & r_i\sin\theta_i \\ 0 & \sin\alpha_i & \cos\alpha_i & d_i \\ 0 & 0 & 0 & 1 \end{bmatrix} \tag{60}$$

The previous matrix can be simplified by following equation A_i matrix. The matrix A_i is composed from 3×3 rotation matrix R_i, 3×1 translation vector P_i, 1×3 perspective vector and scaling factor.

$$A_i = \begin{bmatrix} R_{i(3x3)} & P_{i(3x1)} \\ 0_{(1x3)} & 1 \end{bmatrix} \tag{61}$$

4.3. T matrix

The T matrix can be formulated by Eq. (62). The matrix is a sequence of D-H matrices and is used for obtaining end-effector coordinates. The T matrix can be built from several A matrices depending on the number of manipulator joints.

$$T_0^3 = T_0^1 . T_1^2 . T_2^3 \tag{62}$$

Inside the T matrix is the translation vector P_i, which includes joint coordinates, where the X, Y, and Z positions are P_1, P_2, and P_3, respectively [6].

5. The pseudoinverse method

If the number of independent coordinates n (joint parameters) is larger than the number of reference manipulator endpoint coordinates m (three in Cartesian coordinate system for the point), it shows that a redundancy problem has occurred. In this case, it can exist in infinite combinations of independent coordinates for the only endpoint position. Jacobian matrix J has a size of m rows and n columns ($m \neq n$), i.e., J is a non-square matrix. In general, it cannot be computed inverse matrix from non-square matrix.

In order to solve inverse kinematics task for this case, pseudoinverse of Jacobian matrix (denotes J^+) is used. This method uses singular value decomposition (SVD) of Jacobian matrix to determine J^+.

Every matrix J can be decomposed with the usage of SVD to three matrices Eq. (63):

$$J = U\Sigma V^T \tag{63}$$

Where

J is $m \times n$ matrix.

U is $m \times m$ orthogonal matrix, i.e. $U^{-1} = U^T$.

V is $n \times n$ orthogonal matrix, i.e. $V^{-1} = V^T$.

Σ is $m \times n$ diagonal matrix, which contains singular values of matrix J on its major diagonal.

$$\begin{bmatrix} j_{11} & j_{12} & \cdots & j_{1n} \\ j_{21} & j_{22} & \cdots & j_{2n} \\ \vdots & \vdots & \ddots & \vdots \\ j_{m1} & j_{m2} & \cdots & j_{mn} \end{bmatrix} = \begin{bmatrix} u_{11} & u_{12} & \cdots & u_{1m} \\ u_{21} & u_{22} & \cdots & u_{2m} \\ \vdots & \vdots & \ddots & \vdots \\ u_{m1} & u_{m2} & \cdots & u_{mm} \end{bmatrix} . \begin{bmatrix} \sigma_1 & 0 & \cdots & 0 \\ 0 & \sigma_2 & \cdots & 0 \\ \vdots & \vdots & \ddots & \vdots \\ 0 & 0 & \cdots & \sigma_d \end{bmatrix} . \begin{bmatrix} v_{11} & v_{21} & \cdots & v_{n1} \\ v_{12} & v_{22} & \cdots & v_{n2} \\ \vdots & \vdots & \ddots & \vdots \\ v_{1n} & v_{2n} & \cdots & v_{nn} \end{bmatrix} \tag{64}$$

Where $d = m$ for $m < n$ and $d = n$ for $m > n$, because Σ is a non-square matrix.

To determine matrices U and Σ, we multiply matrix J by its transpose matrix J^T from the right:

$$J.J^T = (U\Sigma V^T).(U\Sigma V^T)^T \tag{65}$$

$$J.J^T = U\Sigma V^T.V\Sigma^T U^T \tag{66}$$

$$J.J^T = U\Sigma\Sigma^T U^T \tag{67}$$

We multiply the above Eq. (67) by matrix U from the right:

$$JJ^T.U = U.\Sigma\Sigma^T.U^T U \tag{68}$$

$$JJ^T.U = U.\Sigma\Sigma^T \tag{69}$$

It leads to eigenvalue problem for JJ^T matrix. U is $m \times m$ square matrix, which contains eigenvectors of JJ^T matrix in its columns and $\Sigma\Sigma^T$ is diagonal matrix of eigenvalues $\lambda_1, \ldots, \lambda_m$.

To determine matrices V and Σ, we multiply matrix J by its transpose matrix J^T from the left:

$$J^T.J = \left(U\Sigma V^T\right)^T.\left(U\Sigma V^T\right) \tag{70}$$

$$J^T.J = V\Sigma^T U^T.U\Sigma V^T \tag{71}$$

$$J^T.J = V\Sigma^T\Sigma V^T \tag{72}$$

We multiply the above Eq. (72) by matrix V from the right:

$$J^T J.V = V.\Sigma^T\Sigma.V^T.V \tag{73}$$

$$J^T J.V = V.\Sigma^T\Sigma \tag{74}$$

It leads to eigenvalue problem for J^TJ matrix. V is $n \times n$ square matrix, which contains eigenvectors of J^TJ matrix in its columns and $\Sigma^T\Sigma$ is diagonal matrix of eigenvalues $\lambda_1, \ldots, \lambda_n$.

Matrices JJ^T and J^TJ are symmetric matrices and they have the same nonzero eigenvalues. Eigenvalues and eigenvectors of the real symmetric matrices are always real numbers and real vectors.

The eigenvalues are equal to square of the singular values: $\lambda_i = \sigma_i^2$, where $i = 1, \ldots, d$. The number of nonzero eigenvalues is $d = m$ for $m < n$ and $d = n$ for $m > n$. The number of zero eigenvalues is $|m - n|$.

When values of matrices U, Σ, and V were computed, we can determine pseudoinverse of Jacobian matrix as follows:

$$J^+ = \left(U.\Sigma.V^T\right)^{-1} \tag{75}$$

$$J^+ = V.\Sigma^+.U^T \tag{76}$$

Where

$$\Sigma^{+} = \begin{bmatrix} \frac{1}{\sigma_1} & 0 & \cdots & 0 \\ 0 & \frac{1}{\sigma_2} & \cdots & 0 \\ \vdots & \vdots & \ddots & \vdots \\ 0 & 0 & \cdots & \frac{1}{\sigma_d} \end{bmatrix} \tag{77}$$

Now, we can solve inverse kinematics task for the cases, when Jacobian matrix is non-square:

$$\Delta Q = J^{+}.\Delta X \tag{78}$$

Pseudoinverse J^{+}, also called Moore-Penrose inverse of Jacobian matrix, gives the best possible solution in the sense of least squares [1].

6. The Jacobian matrix transpose method

We designed Jacobian matrix transpose method simulation [1–3]. The basic idea was written using Eq. (79). We used the transpose of Jacobian matrix, instead of the inverse of Jacobian matrix, in this method. We set $\Delta\theta$ equal to

$$\Delta\theta = \alpha J^{T} \vec{e} \tag{79}$$

Where α is:

$$\alpha = \frac{\left\langle \vec{e}, JJ^{T} \vec{e} \right\rangle}{\left\langle JJ^{T} \vec{e}, JJ^{T} \vec{e} \right\rangle} \tag{80}$$

Whole simulation is described by block diagram shown in **Figure 14**. In the first step, we defined requested error. This error represented difference between reference coordinates and actual coordinates. Error that we consider as unacceptable, we set to 200 μm. This is position repeatability of DOBOT. We calculate the increment of requesting angles $\Delta\theta$ in each iteration. In the first iteration, $\Delta\theta$ is equal to zero.

Figures 15 and **16** represent simulation result of DOBOT movement same as in simulation of Jacobian matrix pseudoinverse. Simulation was split on three parts. First part (solid line in chart) is movement from starting position to position (x, y, z) = (100, 150, 160) mm. Second part (dotted line in chart) is movement from previous position to position (50, 90, 80). And third part (dashed line in chart) is movement to position (150, 180, 140).

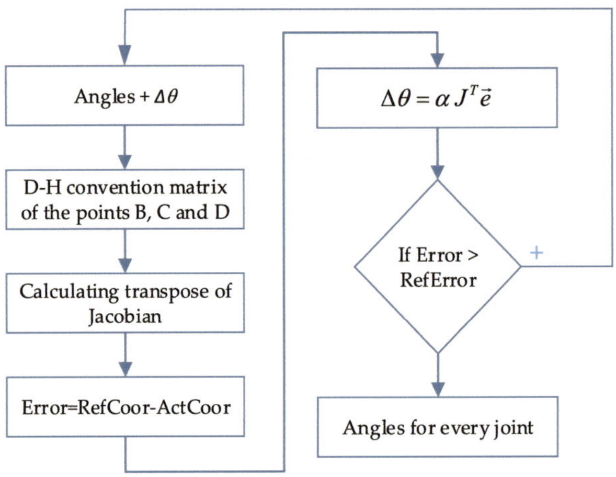

Figure 14. Block diagram of Jacobian matrix transpose method simulation.

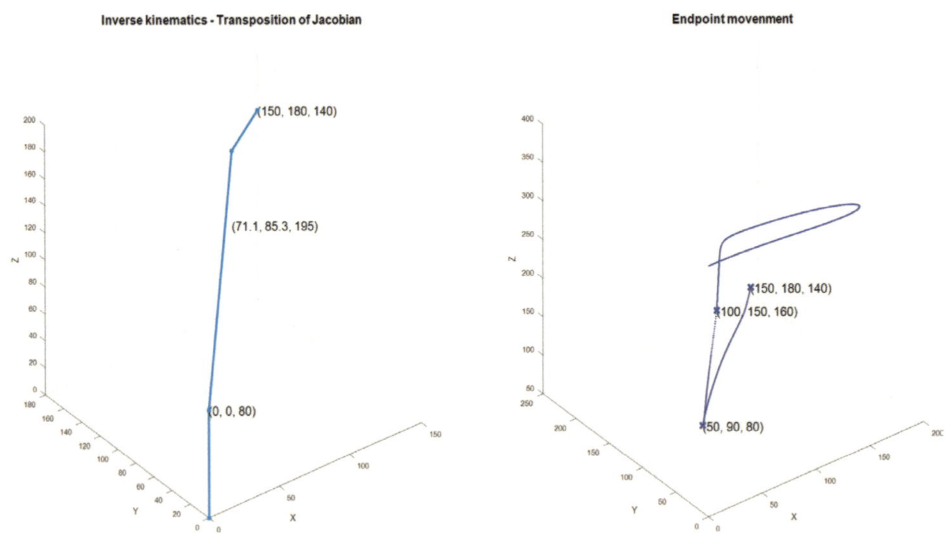

Figure 15. Simulation result of Jacobian matrix transposition in inverse kinematics model of the DOBOT manipulator.

7. Conclusion

As we can see in simulation results from previous subchapters, every method for inverse kinematics has some positives and negatives. Comparison of both methods is shown in **Table 2**. Pseudoinverse method is faster than transposition method, but is harder to implement in a DSP or a microcontroller. In Matlab environment, pseudoinverse method is easily made by the pinv() command. If we want to simplify inverse kinematics and we don't need fast calculating time, it is more readily to use transposition method. In the case of using DOBOT manipulator, it is considered to use the analytical model. In the case of more complicated manipulator, this method is inapplicable.

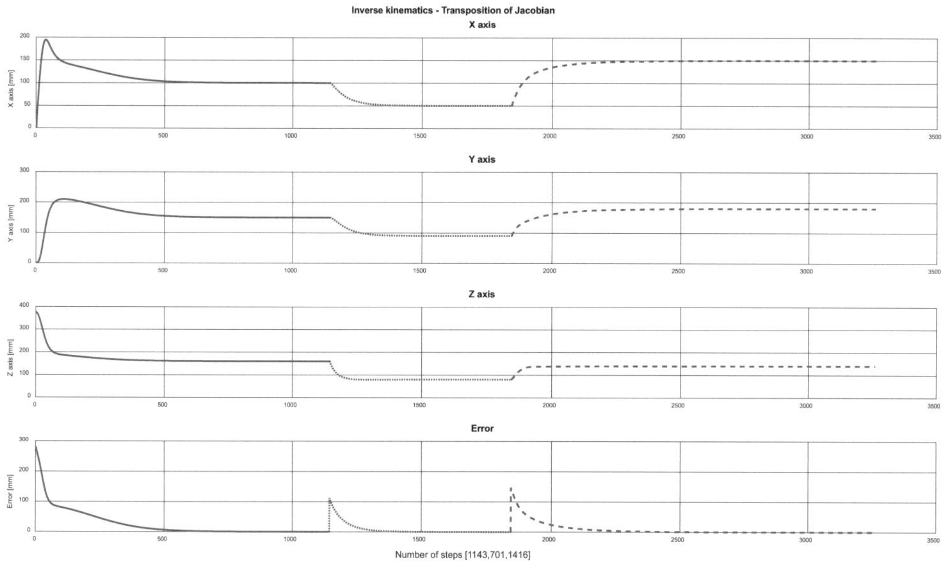

Figure 16. Simulation results of DOBOT axis and total error of coordinates for Transposition method.

Simulation part	Pseudoinverse method (number of iterations)	Transposition method (number of iterations)
Part 1 (solid line)	22	55
Part 2 (dotted line)	5	34
Part 3 (dashed line)	6	68

Table 2. Comparison of pseudoinverse and transposition method.

Comparison of both methods is shown in **Table 2**. As we can see in **Table 2**, the main criteria are number of iterations. Pseudoinverse method is much better, but only for simulation. If we can use this method in real-time application, like dSPACE from MathWorks® or implementation to DSP, we will not achieve such results like in **Table 2**. It is caused by using singular value decomposition (SVD), which is very demanding for a computation performance. In the other case, transposition of Jacobian matrix is much easier for implementation and need lower performance.

In the next research, we considerate the use suitable iterative method, like damped least squares. We also designed several implementation methods of Jacobian matrix transposition to DSP (TMS430, C2000™). It is very important to try more implementation methods for the most possible shortening of the calculation time.

Acknowledgements

The results of this work are supported by Grant No. APVV-15-0571: research of the optimum energy flow control in the electric vehicle system.

Author details

Ondrej Hock and Jozef Šedo*

*Address all correspondence to: jozef.sedo@fel.uniza.sk

Department of Mechatronics and Electronics, Faculty of Electrical Engineering, University of Žilina, Žilina, Slovakia

References

[1] Buss SR. Introduction to inverse kinematics with Jacobian transpose, pseudoinverse and damped least squares methods. IEEE Journal of Robotics and Automation. 2004;**17**:1-19

[2] Balestrino A, De Maria G, Sciavicco L. Robust control of robotic manipulators. In: Proceedings of the 9th IFAC World Congress; 1984;**5**:2435-2440

[3] Wolovich WA, Elliot H. A computational technique for inverse kinematics. In: Proceedings of the 23rd IEEE Conference on Decision and Control; 1984. pp. 1359-1363

[4] Denavit–Hartenberg [Internet]. 2017. Available from: https://en.wikipedia.org/wiki/Denavit%E2%80%93Hartenberg_parameters [Accessed: 03-08-2017]

[5] Desai JP. D-H Convention, Robot and Automation Handbook. USA: CRC Press; 2005 ISBN: 0-8493-1804-1

[6] Rehiara AB. Kinematics of Adept Three Robot Arm. In: Satoru Goto, editor. Robot Arms. 2011. InTech. ISBN: 978-953-307-160-2. Available from: http://www.intechopen.com/books/robot-arms/kinematicsof-adeptthree-robot-arm [Accessed: 01-08-2017]

[7] The MathWorks, Inc., SimMechanics 2User's Guide [Internet], 2007. Available from: https://mecanismos2mm7.files.wordpress.com/2011/09/tutorial-sim-mechanics.pdf [Accessed: 08-04-2017]

[8] Bingul KS. The inverse kinematics solutions of industrial robot manipulators. In: IEEE Conference on Mechatronics; Istanbul, Turkey; 2004. pp. 274-279

[9] Serdar Kucuk, Zafer Bingul. In: Sam Cubero, editor. Robot Kinematics: Forward and Inverse Kinematics, Industrial Robotics: Theory, Modelling and Control. 2006. InTech. ISBN: 3-86611-285-8. Available from: http://www.intechopen.com/books/industrial_robotics_theory_modelling_and_control/robot_kinematics__forward_and_inverse_kinematics [Accessed 01-08-2017]

[10] WebPage [Internet], 2017. Available from: http://www.dobot.cc/ [Accessed: April 04, 2017]

Chapter 5

How to Expand the Workspace of Parallel Robots

Takashi Harada

Additional information is available at the end of the chapter

http://dx.doi.org/10.5772/intechopen.71407

Abstract

In this chapter, methods for expanding the workspace of parallel robots are introduced. Firstly, methods for expanding the translational workspace of the parallel robot are discussed. The parallel robot has multiple solutions of the inverse and forward displacement analysis. By changing its configurations from one solution to another, the parallel robot can expand its translational workspace. However, conventional nonredundant parallel robot encounters singularity during the mode change. Singularity-free mode changes of the parallel robot by redundant actuation are introduced. Next, methods for expanding the rotational workspace of the parallel robot are shown. In order to achieve the large rotation, some mechanical gimmicks by gears, pulleys, and helical joints have been embedded in the moving part. A novel differential screw-nut mechanism for expanding the rotational workspace of the parallel robot is introduced.

Keywords: parallel robot, workspace, mode change, kinematics, singularity, differential, helical joint

1. Introduction

The parallel robot has excellent characteristics such as high speed, high precision, and high rigidity [1]. However, mechanical collisions between limbs and complexly existing singular configurations restrict its workspace. In this chapter, firstly, methods for expanding the translational workspace of the parallel robot are discussed. The parallel robot has multiple solutions of the inverse and forward displacement analysis. By changing its configurations from one solution to another, the parallel robot can expand its translational workspace. However, conventional nonredundant parallel robot encounters singularity during the mode change. Singularity-free mode changes of the parallel robot by redundant actuation are introduced. Next, methods for expanding the rotational workspace of the parallel robot are shown. In order to achieve the large rotation, some mechanical gimmicks by gears, pulleys, and helical joints have been embedded in the moving part. A novel differential screw-nut mechanism for expanding the rotational workspace of the parallel robot is introduced.

2. Expanding the translational workspace

2.1. Mode change of the parallel robot

Schematic model of a planar two-dof (degree-of-freedom) serial robot and a planar two-dof parallel robot is illustrated in **Figures 1** and **2**, respectively. *R* and *P* represent the rotational pair and prismatic (sliding) pair, respectively. Underline indicates that an actuator is located at the pair, namely, $\underline{R}$ and $\underline{P}$ represent active rotational pair and active prismatic pair, and *R* and *P* represent passive ones. The serial robot in **Figure 1** is represented as $\underline{RR}$ mechanism, and the parallel robot in **Figure 2** is represented as 2$\underline{P}RR$ mechanism. In 2$\underline{P}RR$, "2" means that the two $\underline{P}RR$ mechanisms are connected to the output link (end effector or hand) in parallel. In order to control the position of the hand, forward displacement analysis gives the position and orientation of the hand from the displacements of the active pairs (joints), and inverse displacement analysis gives the displacements of the active joints from the position and orientation of the hand. As shown in **Figures 1** and **2(a)**, different solutions for inverse displacement analysis and different position of the joints for a position and orientation of the hand exist. In the robotics, the solution of the inverse displacement analysis is called as the working mode. Changing the positions of the joints between the different working modes, at that time the robot changes its configuration, is named as the working mode change. In the case of the serial robot, there exists only one solution for the forward displacement analysis; namely, when the displacement of the active joints is given, the position and orientation of the robot are fixed. However, in the case of the parallel robot, there exist different solutions for the forward displacement analysis as shown in **Figure 2(b)**. The solution of the forward displacement analysis is named as the assembly mode. Parallel robot can change the position and orientation of the hand with the same displacements of the actuated joints. Note here that the parallel robot in **Figure 2** is designed as free of the mechanical interferences such as limb collisions. Parallel robot can expand the workspace of the hand by the assembly mode change as shown in **Figure 2(b)**.

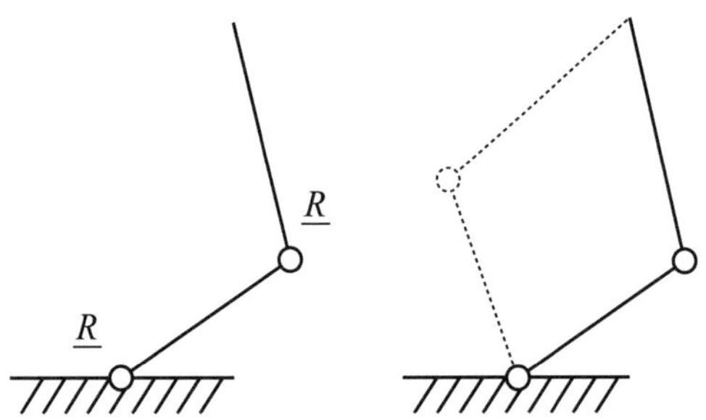

Figure 1. Working mode change of $\underline{RR}$ serial robot.

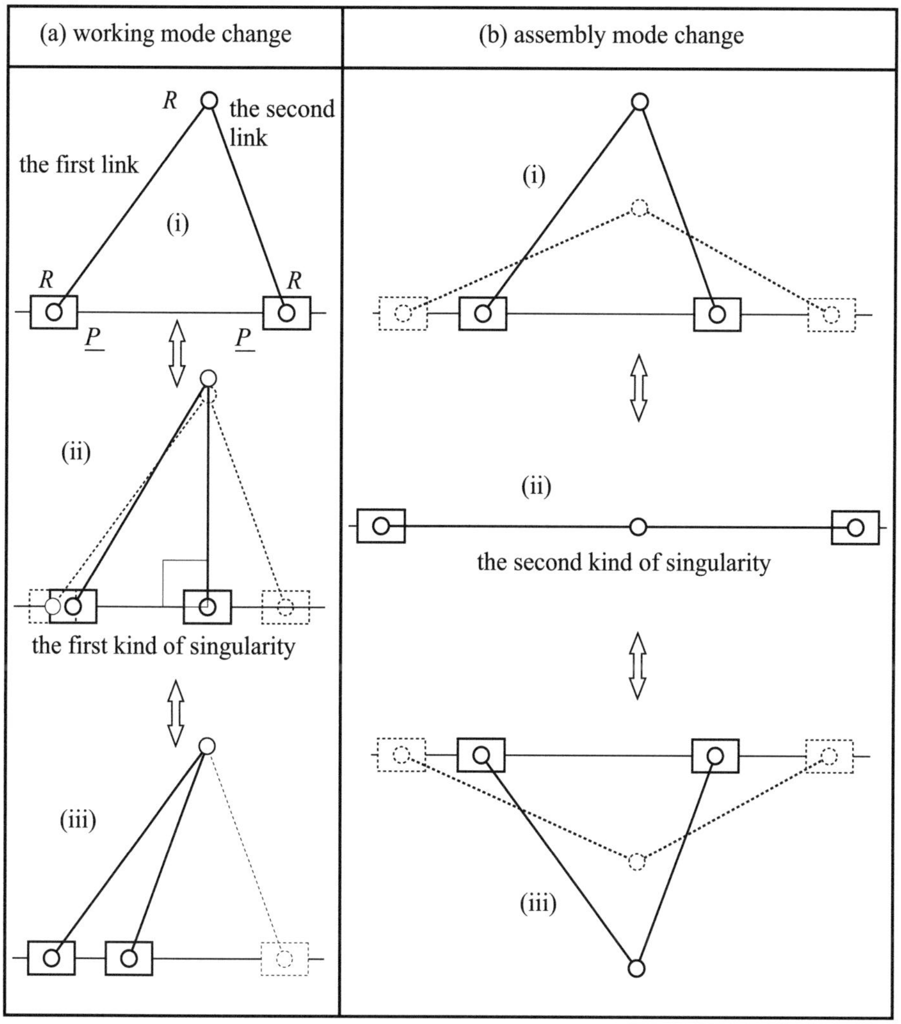

Figure 2. Assembly and working mode change of 2*PRR* parallel robot.

2.2. Kinematics of the parallel robot

As mentioned before, the parallel robot can expand its workspace by the mode change. However, parallel robot encounters the singular configuration between the mode changes. In this section, kinematics and singular configuration of a 2*PRR* parallel robot are shown in **Figure 3**. Coordinate frames Σ_0 and Σ_i $(i = 1, 2)$ are defined for the fixed coordinate and position of each actuator, respectively. Vector bi is defined for the position of the origin Σ_i with respect to the coordinate Σ_0. In the case of the 2*PRR* parallel robot in **Figure 3**, $\mathbf{b}_i = \mathit{0}$ $(i = 1, 2)$ because each coordinate Σ_i is coincident with Σ_0. Unit direction vector and displacement of the linear actuator are defined as $\mathbf{u}_i$ and $\mathbf{q}_i$, respectively. Unit direction vector and length of the rod are defined as $\mathbf{w}_i$ and l_i, respectively. Position of the hand (output link) is given as

$$\mathbf{x} = \mathbf{b}_i + q_i \boldsymbol{u}_i + l_i \boldsymbol{w}_i \tag{1}$$

Equation (1) is the loop closure equation of the parallel robot. Next, the forward displacement analysis and inverse displacement analysis of the parallel robot are going to be derived.

Eq. (1) is deformed as

$$l_i \boldsymbol{w}_i = \mathbf{x} - \mathbf{b}_i - q_i \boldsymbol{u}_i \tag{2}$$

The unit direction vector $\mathbf{w}_i$ is eliminated by raising the both sides of Eq. (2) to the second power as,

$$l_i^2 = (\boldsymbol{x} - \boldsymbol{b}_i - q_i \boldsymbol{u}_i)^T (\mathbf{x} - \mathbf{b}_i - q_i \boldsymbol{u}_i) \tag{3}$$

As shown in **Figure 3**, $\mathbf{u}_i$ and $\mathbf{b}_i$ are given as,

$$\mathbf{u}_i = \begin{bmatrix} 1 \\ 0 \end{bmatrix}, \mathbf{b}_i = \begin{bmatrix} 0 \\ 0 \end{bmatrix} \quad (i = 1, 2) \tag{4}$$

Substituting Eq. (4) into Eq. (3), one obtains a quadratic equation about the displacement of the actuator as,

$$q_i^2 - 2 q_i x + (y^2 - x^2 - l_i^2) = 0 \tag{5}$$

Solutions of the inverse displacement analysis are given by the solution of Eq. (5) as,

$$q_i = x \pm \sqrt{l_i^2 - y^2} \tag{6}$$

Eq. (6) represents that there are two solutions for the inverse displacement analysis. It means that the parallel robot has two working modes. **Figure 2(a)** (i) and (iii) represents the configurations of the parallel robot of the two working modes.

In the forward displacement analysis, positions x and y of the hand are solved by the simultaneous equations about Eq. (1) of each arm as,

$$\begin{cases} (x - q_1)^2 + y^2 = l_1^2 \\ (x - q_2)^2 + y^2 = l_2^2 \end{cases} \tag{7}$$

Each equation of (7) represents that a circle of the central position is $(q_i, 0)$ and the radius is l_i. The solution of Eq. (7) is given as the crossing point of the two circles. There exist two crossing points of the circles; namely, the parallel robot has two assembly modes. **Figure 2(b)** (i) and (iii) represents each configuration of the assembly mode of the parallel robot.

When $y = l_i$, at that time, the value in the square root in Eq. (6) becomes zero and the inverse displacement analysis has a duplication solution. When $y > l_i$, at that time, value in the square root in Eq. (6) becomes negative and the inverse displacement analysis has no solution. When

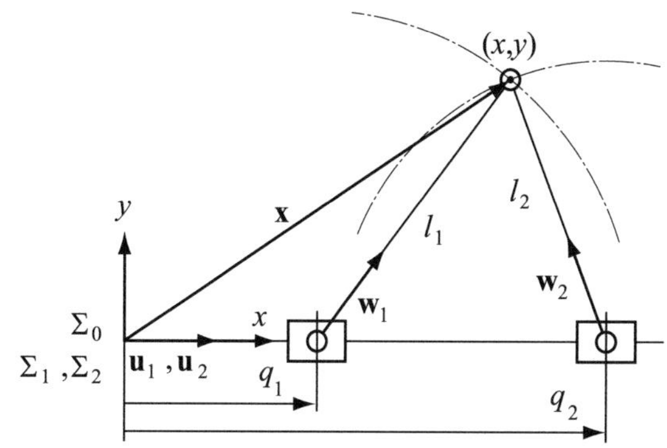

Figure 3. Kinematic model of 2PRR parallel robot.

the distance between the two circles of Eq. (7) equals $(l_i + l_2)$, the forward displacement analysis has a duplication solution. When the distance between the two circles of Eq. (7) becomes larger than $(l_i + l_2)$, the forward displacement analysis has no solution.

2.3. Singularity analysis of the parallel robot

2.3.1. The Jacobian of the parallel robot

Differentiating the both sides of Eq. (1) with respect to time, one obtains

$$\dot{\mathbf{x}} = \dot{q}_i \boldsymbol{u}_i + l_i \dot{\mathbf{w}}_i \tag{8}$$

In Eq. (1), $\mathbf{b}_i$, $\mathbf{u}_i$, and l_i are constant values and their time derivatives become zero. Because $\mathbf{w}_i$ is unit direction vector, one obtains

$$\mathbf{w}_i^T \mathbf{w}_i = 1 \tag{9}$$

Differentiating the both sides of Eq. (9) with respect to time, one obtains

$$\boldsymbol{w}_i^T \dot{\mathbf{w}}_i = 1 \tag{10}$$

Multiplying the both sides of Eq. (8) by $\mathbf{w}_i^T$ and taking into consideration about Eq. (9), one obtains

$$\mathbf{w}_i^T \dot{\mathbf{x}} = \mathbf{w}_i^T \boldsymbol{u}_i \dot{q}_i \tag{11}$$

The displacements of all actuators q_1 and q_2 are defined by vector form as $\mathbf{q} = [q_1 \;\; q_2]^T$; then, Eq. (10) is expressed by matrix form as

$$\mathbf{J}_x \dot{x} = \mathbf{J}_q \dot{q} \tag{12}$$

$$\mathbf{x} = [\dot{x} \ \ \dot{y}]^T, \mathbf{q} = [\dot{q}_1 \ \ \dot{q}_2]^T$$

$$\mathbf{J}_x = \begin{bmatrix} \boldsymbol{w}_1^{\ T} \\ \boldsymbol{w}_2^{\ T} \end{bmatrix}, \mathbf{J}_q = \begin{bmatrix} \boldsymbol{w}_1^{\ T}\boldsymbol{u}_1 & 0 \\ 0 & \boldsymbol{w}_2^{\ T}\boldsymbol{u}_2 \end{bmatrix}$$

$\mathbf{J}_x$ and $\mathbf{J}_q$ are the Jacobian of the parallel robot.

2.3.2. *Second kind of singularity*

When the Jacobian matrix $\mathbf{J}_x$ becomes singular, namely the determinant of the Jacobian becomes zero as

$$\det(J_x) = 0 \tag{13}$$

Then, the parallel robot is in the second kind of singularity [2] or in direct kinematics singularity [3]. The 2*PRR* parallel robot occurs in the second kind of singularity when the direction of the two rods $\mathbf{w}_1$ and $\mathbf{w}_2$ becomes identical.

$$\det\left(\begin{bmatrix} \boldsymbol{w}_1^{\ T} \\ \boldsymbol{w}_2^{\ T} \end{bmatrix}\right) = 0 \tag{14}$$

This case corresponds to that two rods of the parallel robot lay in one line as shown in **Figure 2(b)** (ii). As shown in **Figure 2(b)**, the parallel robot encounters the second kind of singularity during the assembly mode change.

2.3.3. *First kind of singularity*

When the Jacobian matrix $\mathbf{J}_q$ is singular

$$\det(J_q) = 0 \tag{15}$$

Then, the parallel robot is in the first kind of singularity [2] or in inverse kinematics singularity [3]. The 2*PRR* parallel robot occurs in the first kind of singularity when the direction of at least one rod is perpendicular to the direction of the actuator as shown in **Figure 2(a)** (ii).

$$\det\left(\begin{bmatrix} \boldsymbol{w}_1^{\ T}\boldsymbol{u}_1 & 0 \\ 0 & \boldsymbol{w}_2^{\ T}\boldsymbol{u}_2 \end{bmatrix}\right) = 0 \tag{16}$$

As shown if **Figure 2(a)**, the parallel robot encounters the first kind of singularity during the working mode change.

2.3.4. *The third kind of singularity*

The third kind of singularity [2] occurs when both $\mathbf{J}_q$ and $\mathbf{J}_x$ are simultaneously singular. It requires certain conditions of the linkage parameter. In the case of the 2*PRR* parallel robot, when the lengths of the rods are identical, the third kind of singularity occurs as shown in **Figure 4**. The third kind of singularity is referred as the combined singularity [3].

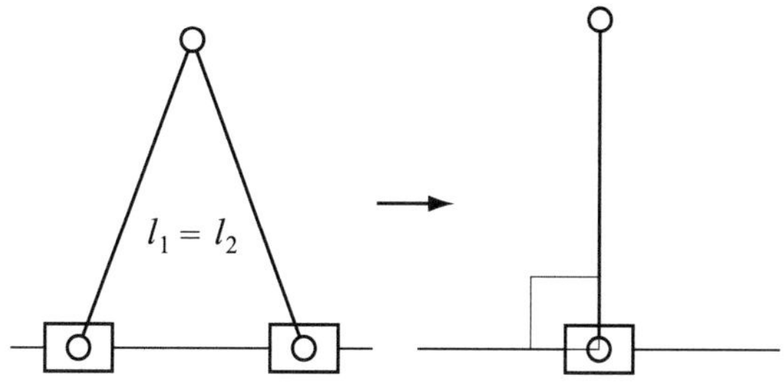

Figure 4. The third kind of singularity of the 2PRR parallel robot.

2.4. Passing the singular configuration during the mode changes

2.4.1. Passing the second kind of singularity during the assembly mode change

2.4.1.1. Using the inertia

As shown in **Figure 5(a)**, if the parallel robot has some speed just before the singular configuration, the robot keeps moving according to the law of inertia; then, the parallel robot passes

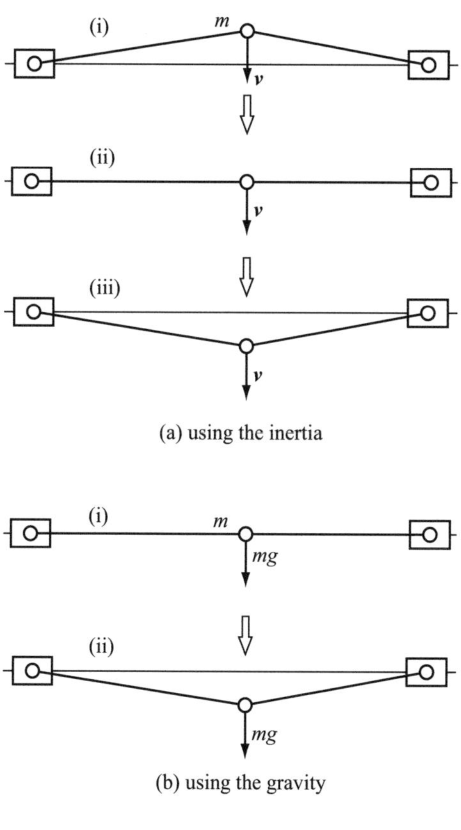

Figure 5. Assembly mode change of nonredundant 2PRR parallel robot.

the second kind of singularity. Note here if the parallel robot stops at the singular configuration, it is impossible to pass the singular configuration by the inertia.

2.4.1.2. *Using the gravity*

As shown in **Figure 5(b)**, when the gravity force acts to the lower direction, the parallel robot can pass the singular configuration even if the parallel robot stops at the singular configuration. However, the robot cannot move to the upper direction against the gravity. When the robot moves on the horizontal plane where the gravity force does not act on the robot, the parallel robot cannot pass the singular configuration.

2.4.1.3. *Using the redundancy*

A robot has actuation redundancy when it is driven by number of actuators greater than the degree of freedom. The actuation redundancy may increase the cost of the robot and complexity of control. However, the actuation redundancy is one of the most effective methods for avoiding the singularity during the mode change.

Figure 6 represents 3*PRR* parallel robot, a planar two-dof parallel robot redundantly actuated three actuators. The Jacobian of the parallel robot is given as

$$\mathbf{J}_x \dot{x} = \mathbf{J}_q \dot{q} \tag{17}$$

$$\mathbf{x} = [\dot{x} \quad \dot{y}]^T, \mathbf{q} = [q_1 \quad q_2 \quad q_3]^T$$

$$\mathbf{J}_x = \begin{bmatrix} \boldsymbol{w}_1^T \\ \boldsymbol{w}_2^T \\ \boldsymbol{w}_3^T \end{bmatrix}, \mathbf{J}_q = \begin{bmatrix} \boldsymbol{w}_1^T \boldsymbol{u}_1 & 0 & 0 \\ 0 & \boldsymbol{w}_2^T \boldsymbol{u}_2 & 0 \\ 0 & 0 & \boldsymbol{w}_2^T \boldsymbol{u}_2 \end{bmatrix}$$

where $\mathbf{J}_x$ is 3×2 matrix of its full rank that equals two. For convenience, singularity analysis is applied to the 3×2 transposed Jacobian matrix of $\mathbf{J}_x^T$. When at least one 2×2 minor of $\mathbf{J}_x^T$ is nonsingular, the rank of the $\mathbf{J}_x$ equals two; namely, the $\mathbf{J}_x$ has full rank. At this time, the robot still works as two-dof parallel robot. For example, when the first rod and the second rod are collinear as shown in **Figure 6** (iii), 2×2 minors of the $\mathbf{J}_x^T$ become

$$\begin{cases} \det([w_1 \quad w_2]) = 0 \\ \det([w_1 \quad w_3]) \neq 0 \\ \det([w_2 \quad w_3]) \neq 0 \end{cases} \tag{18}$$

As shown in Eq. (18), one minor is singular and the other two minors are nonsingular; namely, the $\mathbf{J}_x$ still has the full rank.

Now, the parallel robot loses the redundancy but keeps the nonsingularity. The parallel robot can pass the singular configuration of **Figure 6** (iii). In the same way, the parallel robot can pass the singular configuration of **Figure 6** (ii). Thus, the parallel robot achieves the assembly mode change from **Figure 6** (i) to (iv).

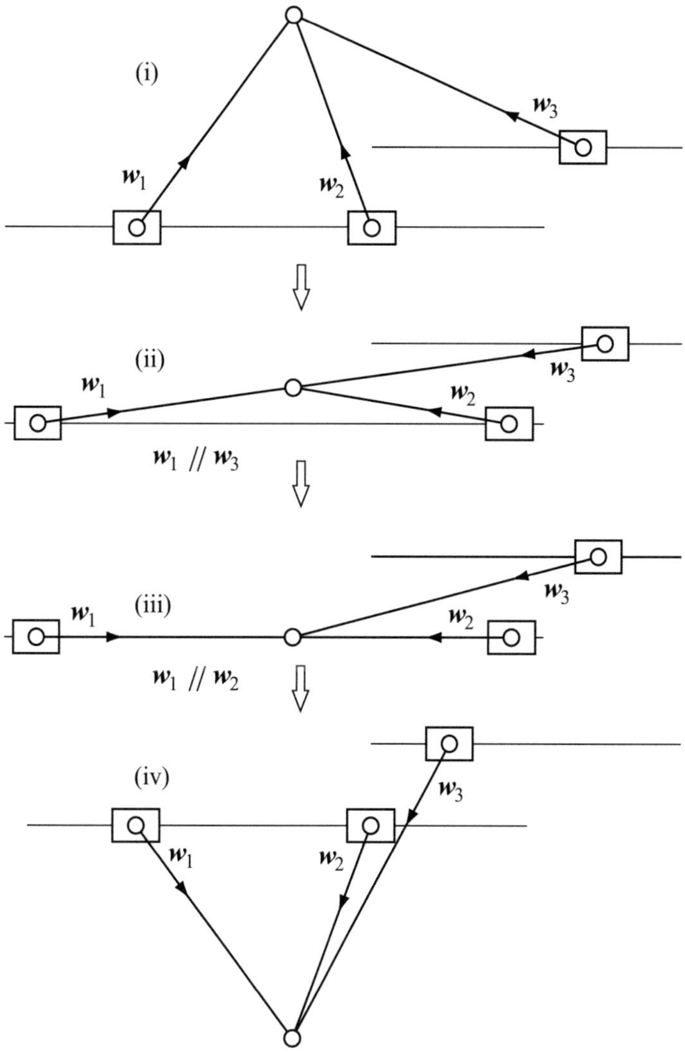

Figure 6. Assembly mode change of redundant 3*PRR* parallel robot.

2.4.2. Passing the first kind of singularity during the working mode change

The parallel robot encounters the first kind of singularity during the working mode change. In the velocity control of a robot, the velocity of the actuator is controlled to trace the desired velocity of the actuator $\dot{\mathbf{q}}_r$, which is given from the desired velocity of the end-point $\dot{\mathbf{x}}_r$ from Eq. (12) as

$$\dot{\mathbf{q}}_r = \left(J_q^{-1} J_x\right) \dot{\mathbf{x}}_r \tag{19}$$

However, $\mathbf{J}_q^{-1}$ cannot be calculated when $\mathbf{J}_q$ is singular at the first kind of singularity. In this case, the desired velocity of the actuator is directly given instead of being indirectly given from Eq. (12).

2.5. Researches on mode change of the parallel robot

Researches on expanding the workspace by the mode change of the parallel robot have been reported for nonredundant two-dof planar robot [4, 5], nonredundant three-dof spatial translational robot [6], redundantly driven three-dof translational and rotational planar robot [7], and redundantly driven three-dof spatial translational robot [8]. Redundant actuation is the actuation in one of the most effective methods for avoiding the singularity. However, the redundancy is not always the answer to avoiding the singularity. Additional ingenuities, for example, path planning for mode change, or asymmetrical design for the robot, are required.

3. Expanding the rotational workspace

3.1. Conventional methods

In this section, methods for expanding the rotational workspace of the parallel robot are introduced. **Figure 7(a)** represents the Stewart Platform [9], six-dof parallel robot with three-dof translations and three-dof rotations. The moving plate is driven by linear actuator embedded six limbs. Each limb is paired by a passive universal joint (*U*) with the base plate and paired by a passive spherical joint (*S*) with the moving plate. The Stewart Platform is categorized into 6*U*<u>*P*</u>*S* parallel mechanism. The Stewart Platform generates high power with a hydraulic linear actuator. Flight simulator and driving simulator for carrying heavy cockpit of aircraft are typical applications of the Stewart Platform. However, the robot has a drawback of small rotation around z axis because of the mechanical interference between the limbs. **Figure 7(b)** represents a six-dof cable-driven parallel robot [10]. This robot also has the same drawback of small rotation around z axis because of the cable interference. Small rotation of parallel robots puts restriction to their applications.

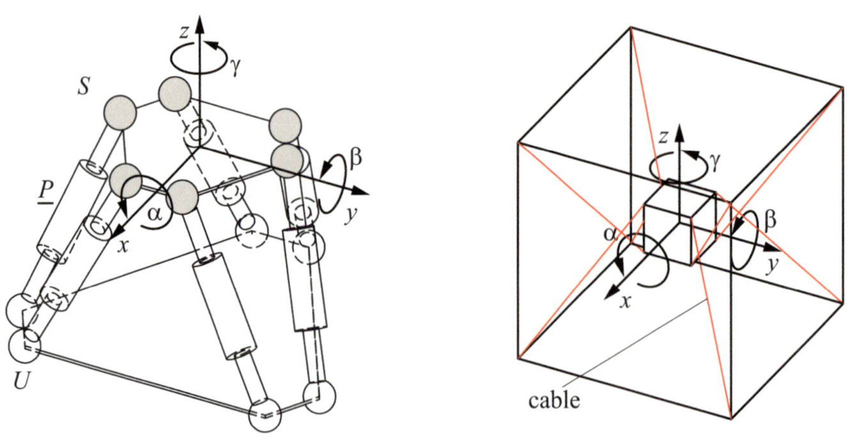

(a) Stewart Platform

(b) Cable driven parallel robot

Figure 7. Conventional six-dof parallel robot.

In order to cope with the problem, Clavel invented a novel parallel robot DELTA [11] as shown in **Figure 8**. The DELTA generates three-dof translation and one-dof rotation around the z-axis.

Three control arms are rotated by three motors located on the base plate. Each arm is connected to the moving part by a parallelogram rod with spherical joints. The gripper on the moving part is connected to the motor on the base plate with a telescopic arm via universal joints (**Figure 9**).

Researches on expanding the rotational workspace of the DELTA like parallel robot have been reported from research group at the Montpelier University. The basic idea is to convert the translational motion to the rotational motion by the mechanical gimmicks such as pulley, rack and pinion, and screw [12]. Crank embedded moving part has been proposed by McGill University for two-limb parallel robot [13] and by Fraunhofer IPA for cable-driven parallel robot [14].

3.2. Differential screw drive system

Recently, differential screw drive systems composed of two screw-nut pairs of different leads have been applied to robot systems [15, 16]. There are four driving methods of the differential drive systems;

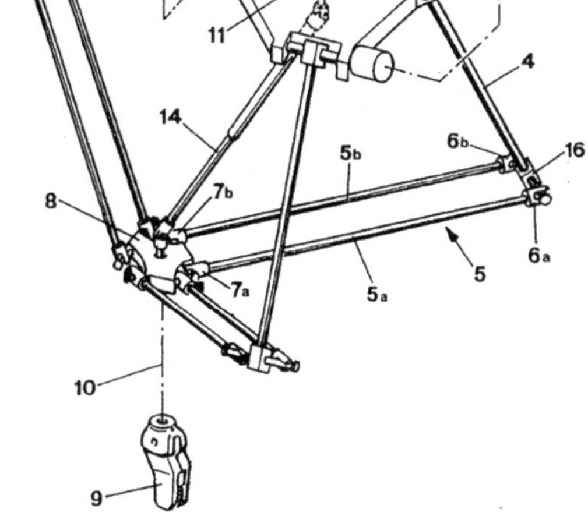

1. base member
2. axes of rotation
3. fixed portion
4. control arms
5. linking bars
6. double articulations in cardan form
7. two double articulations
8. cardan type to the movable member
9. working member (gripper)
10. axis of the rotation of the working member
11. fixed motor
12. managing computer
13. three actuators
14. telescopic arm
15. one end of the control arm
16. the other end of the control arm

Figure 8. The DELTA robot [11].

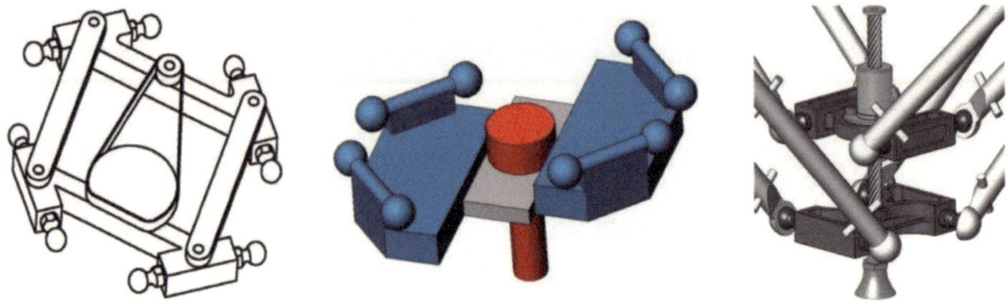

Figure 9. Mechanical gimmicks enlarge the rotational workspace of parallel robot [12].

(a) Rotations of the two nuts are converted to the rotation and translation of the screw pair, which is coaxially arranged with one end interconnected with each other.

(b) Translations of the two nuts are converted to the rotation and translation of the screw pair, which is coaxially arranged with one end interconnected each other.

(c) Rotations of the two screws are converted to the rotation and translation of the nut pair, which is coaxially arranged with one end interconnected each other.

(d) Translations of the two screws are converted to the rotation and translation of the nut pair, which is coaxially arranged with one end interconnected each other.

In this section, kinematics of the differential drive system (b) as shown in **Figure 10** is discussed for enlargement of the rotational workspace of the parallel robot.

Let l_i $(i = 1, 2)$ be the lead of the ith screw nut. n_i $(i = 1, 2)$, z, and γ represent the position of the ith nut, position, and angle of the screw pair, respectively. Relation of these parameters is given as

$$n_i = z - \frac{l_i}{2\pi}\gamma \tag{20}$$

Equation (20) is given in the matrix form as

$$\begin{bmatrix} n_1 \\ n_2 \end{bmatrix} = \begin{bmatrix} 1 & -\frac{l_1}{2\pi} \\ 1 & -\frac{l_2}{2\pi} \end{bmatrix} \begin{bmatrix} z \\ \gamma \end{bmatrix} \tag{21}$$

Equation (21) gives the inverse kinematics of the differential drive system. The forward kinematics of the system is derived by inverting Eq. (21) as

$$\begin{bmatrix} z \\ \gamma \end{bmatrix} = \frac{1}{l_2 - l_1} \begin{bmatrix} l_2 & -l_1 \\ 2\pi & -2\pi \end{bmatrix} \begin{bmatrix} n_1 \\ n_2 \end{bmatrix} \tag{22}$$

If two leads of the screw-nut pair are identical as

$$l_1 = l_2 \tag{23}$$

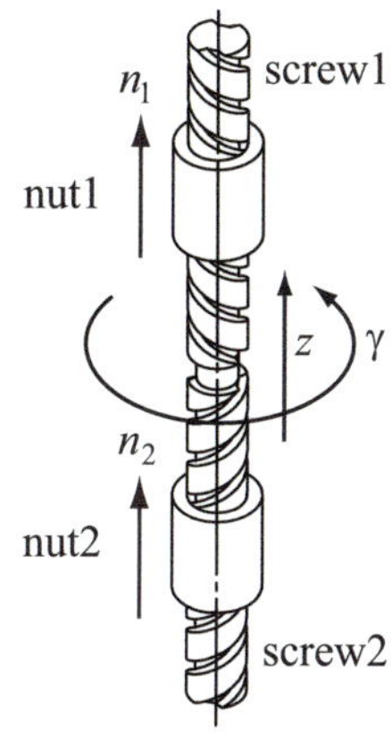

Figure 10. Differential screw drive system (translations of the two nuts).

then the matrix in Eq. (22) becomes singular. It is necessary that the leads of the two screw nuts are different from the differential drive system.

Figure 11 represents a 4-dof parallel robot that the differential drive system (a) embedded in the moving part [15, 16]. *H* represents a passive helical joint composed of screw and nut pair. In **Figure 11**, two *H* pairs have the same lead, but the different helix hands (right and left). In **Figure 9 (c)**, one *H* pair and one *R* pair are embedded in the moving part. This is included in the differential drive system (a), one of the *H* pair has zero lead.

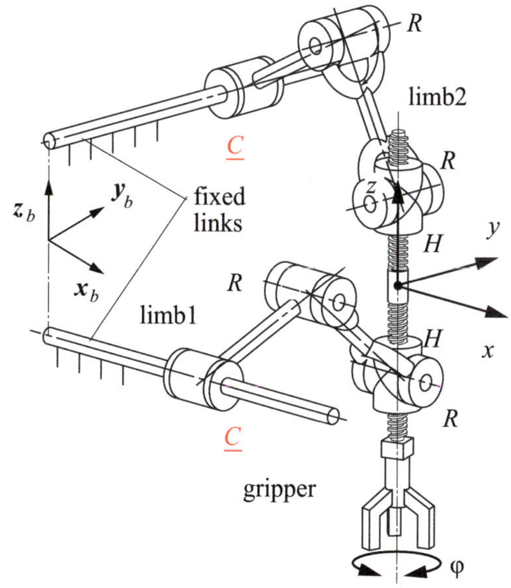

Figure 11. Parallel robot with the differential drive system embedded moving part [15].

4. Conclusion and future works

After commercialization of the DALTA robot, a lot of parallel robots can be found in machine factory and other industries. Recently, one can get the kit of a 3D printer machine by linear DELTA mechanism under 500 USD. It may be said that the first-generation parallel robot such as DELTA and Stewart Platform is getting to reach the mature stage. In this chapter, expanding the workspace of the parallel robot was introduced. The translational workspace was expanded by singularity-free mode change using the actuation redundancy. The differential drive mechanism converts the translational motion to the rotational motion, which expands the rotational workspace of the parallel robot. These parallel robots are expected as the next-generation robot.

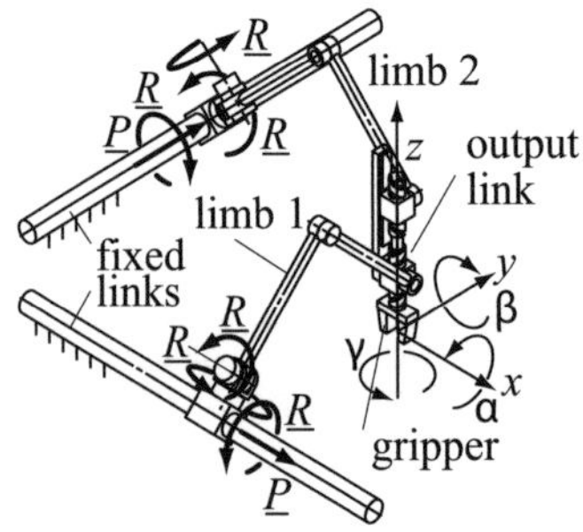

(a) 6-dof parallel robot

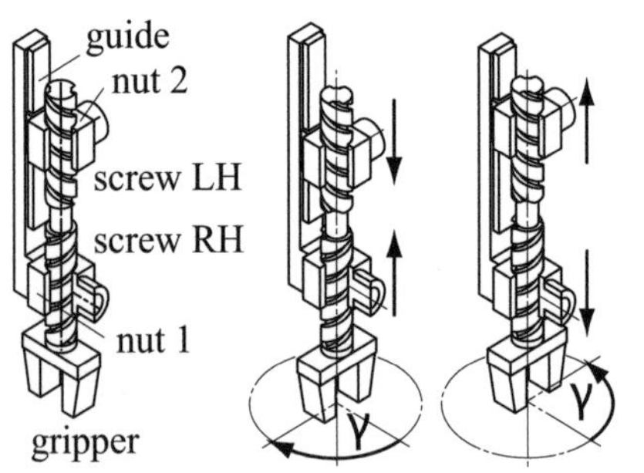

(b) differential screw embedded output link

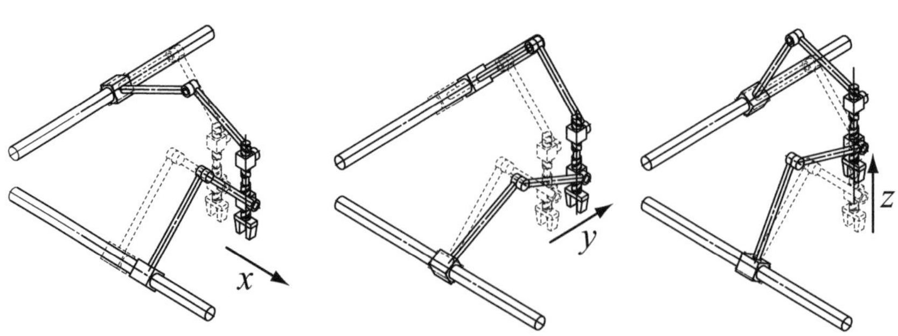

Figure 12. Two-limb six-dof parallel robot with the differential drive embedded output link [18].

http://dx.doi.org/10.5772/intechopen.71407

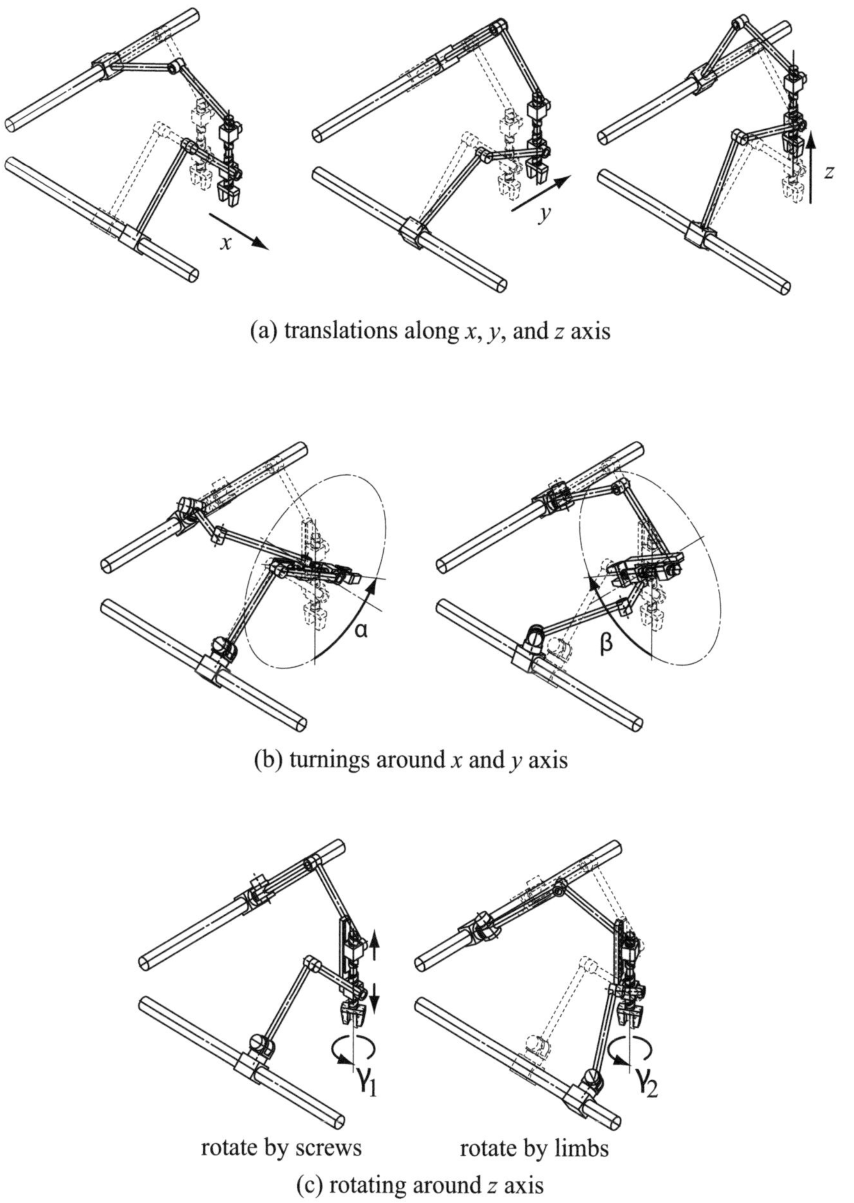

Figure 13. Motion of the two-limb six-dof parallel robot.

At the end of this chapter, an example of the next-generation parallel robot is introduced. **Figure 12(a)** represents a novel two-limb six-dof parallel robot [17]. As shown in **Figure 12(b)**, a differential screw mechanism is embedded in the output link for enlarging the rotational workspace of the gripper. This parallel robot extends its degree of freedom from four (three translations and one rotation as shown in **Figure 11**) to six (three translations and three rotations) as shown in **Figure 13**. We are working on the kinematic analysis and detailed mechanical design of the parallel robot [18].

Acknowledgements

This work was supported by JSPS KAKENHI Grant Number 15K05918 and 24560314.

Author details

Takashi Harada

Address all correspondence to: harada@mech.kindai.ac.jp

Department of Mechanical Engineering, Faculty of Science and Engineering, Kindai University, Higashiosaka, Osaka, Japan

References

[1] Merlet JP. Parallel Robots. 2nd ed. Netherlands: Springer; 2006. 402 p. DOI: 10.1007/1-4020-4133-0

[2] Gosselin C, Angeles J. Singularity analysis of closed-loop kinematic chains. IEEE Transactions on Robotics and Automation. 1990;**6**(3):281-290. DOI: 10.1109/70.56660

[3] Tsai LW. Robot Analysis. 1st ed. New York: Wiley-Interscience; 1998. 520 p. ISSN: 978-0471325932

[4] Hesselbach J, Woffmeister H, Loohß T, Krefft M, Armbrecht C. Parallel kinematic concept for stationary high performance cutting in wood machining. Production Engineering. 2007;**1**(2):205-212. DOI: 10.1007/s11740-007-0033-9

[5] Campos L, Bourbonnais F, Bonev I and Bigras P. Development of a five-bar parallel robot with large workspace. In: Proceedings of the ASME 2010 International Design Engineering Technical Conferences & Computers and Information in Engineering Conference; 15-18 August 2010; Montreal. 2010. pp. 917-922. DOI: 10.1115/DETC2010-28962

[6] Budde C, Last P and Hesselbach J. Development of a triglide-robot with enlarged workspace. In: Proceedings of IEEE International Conference on Robotics and Automation; 10-14 April 2007; Roma. 2007. pp. 543-548. DOI: 10.1109/ROBOT.2007.363043

[7] Harada T. Mode changes of a planar 3 DOF redundantly actuated parallel robot. International Journal of Materials, Mechanics and Manufacturing. 2016;**4**(2):123-126. DOI: 10.7763/IJMMM.2016.V4.238

[8] Harada T. Mode changes of redundantly actuated asymmetric parallel mechanism. Journal of Mechanical Engineering Science. 2015;**230**(3):454-462. DOI: 10.1177/0954406215588479

[9] Steward D. A platform with six degrees of freedom. Proceedings of the Institution of Mechanical Engineers. 1965;**180**(1):371-386. DOI: 10.1243/PIME_PROC_1965_180_029_02

[10] Bruckmann T, Pott A, editors. Cable-Driven Parallel Robots. 1st ed. New York: Springer; 2013. 454 p. DOI: 10.1007/978-3-642-31988-4

[11] Clavel R. Device for the Movement and Positioning of an Element in Space. US Patent No. 4,976,582. 1989

[12] Company O, Pierrot F, Krut S, Nabat V. Simplified dynamic modelling and improvement of a four-degree-of-freedom pick-and-place manipulator with articulated moving platform. Journal of Systems and Control Engineering. 2009;**223**(1):13-29. DOI: 10.1243/09596518JSCE616

[13] Gauthier J, Angeles J, Nokleby SB, Morozov A. The kinetostatic conditioning of two-limb schonflies motion generators. Transactions of the ASME Journal of Mechanisms and Robotics. 2008;**1**(1):011010. DOI: 10.1115/1.2960544

[14] Miermeister P, Pott A. Design of cable-driven parallel robots with multiple platforms and endless rotating axes. In: Kecskeméthy A, Flores FG, editors. Interdisciplinary Applications of Kinematics. 1st ed. Lima: Springer; 2013. p. 21-29. DOI: 10.1007/978-3-319-10723-3_3

[15] Harada T, Angeles J. Kinematics and singularity analysis of a CRRHHRRC Schönflies motion generator. Transactions of the Canadian Society for Mechanical Engineering. 2014;**38**(2):173-183. ISSN: 0315-8977

[16] Harada T, Friedlaender T and Angeles J. The development of an innovative two-DOF cylindrical drive: Design, Analysis and preliminary tests. In: Proceedings of IEEE International Conference on Robotics and Automation; 31 May-7 June 2014; Hong Kong. 2014. pp. 6338-6344. DOI: 10.1109/ICRA.2014.6907794

[17] Harada T and Makino T. Design of a novel 6 DOF parallel robot for haptic device. In: 12th International Conference on Ubiquitous Healthcare; October 30–November 2; Osaka, Japan. 2015

[18] Harada T and Angeles J. From the McGill Pepper-Mill Carrier to the Kindai Atarigi Carrier. In: Proceedings of the 2017 IEEE International Conference on Robotics and Biomimetics; December 5-8; Macau SAR, China. 2017. Printing

Chapter 6

Kinematic and Biodynamic Model of the Long Jump Technique

Milan Čoh, Milan Žvan and Otmar Kugovnik

Additional information is available at the end of the chapter

http://dx.doi.org/10.5772/intechopen.71418

Abstract

The main aim of the study was to determine the kinematic model for long jump and define the kinematic and dynamic parameters of an elite long jumper's technique. The theoretical model was based on real data where the jumper was defined with a joint mass point. In view of certain previous similar studies, our study identified kinematic and dynamic parameters directly without using the inverse mechanics method. The analysis was made on two jumps of the top level athlete G.C., who won the bronze medallion in long jump at the World Championships in Seville. The kinematic parameters of the take-off, flight and landing were measured with a 3-D video ARIEL system (Ariel Dynamics Inc., USA). The dynamic characteristics of take-off in the X, Y and Z axes were registered with a force-platform (KISTLER-9287), which was installed immediately prior the take-off board. The take-off efficiency was defined best by the following parameters: horizontal velocity, VXTO—8.10 m s^{-1}; vertical velocity, VYTO—3.90 m s^{-1}; angle of projection, PATO—24.1°; duration of compression phase, TDMKF—84 ms, duration of lift phase, MKFTO—43 ms and maximal force in Y-vertical axis, FYMAX—5132 N. An important factor of a rational technique of long jump is also the landing, which is defined by the landing distance and fall-back distance. The efficiency of the landing depended on the landing distance L3—0.63 m and fall-back distance LFB, which amounted to 0.15 m.

Keywords: long jump, technique, kinematics, dynamics, model, top sport

1. Introduction

The long jump consists of four interconnected phases: approach, take-off, flight and landing. According to some existent studies [2, 6, 9, 11], the approach and take-off are the most important factors that affect the result. The fundamental problem of long jump, from the biomechanical point of view, is the transformation of horizontal velocity to a resultant of the vertical and horizontal velocities in the take-off phase. It is very important that the athlete realises the greatest possible vertical velocity with the smallest possible loss of horizontal velocity. The jumping distance is defined, according to the theoretical model after Ballreich and Bruggemann, [1], by the take-off distance L1, flight distance L2 and landing distance L3. The CM flight parabola is defined by the following parameters: relative height of CM at take-off, resultant of the vertical and horizontal velocity of take-off, take-off angle and air resistance [1, 5, 6, 7]. Most of the biomechanical studies of long jump technique till now dealt with studying the kinematic characteristics with high-frequency film or video cameras. However, there are very few studies on top long jumpers with emphasis on the dynamic characteristics of take-off, which is the most important generator of the long jump result. The main purpose of our study was therefore a complex analysis of both the kinematic model of long jump and the dynamic model of take-off.

2. Methods and procedures

2.1. Kinematic model

The analysis was performed on two jumps of one of the world's best long jumpers G.C. (GC—body height 178.5 cm, body mass 69 kg, age 24 years, personal best 8.40 m). The length of the first analysed jump was 7.93 m and the second 8.25 m.

Kinematic parameters were obtained with a 3-D kinematic system ARIEL (Ariel Dynamics Inc., USA) with four synchronised cameras (Sony—DVCAM DSR 300 PK), shooting at 100 Hz. The first two cameras, set a 90° angle to the filmed object, were used to analyse the last two

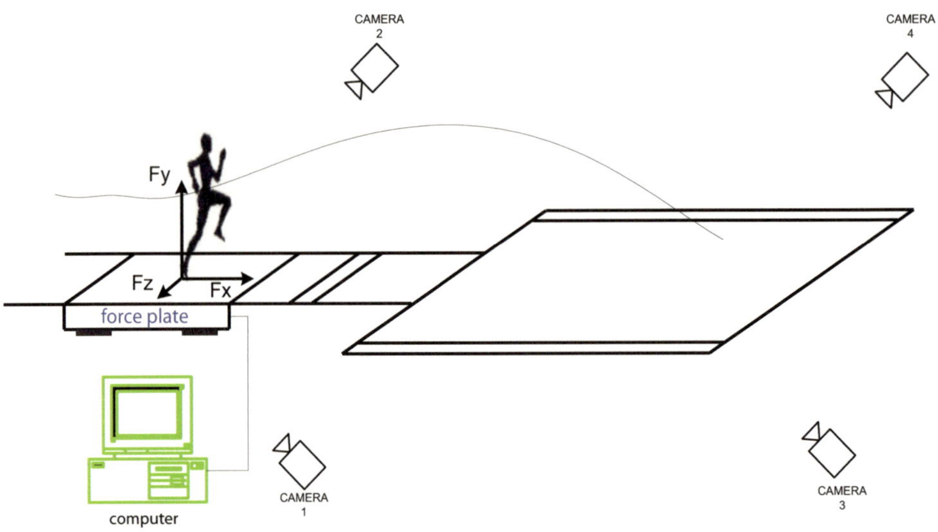

Figure 1. Measurement system for kinematic and dynamic parameters of the take-off in long jump.

http://dx.doi.org/10.5772/intechopen.71418

strides of the approach and the take-off, with the other two the flight and landing (**Figure 1**). In the kinematic analysis, a 15-segment model was digitised. The model segments represent parts of the body, connected with point joints. The masses and centres of gravity of the segments and the common body centre of gravity (CM) were computed according to the anthropometric model of Dempster [1]. All kinematic parameters were filtered with a seventh-level Butterworth filter.

2.2. Dynamic model

The dynamic parameters of take-off were registered with a force-platform Kistler 9287, area 900 × 600 mm, covered with a tartan surface and installed before the take-off board on the approach track of long jump. Forces were measured in three directions: X—horizontal, Y—vertical and Z—lateral (**Figure 2**).

The jump length (L) was measured from the front impression of the foot of the take-off leg on the force-platform to the point of contact of the feet in the sand of the landing pit. The programme package Matlab (Mathworks Inc., USA) was used to analyse the measured forces. The registration frequency on the force-platform was 2000 Hz.

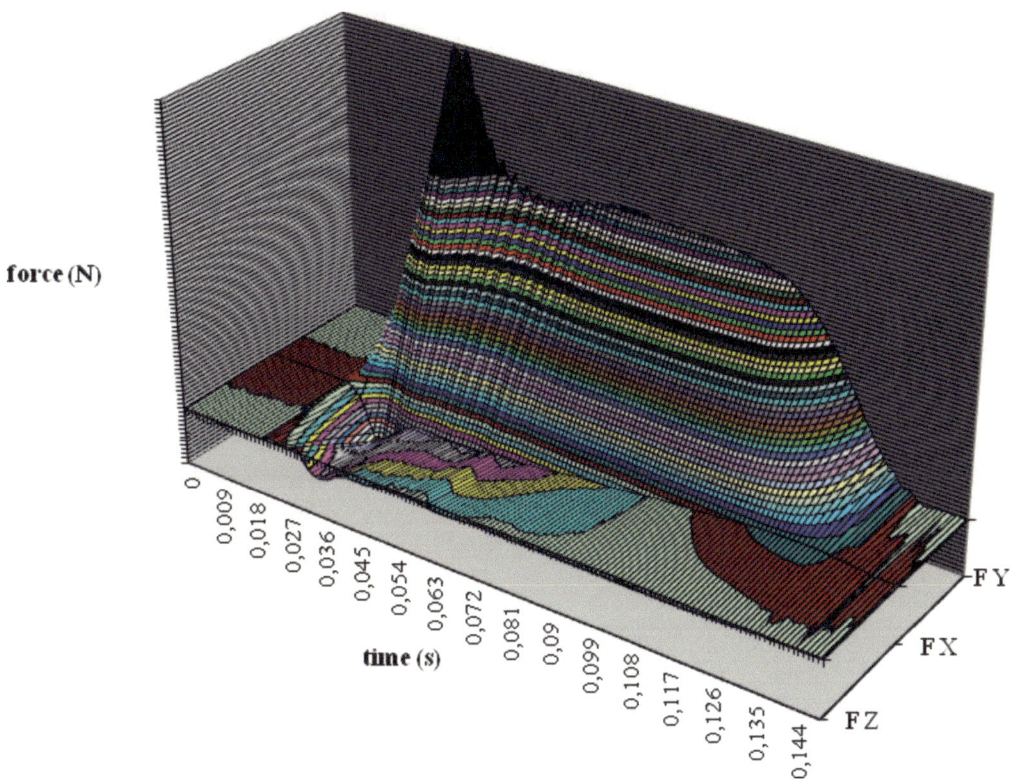

Figure 2. Take-off phase 3D-surface contour plot.

2.3. Optimising angle of projection at take-off action

We ask ourselves how to determine the best angle of projection at take-off for long jumpers. In our case, we model jumpers as mass points and observe their trajectories for a case where there is no air friction. We already know that energy losses have to be taken into consideration because jumper always loses some energy in a take-off phase, when he/she is transforming his/her horizontal velocity before he/she jumps into combination of horizontal and vertical velocity.

Let us first look at the example when height of centre of mass at take-off and height of centre of mass when he/she first touches the ground are the same. Let us denote velocity before jump with v_0 and velocity after take-off with v_1. Angle of projection shall be denoted with φ and jumpers mass with m. All mechanical energy before take-off is $W_0 = \frac{1}{2} m v_0{}^2$. This is jumpers kinetic energy. δW_0 denotes the amount of energy that is lost at the time of take-off as heat and sound [16], and of energy before take-off transforms into vertical kinetic energy. We denote this part as $\Delta W = \frac{1}{2} m v_1{}^2 (\sin\varphi)^2$. We can therefore write:

$$W_1 = W_0 - \delta W - \Delta W \tag{1}$$

which means

$$\frac{1}{2} m v_1{}^2 = \frac{1}{2} m v_0{}^2 - \frac{\delta}{2} m v_0{}^2 - \frac{1}{2} m v_1{}^2 (\sin\varphi)^2. \tag{2}$$

From this we get

$$v_1 = \sqrt{\frac{1-\delta}{1+\sin\varphi^2}}\, v_0. \tag{3}$$

The length of a jump of a mass point, which in our case is the centre of mass, is

$$D = D(\varphi) = \frac{v_1{}^2 \sin 2\varphi}{g} = \frac{\sin 2\varphi}{1+(\sin\varphi)^2}\, p, \tag{4}$$

where

$$p = \frac{(1-\delta) v_0{}^2}{g}. \tag{5}$$

To find the optimal value of φ, we need to find the extreme of function $D = D(\varphi)$. We do so by finding the zeros of the first derivate of function $D = D(\varphi)$.

$$\frac{\partial D}{\partial \varphi} = \frac{3\cos 2\varphi - 1}{(1+(\sin\varphi)^2)^2}\, p. \tag{6}$$

We must also take a look at the second derivate to establish if angle that we compute is maximum or minimum

$$\frac{\partial^2 D}{\partial \varphi^2} = \frac{4(3 \sin 4\varphi + 14 \sin 2\varphi)}{(\cos 2\varphi - 3)^2}. \tag{7}$$

Zeros of this function coincide with zeros of $f(\varphi) = 3 \cos 2\varphi - 1$. From this equation we get the optimal angle of projection

$$\varphi_{opt} = \frac{1}{2} \cos^{-1} \frac{1}{3} = 35.26°. \tag{8}$$

With this we have established that optimal angle is independent of values v_0 and δ. The fact $\frac{\partial^2 D}{\partial \varphi^2}(\varphi_{opt}) < 0$ also holds true, which brings us to the conclusion that φ_{opt} in fact is the optimal angle for longest jump, that is, is the maximum of function $D = D(\varphi)$.

Of course height of the centre of mass of the jumper at take-off is bigger than the height of the centre of mass when the jumper first touches the ground because he/she pulls his/her legs closer to their bodies when landing (**Figure 2**—kinematic and biodynamic model of the long jump technique). In this case, the length of the jump is defined as $L = D + d$, where point $(D, -h)$ is defined as an intersection of curves:

$$y(x) = x \tan \varphi - \frac{g x^2}{2 h v_1^2 (\cos \varphi)^2} \tag{9}$$

and $y(x) = -h$, where h denotes the height difference of centre of mass before and after the jump. Computing this intersection gives us

$$D(\varphi, h, v_1) = \frac{v_1^2 \sin 2\varphi}{2g} \left(1 + \sqrt{1 + \frac{2gh}{v_1^2 (\sin \varphi)^2}}\right) \tag{10}$$

which means

$$L(\varphi, h, v_1, d) = \frac{v_1^2 \sin 2\varphi}{2g} \left(1 + \sqrt{1 + \frac{2gh}{v_1^2 (\sin \varphi)^2}}\right) + d. \tag{11}$$

Taking into consideration that $v_1 = \sqrt{\frac{1-\delta}{1+\sin \varphi^2}}\, v_0$ and $p = \frac{(1-\delta) v_0^2}{g}$ we get

$$L(\varphi, p, h, d) = \frac{p}{2} \frac{\sin 2\varphi}{(1+(\sin \varphi)^2)^2} \left(1 + \sqrt{1 + 2 \frac{1+(\sin \varphi)^2}{(\sin \varphi)^2} \frac{h}{p}}\right) + d. \tag{12}$$

In the same way, as we did before, we can compute the maximum value of function $L = L(\varphi, p, h, d)$ by finding the zeros of the function $\frac{\partial L}{\partial \varphi}$. With a lot of algebra and using Wolfram alpha we can show that

$$\varphi_{opt} = 2\tan^{-1}\left(\sqrt{\frac{16h}{p} + 5 - \frac{2\sqrt{2}\sqrt{3p^2 + 20ph + 32h^2}}{p}}\right). \quad (13)$$

In McFarland's research [10], δ is estimated to be 0.1.

3. Main body

3.1. Kinematic analysis

Take-off is one of the key generators of a successful long jump. The take-off model is defined by kinematic (**Table 1** and **Figure 3**) and dynamic parameters. These are interrelated, in accordance with biomechanical laws. In this phase, the jumper must optimally transform the horizontal velocity developed in the approach into speed of take-off.

The take-off speed is the resultant of horizontal velocity at TO and vertical velocity at TO and is one of the most important predictors of an effective jump length. Hay et al. [6], Nixdorf and Bruggemann [14], found that the correlation between the take-off speed and end result is between 0.74 and 0.83. Horizontal velocity at TD for the athlete GC is 9.46 m s^{-1}. In take-off (TD-TO), a reduction of the horizontal velocity of 1.36 m s^{-1} occurs, representing 14.3%. Its

Parameter		Unit	Result
Touchdown			
Length of last stride	(S1)	m	2.05
Touchdown distance	(LTD)	m	0.63
Height of CM at TD	(HTD)	m	0.95
Horizontal velocity at TD	(VXTD)	m s^{-1}	9.46
Vertical velocity at TD	(VYTD)	m s^{-1}	−0.26
Speed at TD	(STD)	m s^{-1}	9.47
Maximum knee flexion	(MKF)	deg.	148.0
Height of CM at max. knee flexion	(HMA)	m	1.01
Take-off			
Take-off distance	(L1)	m	0.29
Height of CM at take-off	(HTO)	m	1.23
Horizontal velocity at TO	(VXTO)	m s^{-1}	8.10
Vertical velocity at TO	(VYTO)	m s^{-1}	3.90
Speed at TO	(STO)	m s^{-1}	8.99
Angular velocity of thigh at TO	(AVTO)	deg s^{-1}	790
Angle of projection at TO	(PATO)	deg.	24.1

Table 1. Kinematic parameters of take-off action.

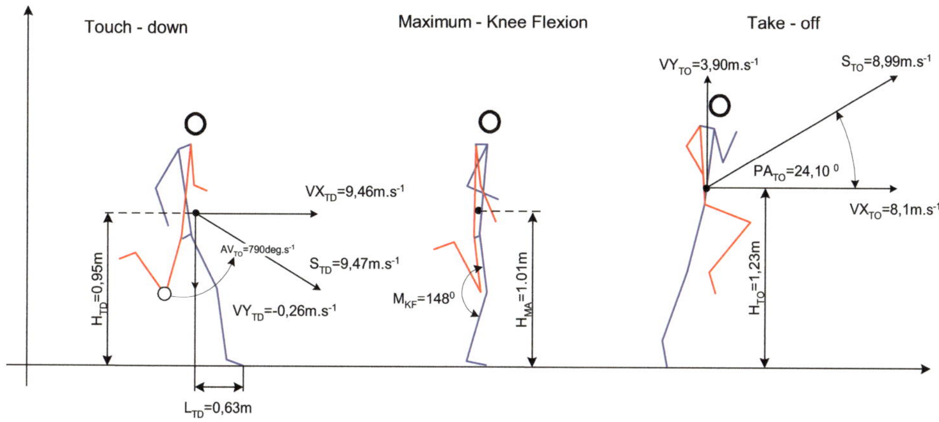

Figure 3. Kinematic model of take-off (G.C., distance 8.25 m).

decrease is connected with the increase of vertical velocity at TO, amounting to 3.90 m s^{-1} (**Figure 3**). We can conclude that the loss in horizontal velocity is proportional to the increase in vertical velocity. The vertical velocity at touchdown is directed downwards and has a negative sign (VYTD = −0.26 m s^{-1}).

The realisation of vertical velocity is directly connected with the magnitude of the projection of CM on the surface at touchdown (touchdown distance = 0.63 m). One of the key biomechanical problems of take-off is how to ensure the greatest possible vertical velocity, while keeping the decrease in horizontal velocity to a minimum. The ratio of the take-off velocity components VXTO:VYTO for the athlete GC is 2.08:1. The results of some similar studies [1, 5, 6, 9, 14] show us that this athlete has an optimal vertical velocity at take-off and a horizontal velocity (VXTO) that is a little too low to allow even longer jumps. The consequence of a relatively high vertical velocity at take-off is also the magnitude of the angle of projection at take-off PATO = 24.10 deg. This magnitude of the angle of projection at take-off later defines the CM parabola of the jumper.

The realisation of a high vertical velocity can be connected with an efficient elevation of CM at take-off. The difference between the lowest position of CM at touchdown and at take-off is 28 cm. This is augmented also with a favourable ratio between the length of the last two strides (2.35 m:2.05 m) and the lowering of CM for 11 cm in the last-but-one stride. The average lowering of CM of the finalists of the 1997 World Championship in Athens was 8 cm [12]. This lowering of CM in the last stride but one increases the vertical acceleration distance in the take-off phase.

A very important factor contributing to the take-off efficiency is the angular velocity of the swing leg (AGTO), 790 deg s^{-1} for the athlete GC. This velocity is not constant, it changes. It is highest in the middle part of the swing amplitude, the lowest at the end of the swing, when the movement of the swing leg is blocked. The consequence of this is the transfer of the force of the mass inertia of this segment (the leg represents, according to the anthropometric model of Dempster, 16.1% of the entire body mass) to the common body centre of gravity of the jumper.

Hay et al. [6] states that the contribution of the swing leg to the entire take-off impulse is between 17 and 20%.

Combining this information with the measured data:

$v_0 = 9.46\frac{m}{s}$, $v_1 = 8.99\frac{m}{s}$, $h = 0.31m$, and $d = 0.63$ we can compute $\varphi_{opt} = 33.39°$ and $L = L(\varphi, p, h, d) = 6.86m$, while the measured distance of the jump was $8.25m$.

Because, we have all the needed data, it is interesting to see how initial velocity before the take-off v_0 affects the optimal angle of projection φ_{opt} computed above. The values for initial velocity were chosen from the interval $[8,11]$ because most elite long jumpers have their initial velocity from this interval (**Figure 4**).

Similarly, we can investigate the relation between coefficient of efficiency δ and the length of the jump (**Figure 5**).

Next graph (**Figure 6**) is in fact actually another graph of correlation between initial velocity before take-off and optimal angle of projection with the difference that this graph shows how the length of jump depends on initial angle of projection for different initial velocities. The optimal angle can be seen as a value of an angle where maximum is achieved.

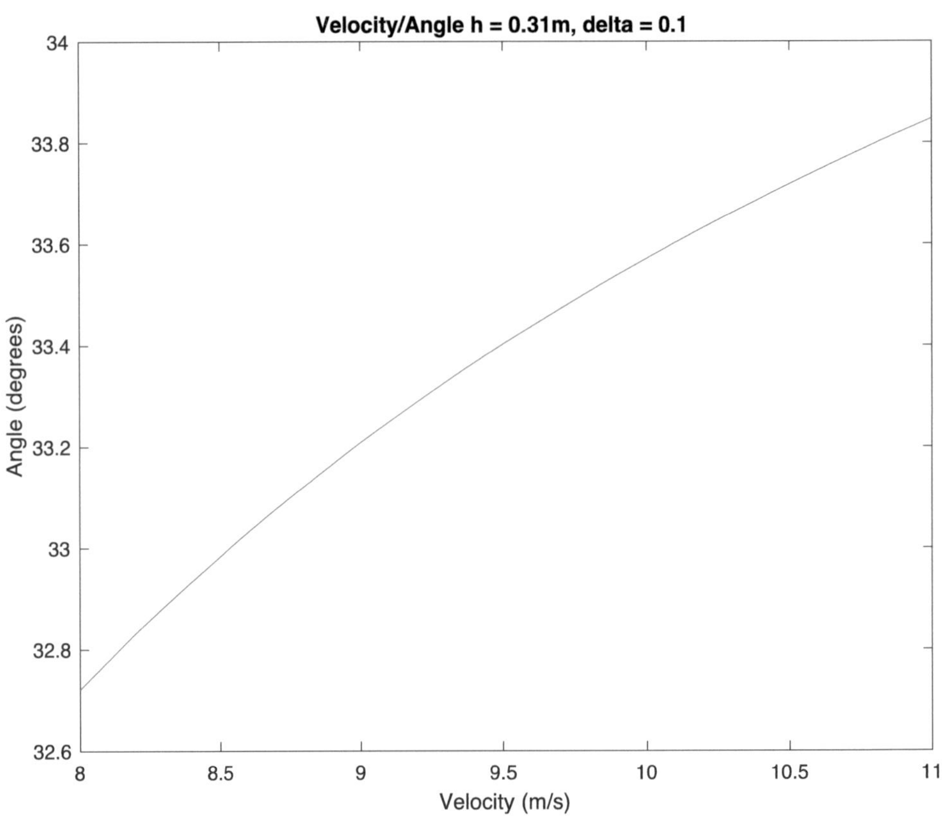

Figure 4. Effect of initial velocity on optimal angle of projection.

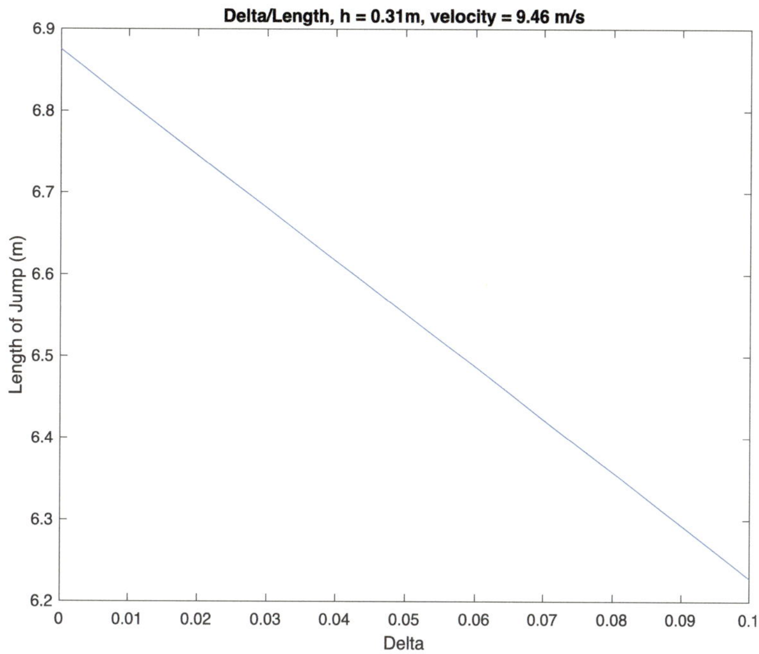

Figure 5. Effect of efficiency coefficient on length of the jump.

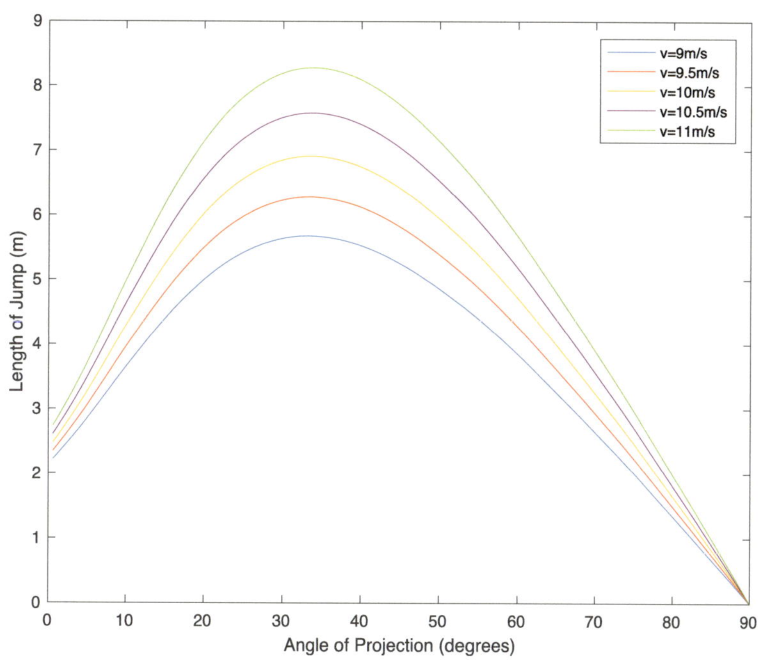

Figure 6. Relation between length of a jump and initial angle of projection.

3.2. Dynamic analysis of take-off action

The dynamic parameters of take-off (**Table 2** and **Figure 7**) were measured directly with a force-platform. In this way, we measured the forces developed by the jumper at take-off in a realistic situation, which is a rarity in studies of this kind. The period of the contact phase was 127 ms. Here, the ratio between the period of compression (84 ms) and the period of lift (43 ms) is important—this was 66%:34% for our athlete, which is a good indicator of an efficient take-off [9, 11], from the point of view of the jump dynamics.

Parameter		Unit	Result
Contact time	(TD-TO)	ms	127
Period of compression phase	(TD-MKF)	ms	84
Period of lift phase	(MKF-TO)	ms	43
Max. force in X—horizontal axis	(FXMAX)	N	4581
Max. force in Y—vertical axis	(FYMAX)	N	5132
Max. force in Z—lateral axis	(FZMAX)	N	1396
Force impulse in compression phase	(FIMCP)	N s	−101.9
Force impulse in lift phase	(FIMLP)	N s	8.1
Total force impulse	(TFIMP)	N s	328.8

Table 2. Dynamic parameters of take-off action.

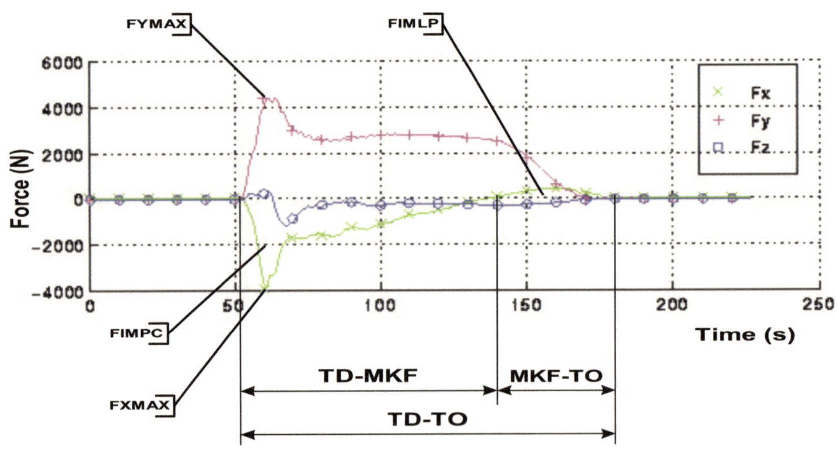

TD-TO	=	0,127	s	MZMAX	=	1395,90	N
TD-MKF	=	0,084	s	TFIMP	=	328,802	N/s
MKF-TO	=	0,043	s	FIMLP	=	8,106	N/s
FXMAX	=	4580,94	N	FIMPC	=	-101,920	N/s
MYMAX	=	5131,57	N				

Figure 7. Force diagram of the take-off phase.

The compression phase by definition lasts from the instant the foot of the take-off leg is placed on the ground till the moment of maximal amortisation in the knee of the take-off leg (TD-MKF). The lift phase then follows till take-off (MFK-TO). In the compression phase, our athlete developed a maximal force of 4581 N in the horizontal direction and 5132 N in the vertical direction. The muscular regime is eccentric in the compression phase period and concentric in the lift phase period. The long jump take-off is a typical example of a two-phase eccentric-concentric muscular contraction, whose efficiency depends mainly on two physiological factors. The first in the switching time—the utilisation of the elastic energy potential saved in the muscles, connecting tissue and tendons. This elastic potential is available only for a certain period of time. This time is defined by the life-time of the cross-bridges between the actin and the myosin in the sarcomeres, lasting from 15 to 100 ms. If the concentric contraction does not follow the eccentric one quickly enough, there is no transfer of the elastic energy from the first to the second phase. Studies have shown that elastic energy is the main force generator in the eccentric (compression) phase and chemical energy of the muscles in the concentric (lift) phase. An efficient integration of elastic and chemical energy in the muscles and tendons can cause a 40% greater resultant force [8, 15, 17].

The second factor of an economic eccentric-concentric contraction is the ability of the muscles to resist rapid extension—stiffness. Stiffness, as a neural mechanism, depends mostly on the pre-activation of the muscles and the action of reflexes: miotatic and Golgi tendon reflex [3, 4, 8, 13]. In light of the biomechanical characteristics, the short-range elastic stiffness is a characteristic for long jump take-off, where it is a matter of an immediate mechanic response of the activated muscle to eccentric contraction. In our experiment, we found the eccentric contraction time to be 84 ms and the athlete develops in this phase a maximal vertical ground reaction force of 5132 N. This force is almost 7.5 times the body mass of the athlete. An important kinematic criterion of the muscles' efficiency while resisting stretching (lowering of CM) is the angle of flexion in the take-off leg's knee—MKF = 148 deg. This angle must be sufficiently large. A marked lowering of CM in the MKF phase results in prolonging the compression phase period and through this an inefficient integration of the eccentric and concentric action of the muscles.

The results in the study of Lees et al. [9] show that force impulse in the compression phase is the primary indicator of vertical velocity in the lift phase. A study by Ballreich and Bruggemann [1] namely showed a much higher correlation between vertical velocity of take-off with effective length of the jump (r = 0.89), than with horizontal velocity of take-off (r = 0.21). The force impulse in the compression phase for the athlete GC was −101.9 N s.

3.3. Neuromuscular mechanisms of explosive power

In many sport disciplines, power is a major biomotor ability used in the prediction of results. To some extent, other biomotor abilities are also associated with power. It is of little wonder that many kinesiological studies delve into the subject of power, investigating its structure, the training methodology, the application of new methods and diagnostic procedures. In modern kinesiological science, power is undoubtedly one of the most meticulously researched biomotor abilities, yet many questions in this field remain unanswered. The classification of power is based on different criteria. According to the criterion of action, authors [2, 4, 7, 8, 13] distinguish among peak power, explosive power and endurance. Another criterion is that of neuromuscular activity, where power manifests itself in the form of isometric contraction as well as concentric, eccentric or eccentric-concentric contraction. Isometric contraction occurs in conditions where muscular force equals an external force, which is why there is no movement between two points of the muscle. Eccentric contraction occurs in conditions where the external loading is greater than the force of the activated muscles. In real motor situations, an eccentric-concentric muscle contraction is the most common type and manifests itself in take-off power. Take-off power is a special type of explosive power in eccentric-concentric conditions and is most often seen in cyclic, acyclic and combined movement structures. Its main characteristic is the utilisation of elastic energy in the eccentric-concentric cycle of a muscle contraction. The contribution of the elastic properties of the muscle-tendon complex depends on the velocity of the transition. The transition must be as fast as possible and should not exceed an interval of 260 ms [7]. This type of muscle contraction uses less chemical energy to perform mechanical work compared to a concentric contraction alone, thus speeding up the movement. If a concentric phase of contraction follows an eccentric phase quickly enough, the elastic elements release the accumulated energy in kinetic and mechanical work at the beginning of the concentric phase, thereby increasing the muscular force [15].

The movement structures that occur in specific sport situations are associated with different inputs of eccentric and concentric muscle contractions. Due to the inter-segmentary transmission of energy and the optimisation of the take-off action, the thigh muscles' activity is important in vertical jumps in conditions of a concentric or eccentric-concentric muscle contraction [2, 13, 16]. With vertical jumps, the muscles engage in the take-off action following the proximal-distal principle of muscle activation. In the first phase of the jump when the vertical velocity of the body's centre of gravity increases, the extensors of the trunk and hip are the most active muscles. The key role in this phase of the take-off action is played by m. gluteus maximus, which is able to develop great force due to the relatively low angular velocity of the hips. The thigh muscles generate the peak activation at the beginning of the hip extension [4, 6]. Electrical activity in a muscle results from a change in the electrical potential of the muscle membrane. It is measured by means of electrical potential that can be detected on the surface of the skin. For this purpose, surface electrodes are fastened above the muscle whose EMG signal is to be measured. Silver-silverchloride (Ag-AgCl) bipolar surface electrodes are generally used. The skin at the recording site where the electrodes are applied must be carefully prepared, cleansed and treated. Take-off power in conditions where the active muscles first extend (eccentric contraction) and then shorten (concentric contraction) is measured by means of a countermovement vertical jump and a drop jump.

3.4. Kinematic analysis of flight phase

On the basis of the kinematic parameters (**Table 3** and **Figure 8**), we see different contributions of the individual component distances to the final result. For the athlete GC, the flight distance (L2) has the greatest absolute and relative share in the total jump length—88%, then landing distance (L3)—7.7% and the smallest take-off distance (L1)—3.5%. The flight distance is defined as the horizontal distance between the point CM at take-off and the same point at the moment of contact with the sand.

http://dx.doi.org/10.5772/intechopen.71418

Parameter		Unit	Result
Distance of jump	(L)	m	8.25
Official distance	(OD)	m	8.10
Flight distance	(L2)	m	7.33
Max. height of CM in flight	(HMF)	m	1.88

Table 3. Kinematic parameters of the flight phase.

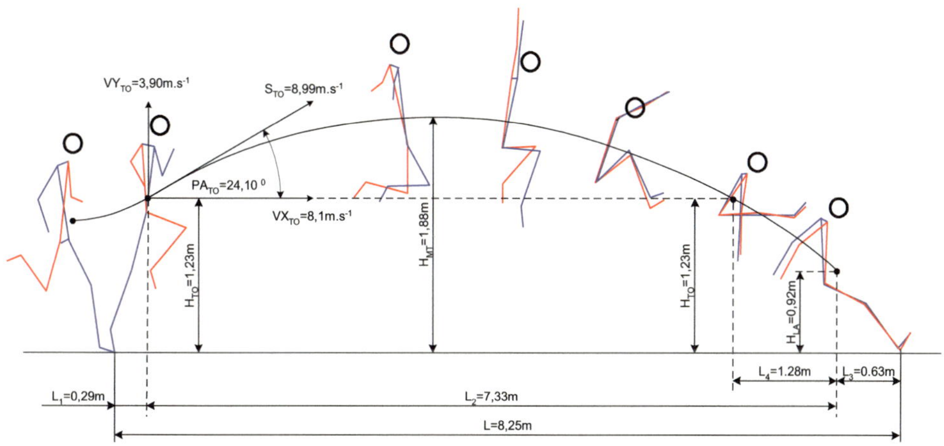

Figure 8. Kinematic model of flight and landing (GC, distance 8.25 m).

The small partial share of the take-off distance of the athlete GC is the consequence of the take-off angle, which amounted to 71.9 deg and the ratio between the vertical (VYTO) and horizontal component of velocity (VXTO) at take-off. The relatively large share of the landing distance (L3) can be ascribed to a very economic technique of landing. In the flight phase, the hang-style long jump is characteristic for the athlete GC, which he utilises very effectively as preparation for landing. A high parabola of CM flight can be seen for this athlete, which is the consequence of a high take-off angle (PATO = 24.1 deg). The highest point of the parabola is 65 cm above the height of CM at take-off. The characteristics of the flight phase are of course also influenced by other factors, especially air resistance, which was not the object of study in this research.

3.5. Kinematic analysis of landing phase

An economic technique of landing is defined by: landing distance (L3), height of CM at the moment of contact of the feet with the sand (HLA) and landing flight distance (L4)—**Table 4** and **Figure 8**.

One of the most important variables in landing is the horizontal landing distance, defined by the projection of point CM and the initial contact with the sand minus the distance lost by the athlete by falling backward or otherwise breaking the sand at a point closer to the board than the initial point of contact (LFB). The index of technique's economy is: IR = L3 − LFB. Hay and Nohara [7]

Parameter		Unit	Result
Landing distance	(L3)	m	0.63
Height of CM at landing	(HLA)	m	0.92
Landing flight distance	(L4)	m	1.28
Fall-back distance	(LFB)	m	0.15

Table 4. Kinematic parameters of landing.

found on a representative sample of 12 best American long jumpers that the average fall-back distance is 11 cm, for our athlete this distance is 15 cm. This leads us to conclude that his technique is economic and efficient. Mendoza [11] also finds that values of landing distance (L3) between 57 and 65 cm point to a very economic landing. The athlete can contribute an important part to the competitive result with good landing technique, with this phase being dependent on specific motor abilities, especially power of the hip flexors and belly muscles; morphologic characteristics, coordinative abilities of the athlete and also the quality of the landing area.

4. Conclusion

The results of kinematic and dynamic analysis showed that the efficiency of long jump is defined mostly by the following take-off parameters: horizontal velocity at TO, vertical velocity at TO, speed at TO, angle of projection at TO, maximal force X—horizontal and Y—vertical axis, force impulse in compression and lift phase. An important factor of a rational technique of long jump is also the landing, which is defined by the landing distance and fall-back distance. This study was conducted with only one elite long jumper (third place at the Universiade in Fukuoka, third place at the Seville World Athletics Championship) and offers an insight into the dynamic and kinematic parameters required for top results in this athletic discipline. Biomechanical diagnostics highlights the objective parameters of the technique on which the training process must be based. The study results hold both theoretical and practical value in applied biomechanics and sport practice, mainly in terms of selecting the optimal training methods.

Author details

Milan Čoh*, Milan Žvan and Otmar Kugovnik

*Address all correspondence to: milan.coh@fsp.uni-lj.si

Faculty of Sport, University of Ljubljana, Ljubljana, Slovenia

References

[1] Ballreich G, Bruggemann G. Biomechanik des Weitsprungs. Biomechanik der Leichtathletik. Stuttgart: Ferdinant Enke Verlag; 1986. pp. 28-47

[2] Čoh M. Biomechanical diagnostic methods in athletic training. Faculty of Sport, University of Ljubljana; 2008

[3] Dempster W. Space Requirements of the Seated Operator. Wright-Patterson Air Force Base, Ohio: WADC Technical Report; 1955. pp. 55-159

[4] Gollhofer A, Kyrolainen H. Neuromuscular control of the human leg extensor muscles in jump exercises under various stretch-load conditions. IJSM. 1991;**12**:34-40

[5] Grahman-Smith P, Lees A. A comparison of the information quality between cinematography and videography long jump technique analysis. Biology of Sport. 1997;**14**(3): 213-225

[6] Hay J, Miller J, Canterna R. The techniques of elite male long jumpers. Journal of Biomechanics. 1986;**19**:855-866

[7] Hay J, Nohara H. Techniques used by elite long jumpers in preparation for take-off. Journal of Biomechanics. 1990;**23**:229-239

[8] Komi P. Stretch-shortening cycle: A powerful model to study normal and fatigued muscle. Journal of Biomechanics. 2000;**33**:1197-1206

[9] Lees A, Smith G, Fowler N. A biomechanical analysis of last stride, touchdown, and take off characteristics of the Men's long jump. Journal of Applied Biomechanics. 1994;**10**(1): 61-78

[10] McFarland E. Encyclopedia of Sport Science. MacMillen; 1997. p. 185

[11] Mendoza L. Individuelle Optimierung der Landeweite beim Weitsprung mit Hilfe der Computersimulation. Leistungssport. 1989;**6**:35-40

[12] Muller H, Hommel H. Biomechanical research project at the VIth World Championships in Athletics, Athens 1997: Preliminary report. New Studies in Athletics. 1997;**2–3**:43-73

[13] Nicol C, Avela J, Komi P. The stretch-shortening cycle. Sports Medicine. 2006;**36**(11): 977-999

[14] Nixdorf E, Bruggemann P. Biomechanical analysis of the long jump. In: Scientific Research Project at the Games of the XXIVth Olympiad—Seoul 1988; International Athletic Foundation; 1990. pp. 263-302

[15] Tihanyi J. Die physiologische und mechanische Grundprinzipien des Krafttraining. Leistungssport. 1987;**17**(2):38-44

[16] Tom A. The mathematics and physics of pole vault. Theta. 1996;**10**(2):14-18

[17] Zatsiorsky V, Kraemer W. Science and Practice of Strength Training. 2006. Available from: HumanKinetics.com

Chapter 7

Kinematic Model for Project Scheduling with Constrained Resources Under Uncertainties

Giuliani Paulineli Garbi and
Francisco José Grandinetti

Additional information is available at the end of the chapter

http://dx.doi.org/10.5772/intechopen.71421

Abstract

Project management practitioners and researchers recognize that the project scheduling efforts are made based on information with many uncertainties and in an environment with constrained resources. This chapter presents the kinematic model named as Coupled Estimate Technique for project scheduling with constrained resources under uncertainties. The Coupled Estimate Technique provides tools of analytical analysis, given that the modelled duration depends on the planned duration and on the resource variability (aleatory uncertainty), as well as the modelled resource depends on the planned resource and on the duration variability (aleatory uncertainty), and also provides tools of graphical analysis, given that the durations and resources of activities, work packages or phases of the project are represented in the bidimensional graphics. In developing the mathematical formulation of the Coupled Estimate Technique, the project precedence diagram was considered as a kinematic chain of robotic manipulators, which may be in chain configuration open (serial), closed (parallel) and/or hybrid. This chapter describes the resource-constrained project scheduling problem (RCPSP) under uncertainties, identifies the limitations and opportunities in the previous work on planning under uncertainties and presents the fundamentals and method of the kinematic model for project scheduling with constrained resources under uncertainties along with a short example of implementation.

Keywords: project scheduling, kinematic model, Coupled Estimate Technique, kinematic chain, robot manipulators, uncertainties and constrained resources

1. Introduction

Project scheduling problems have been well studied in the literature since the 1950s. According to Ref. [1], project scheduling is an important process in project management. The project scheduling literature largely concentrates on the generation of a precedence-feasible and resource-feasible schedule that optimizes scheduling objective(s) (most often the project duration) and that serves as a baseline schedule for executing the project. Baseline schedules serve as a basis on which to allocate resources to different activities in order to optimize some measure of performance and for planning external activities such as material procurement, preventive maintenance and delivery of orders to external or internal customers [2].

However, project scheduling is a difficult process due to scarce resources as well as precedence relations between activities [3]. Admissible schedules must obey constraints such as precedence relations (a task cannot start unless some other tasks have been completed) and resource restrictions (labour and machines are examples of scarce resources with limited capacities) [4]. Thus, the resource-constrained project scheduling problem (RCPSP) consists of project activities subject to many kinds of uncertainties that must be scheduled in accordance with precedence and resource (renewable, non-renewable and doubly constrained) availabilities such that the total duration or makespan of a project is minimized [5].

Most of the variants and extensions of the RCPSP may be summarized and classified within multiple modes, generalized time lags and objective functions, resulting in highly complex optimization problems [1]. The RCPSP has become a well-known standard problem in the context of project scheduling, and numerous researchers have developed both exact and heuristic scheduling procedures. Due to the complex nature of the problem, only a small number of exact algorithms have been presented in the literature, and many heuristic solution algorithms have been presented in the literature [3].

Therefore, this chapter presents the development and implementation of the kinematic model named as Coupled Estimate Technique for project scheduling with constrained resources under uncertainties. In the Coupled Estimate Technique, the modelled duration depends on the planned duration and on the resource variability (aleatory uncertainty), and the modelled resource depends on the planned resource and on the duration variability (aleatory uncertainty).

1.1. Tailoring of the robotic manipulator kinematic concepts to the Coupled Estimate Technique

The development of the Coupled Estimate Technique was based on robotic manipulator kinematic concepts. Kinematics is the science of motion that treats motion without regard to the forces/moments that cause the motion [17]. Robotic manipulator or industrial robot consists of several rigid links connected in series (open kinematic chain) by revolute or prismatic joints; one end of the chain is attached to a supporting base, while the other end is free and equipped with a tool (end-effector) to manipulate objects or perform assembly tasks; and the end-effector could be a gripper, a welding torch, an electromagnet or another device [13].

Robotic manipulator kinematics deals with the analytical description of the spatial displacement of the end-effector with respect to a fixed reference coordinate system, in particular the relations between the joint variables and the position and orientation of the end-effector [15]. There are two (direct and inverse) fundamental problems in robotic manipulator kinematics.

Given a set of joint variables, the direct kinematic problem is performed to find the position and orientation of the end-effector relative to the fixed reference coordinate system. Given the position and orientation of the end-effector relative to the fixed reference coordinate system, the inverse kinematic problem is performed to find all possible sets of joint variables that could be used to attain this given position and orientation. **Figure 1** illustrates a robotic manipulator or industrial robot.

In analogous way, the Coupled Estimate Technique deals with the analytical study of the geometry of project activities of a precedence diagram with respect to a fixed reference two-dimensional coordinate system (0, T, R) where the abscissa is the duration axis (T) and the ordinate is the resource axis (R). Thus, the precedence diagram may be considered as a kinematic chain (open, closed and/or hybrid), where one end of the chain is attached to a fixed reference two-dimensional coordinate system and represents the begin project activity, while the other end is free and represents the end project activity, as well as the rigid links represent the precedences between project activities and joints are represented by the project activities. **Figure 2** presents a precedence diagram analogous to a robotic manipulator in open kinematic chain.

1.2. The main effects of the uncertainties in project scheduling

The majority of research efforts related to RCPSP assume complete information about the scheduling problem to be solved and a static and deterministic environment within which the precomputed baseline schedule is executed [6]. However, in the real world, project activities are subject to considerable uncertainties, stemming from various sources, which are gradually resolved during project execution.

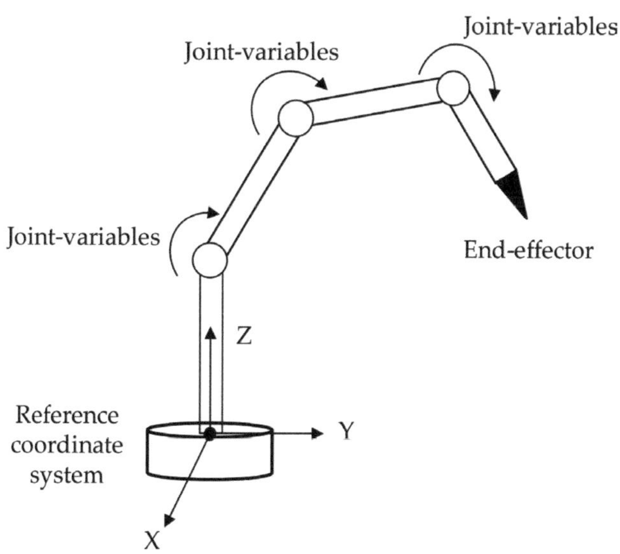

Figure 1. Robotic manipulator or industrial robot.

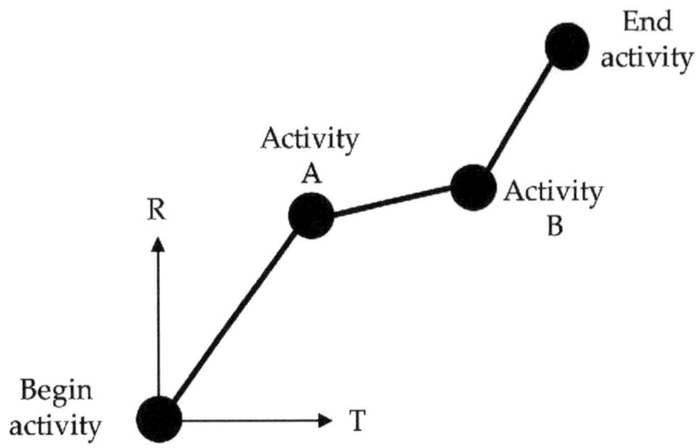

Figure 2. Precedence diagram in open kinematic chain.

The presence of uncertainties in project management is recognized by practitioners and researchers as an important cause underlying the high project failure rate [5]. Project-related uncertainties can lead to numerous schedule disruptions. These uncertainties may stem from the following sources [7]: activities may take more or less time than originally estimated; resources may become unavailable; material may arrive behind schedule; ready times and due dates may have to be changed; new activities may have to be incorporated; activities may have to be dropped due to changes in the project scope; and weather conditions may cause severe delays.

This chapter makes a distinction between uncertainty and risk, assuming that there are degrees of knowledge about the estimations of project duration and resources; the uncertainty may be quantifiable but without probability of occurrence, while the risk is quantifiable and with probability of occurrence [8]. Uncertainty may be aleatory due to the intrinsic variability of projects or epistemic due to the lack of relevant knowledge related to carrying out the project [9]. Thus, quantifying uncertainty is relevant to project scheduling because it sheds light on the knowledge gaps and ambiguities that affect the ability to understand the consequences of uncertainties in project objectives.

1.3. Common approaches for RCPSP under uncertainties

From a modelling viewpoint, there are four approaches for dealing with uncertainty quantification in a scheduling environment where the evolution structure of the precedence network is deterministic: reactive scheduling, stochastic scheduling, scheduling under fuzziness and proactive (robust) scheduling.

1.3.1. Reactive scheduling approach

The reactive (predictive) scheduling models do not try to cope with uncertainty quantification when creating the baseline schedule. Basically, efforts are largely concentrated on repairing, revising or re-optimizing the baseline schedule when an unexpected event occurs [2].

http://dx.doi.org/10.5772/intechopen.71421

The reactive effort may be based on very simple techniques aimed at a quick schedule restoration; a typical example is the well-known right-shift rule. This rule will move forward in time all the activities that are affected by the delays; often, the precedence diagram must be tailored in order to rearrange the changed activities. The reactive effort also may be based on as full rescheduling of the precedence diagram [4].

1.3.2. *Stochastic scheduling approach*

Most of the literature on the RCPSP assumes that activity durations are deterministic; however, activity durations are often uncertain. These uncertainties may be due to different sources, such as estimation errors, late delivery of resources, unforeseen weather conditions, unpredictable incidents (machine or worker) and others [6].

The stochastic RCPSP acknowledges that activity durations are not deterministic, i.e. the activity durations are modelled as stochastic variables. Therefore, the stochastic RCPSP aims to schedule project activities with uncertainty quantification of durations in order to minimize the expected project duration subject to precedence and constrained resources [5]. The stochastic project scheduling models view the project scheduling problem as a multistage decision process; the complete schedule (containing all activities) is constructed gradually as time progresses by means of a scheduling policy, exploiting the available information about the uncertainty of activity durations [1].

1.3.3. *Fuzzy scheduling approach*

Generally, the uncertainty of activity duration in project scheduling was handled by stochastic approaches using a probabilistic-based method. This kind of uncertainty in project duration is associated with randomness. However, for projects never be carried out previously, it is infeasible to determine the probability distribution of activity duration [2].

Therefore, the fuzzy project scheduling approach is used when the probability distributions for the activity durations are unknown due to a lack of historical data and, thus, the activity durations have to be estimated by human experts. In the situations involving the imprecision instead of uncertainty, the project scheduling literature recommends the use of fuzzy numbers for modelling activity durations. Instead of probability distributions, these quantities make use of membership functions, based on possibility theory [3].

1.3.4. *Proactive (robust) scheduling approach*

Traditionally, the robust schedule may absorb some level of unexpected events (machine breakdowns, staffing problems, unexpected arrival of new orders, early or late arrival of raw material and uncertainties in the duration of processing times) [2]. Thus, in order to minimize the impacts of uncertainties and the need of new scheduling or rescheduling, proactive scheduling approach aims at the generation of a robust baseline schedule that incorporates a certain degree of anticipation of potential variability or of potential disruptions according to the objective values (makespan) [3].

The proactive (robust) project scheduling model has prospered widely in the field of machine scheduling. Redundancy-based techniques related to durations and resources have already found their way to the field of project scheduling. The critical chain project management (CCPM) and Success Driven Project Management (SDPM) methods are becoming increasingly popular among project management practitioners [1].

1.4. Main differences between the Coupled Estimate Technique and the common approaches for RCPSP under uncertainties

This chapter presents a kinematic model named as Coupled Estimate Technique for project scheduling with constrained resources under uncertainties. This technique considers precedence, duration, resources and uncertainties related to project activities in order to analytically model the outcomes of project-related events or conditions (uncertainty) with the potential of favourable but mainly adverse consequences on project objectives (duration and resources). This approach can be used to quantify uncertainties; thus, it can help to solve project scheduling problems related to the following limitations and disadvantages identified in the literature review:

- The literature on project scheduling under risk and uncertainty was clearly conceived in a machine scheduling environment [10], whereas this work presents a project scheduling model from the viewpoint of project management.
- Projects are often subject to considerable uncertainty during their execution, but most of the research on project scheduling deals with only one source of uncertainty most often the duration of activities [11]. With the kinematic model detailed herein, the project activities have a set of attributes represented by the duration and resources as well as the uncertainties related to duration and resources.
- Traditional project scheduling is represented mainly by an activity-on-node network [12]. However, this representation is insufficient and inadequate when the activities of a project have a set of attributes. In the kinematic model, project scheduling is represented by IDEF0 (Integrated Computer-Aided Manufacturing Definition for Functional Modelling).
- Project scheduling methods are focused mainly on the basic RCPSP model [3]. In the kinematic model presented here, project scheduling is considered as a kinematic chain (open (serial), closed (parallel) and/or hybrid) formed by a set of rigid links (precedence of activities) that are connected by joints (project activities) with one fixed extremity (activity that represents the beginning of the project) and one free extremity (activity that represents the end of the project), which are represented by a homogeneous transformation matrix.
- The literature on project scheduling states that the generation of proactive (robust) multi-resource baseline schedules in combination with efficient and effective reactive schedule repair mechanisms constitutes a viable area of future research [10]. The kinematic model presented in this chapter provides this combination through the direct and inverse kinematic models. It provides evidence of the influences stemming from uncertainties in project activities, enabling the balancing of durations and resources between project activities and between projects.

http://dx.doi.org/10.5772/intechopen.71421

2. Kinematic fundamentals for the robotic manipulators and for the model proposed by this chapter

This topic aims to present the main fundamentals related to the kinematic problems for the robotic manipulators or industrial robots, as well as to present the main fundamentals related to the kinematic model proposed by this chapter for project scheduling with constrained resources under uncertainties.

2.1. Kinematic problems for robotic manipulators

In order to describe and represent the spatial geometry of the links of a robotic manipulator with to a fixed reference coordinate system, Denavit and Hartenberg (D-H) proposed a systematic approach utilizing matrix algebra [17]. This systematic approach reduces the direct kinematic problem to finding an equivalent 4x4 homogenous transformation matrix that describes the spatial relationship between two adjacent rigid links with respect to a fixed reference coordinate system [13].

In the inverse kinematic problem, given the position and orientation of the end-effector, we would wish to find the corresponding joint variables of the robotic manipulator [15]. The inverse kinematic problem is not as simple as the direct kinematic problem, because the kinematic equations are nonlinear and their solution is not easy (or even possible) in a close form. The inverse kinematic problem may be solved by various methods, such as inverse transform of Paul, screw algebra of Kohli and Soni, dual matrices of Denavit, dual quaternion of Yang and Freudenstein, iterative of Uicker, geometric approaches of Lee and Ziegler and others. **Figure 3** represents a synthesis to the direct and inverse kinematic problem [16].

2.2. Kinematic models for Coupled Estimate Technique

Basically, the project consists of a set of activities that must be performed in order to complete the project. The kinematic model technique for project scheduling with constrained resources under uncertainties named as Coupled Estimate Technique deals with the movements in the

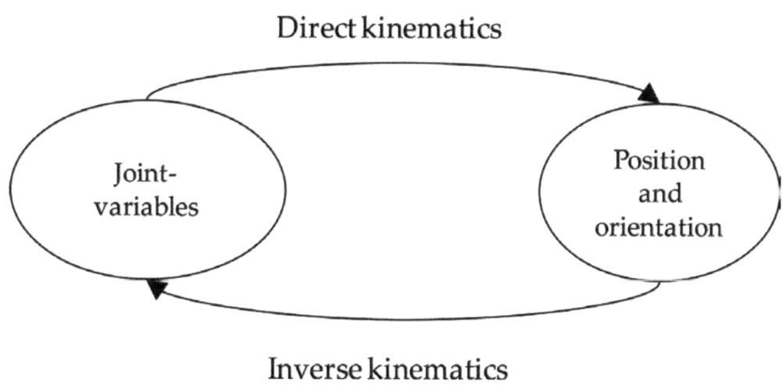

Figure 3. Synthesis to the direct and inverse kinematic problem for robotic manipulators.

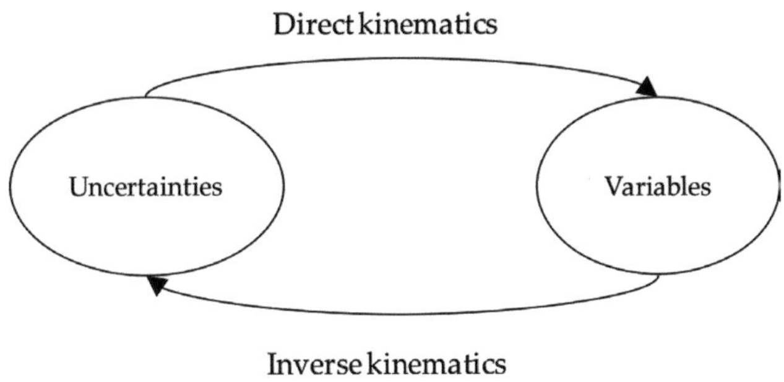

Figure 4. Kinematic model for Coupled Estimate Technique.

scheduled activities without consideration of the causes of the movements. Each project activity has a set of attributes represented by activity estimates (duration, resource and precedence), activity uncertainties (duration, resources and critical factor) and activity variables (duration and resource).

The outcomes of activities depend of their attributes, and the analytical kinematic model provides a metaschedule to the project. To model a project activity with kinematic model, it is assumed that the activity estimates are the constants, the kinematic equations are the orientation, and the modelled outcomes depend of the uncertainties and variables. Thus, the kinematic model may be direct or inverse type according to **Figure 4**.

Therefore, given the activity uncertainties determined by the project team, the direct kinematic model or the CET may be used to model the activity variables. Moreover, given the activity variables determined by the project team, using the inverse kinematics or the CET, the activity uncertainties may be modelled. It is important to emphasize that the estimates of the activity attributes must be part of organizational policy used by the team during the project planning. **Figure 5** illustrates the kinematic model or the CET with the IDEF0 language.

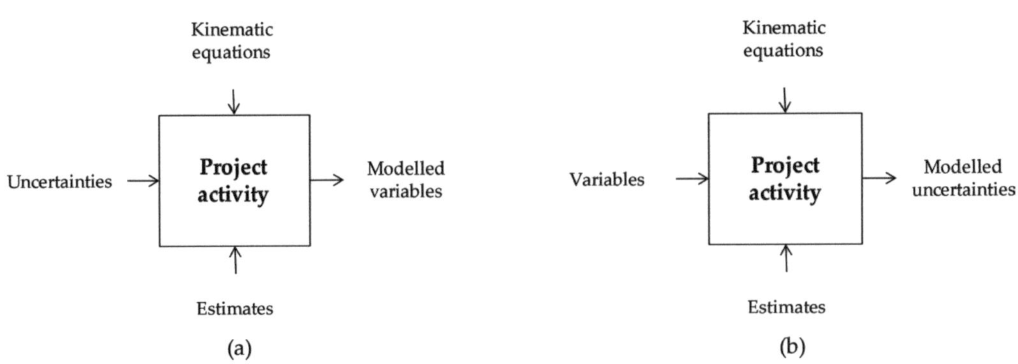

Figure 5. (a) Direct kinematic model and (b) inverse kinematic model.

3. Coupled Estimate Technique concepts

Coupled Estimative Technique (CET) is a project scheduling artefact to model the project activity with constrained resources under uncertainties. This section presents the main aspects of the CET that provide the mathematical formulation involving the activity estimates (duration, resource and precedence), activity uncertainties (duration, resources and critical factor) and activity variables (duration and resource).

Let each project activity be represented in the two-dimensional Cartesian coordinate system (0, T, R) where the abscissa is the duration axis (T) and the ordinate is the resource axis (R). The value set of the duration and resource for project activity lies between the estimated ordered pair (estimated duration (te) and estimated resource (re)) and the modelled ordered pair (modelled duration (tm) and modelled resource (rm)).

Estimated ordered pair is represented in the estimated coordinate system (0, Te, Re), and the modelled ordered pair is represented in the modelled coordinate system (0, Tm, Rm). The rotation alpha expresses the uncertainty in the estimation of the activity duration, and the rotation beta expresses the uncertainty in the estimation of activity resource. **Figure 6** shows the graphical representation of one project activity.

The mathematical formulation of the CET used to model the project activities is obtained by algebraic operations with homogeneous transformation matrices (4 × 4). They map a project activity from estimated coordinate system (estimated ordered pair) to modelled coordinate system (modelled ordered pair) using the homogeneous coordinate notation. The homogeneous transformation matrix (H) consists of four submatrices [13], according to Eq. (1):

$$H = \begin{vmatrix} \text{Rotation} & \text{Position} \\ \text{Perspective} & \text{Scaling} \end{vmatrix} = \begin{vmatrix} R_{3x3} & P_{3x1} \\ f_{1x3} & 1x1 \end{vmatrix} \tag{1}$$

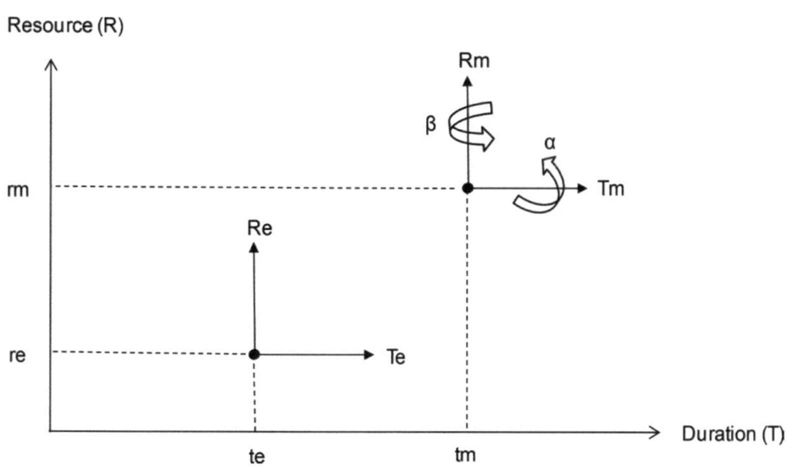

Figure 6. Graphical representation of one project activity.

The rotation submatrix (R3 × 3) transforms one project activity expressed in the estimated coordinate system to the modelled coordinate system. For project scheduling, the rotation submatrices are used to produce the effect of the uncertainty in the estimation of the activity duration by rotation alpha at the duration axis and the effect of the uncertainty in the estimation of activity resource by rotation beta at the resource axis. According to [14, 15], the basic rotation matrices are shown in Eq. (2), rotation alpha at the duration axis and rotation beta at the resource axis:

$$R_{T,\alpha} = \begin{bmatrix} 1 & 0 & 0 \\ 0 & \cos\alpha & -\sin\alpha \\ 0 & \sin\alpha & \cos\alpha \end{bmatrix}; R_{R,\beta} = \begin{bmatrix} \cos\beta & 0 & \sin\beta \\ 0 & 1 & 0 \\ -\sin\beta & 0 & \cos\beta \end{bmatrix} \tag{2}$$

As the duration and resource of the project activity vary between estimated and modelled values, the position submatrix (P3 × 1) represents the coordinates of the estimated ordered pair. Therefore, the first element of the position submatrix is the duration translation or the estimated duration (te), the second element of the position submatrix is the resource translation or the estimated resource (re), and as the project activity is represented in a geometric plane, the third element is null. The position submatrix (P3 × 1) is shown in Eq. (3):

$$P = \begin{bmatrix} te \\ re \\ 0 \end{bmatrix} \tag{3}$$

The perspective transformation submatrix (1 × 3) is useful for computer vision and the calibration of camera models [16]. For the mathematical formulation of the CET, the elements of this submatrix are set to zero to indicate null perspective transformation. The fourth diagonal element is the global scaling factor (1 × 1), which is set to 1 in this work. Values different from 1 are commonly used in computer graphics as a universal scale factor taking on any positive values [17]. The project activity is mathematically represented through homogeneous transformation matrices (4 × 4) which are arranged in four steps to obtain the continuous duration and cost functions as described in **Table 1**.

The first step is performed through the product between the homogeneous matrices ($H_{T,\alpha}$) and ($H_{te;re}$), where ($H_{T,\alpha}$) means a rotation alpha on the duration axis and ($H_{te;re}$) means a translation in the duration and resource axes. The algebraic operation of Eq. (4) corresponds to the effect of the uncertainty in the estimation of the activity duration in estimated resource of the project activity:

Step	Description
1	To represent the effect of the uncertainty in the estimation of the activity duration in estimated resource of the project activity.
2	To represent the effect of the uncertainty in the estimation of the activity resource in estimated duration of the project activity.
3	To represent the joint of the uncertainties (duration and resource) in the estimations (duration and resource) of the project activity.
4	To argue about the range of the uncertainties as well as the critical factor of the project activity.

Table 1. Steps to Develop the Homogeneous Transformation Matrix of the CET.

$$\begin{bmatrix} 1 & 0 & 0 & 0 \\ 0 & \cos\alpha & -\sin\alpha & 0 \\ 0 & \sin\alpha & \cos\alpha & 0 \\ 0 & 0 & 0 & 1 \end{bmatrix} \cdot \begin{bmatrix} 1 & 0 & 0 & te \\ 0 & 1 & 0 & ce \\ 0 & 0 & 1 & 0 \\ 0 & 0 & 0 & 1 \end{bmatrix} = \begin{bmatrix} 1 & 0 & 0 & te \\ 0 & \cos\alpha & -\sin\alpha & ce\,.\cos\alpha \\ 0 & \sin\alpha & \cos\alpha & ce\,.\sin\alpha \\ 0 & 0 & 0 & 1 \end{bmatrix} \quad (4)$$

The first element of the position submatrix (P3 × 1) in Eq. (4) represents the activity duration; the second represents the estimated resource varying according to the uncertainty in the estimation of the activity duration. As the position submatrix (P3 × 1) is a spatial geometric representation (abscissa, ordinate and cote), the third element may be disregarded because the project activity is a plane geometric representation (abscissa and ordinate).

The second step listed in **Table 1** performs the product between the homogeneous matrices ($H_{T,\beta}$) and ($H_{te;\,re}$). The former is the homogeneous matrix composed by a rotation beta on the resource axis, and ($H_{te;\,re}$) means a translation in the duration and resource axes. The algebraic operation of Eq. (5) corresponds to the effect of the uncertainty in the estimation of the activity resource in estimated duration of the project activity:

$$\begin{bmatrix} \cos\beta & 0 & \sin\beta & 0 \\ 0 & 1 & 0 & 0 \\ -\sin\beta & 0 & \cos\beta & 0 \\ 0 & 0 & 0 & 1 \end{bmatrix} \cdot \begin{bmatrix} 1 & 0 & 0 & te \\ 0 & 1 & 0 & re \\ 0 & 0 & 1 & 0 \\ 0 & 0 & 0 & 1 \end{bmatrix} = \begin{bmatrix} \cos\beta & 0 & \sin\beta & te\,.\cos\beta \\ 0 & 1 & 0 & re \\ -\sin\beta & 0 & \cos\beta & -te\,.\sin\beta \\ 0 & 0 & 0 & 1 \end{bmatrix} \quad (5)$$

The first element of the submatrix (P3 × 1) in Eq. (5) represents the activity duration varying according to the uncertainty in the estimation of the activity resource; the second, the activity resource. The third element of the submatrix (P3 × 1) in Eq. (5) may be discarded because the project activity is mapped at the geometric plane. The negative signal arises because the projection of the duration translation is at the negative semi axis of the cote coordinate.

The third step prescribed in **Table 1** performs the sum between Eqs. (4) and (5). The summation shown in Eq. (6) represents the Coupled Estimate Technique (CET) overall homogeneous matrix that provides the joint effect of the uncertainties (duration and resource) in the estimations (duration and resource) of the project activity:

$$\begin{bmatrix} 1+\cos\beta & 0 & \sin\beta & te\,.(1+\cos\beta) \\ 0 & 1+\cos\alpha & -\sin\alpha & ce\,.(1+\cos\alpha) \\ -\sin\beta & \sin\alpha & \cos\beta+\cos\beta & ce\,.\sin\alpha - te\,.\sin\beta \\ 0 & 0 & 0 & 2 \end{bmatrix} \quad (6)$$

The first element of the position submatrix (P3 × 1) in Eq. (6) represents the modelled duration in function of the estimated duration and uncertainty in the estimation of the activity resource. The second element represents the modelled resource in function of the estimated resource and uncertainty in the estimation of the activity duration. As the project activity is mapped at the geometric plane, the third element may be disregarded. However, it might be used to represent other project goals, e.g. the quantification of the activity performance or quality.

Modelled variables	Equations
Duration	tm = θ. te. (1 + cos β)
Resource	rm = θ. re. (1 + cos α)

Table 2. Mathematical formulation of the CET.

And finally, the fourth step in **Table 1** performs some orientations about the range of the uncertainties (duration and resource) and the critical factor of the project activity:

- The uncertainties must range between 0° (highest degree of uncertainty) and 89° (lowest degree of uncertainty). When alpha and beta equal to 0°, the modelled duration equals to the double of estimated duration (2te), and the modelled resource equals to the double of estimated resource (2re).
- When alpha and beta equal to 90°, there are no uncertainties or certainties. Therefore, the modelled duration equals to the estimated duration (te), and the modelled resource equals to the estimated resource (re).
- The certainties must range between 91° (lowest degree of certainty) and 180° (highest degree of certainty). When alpha and beta equal to 180°, the duration and resource modelled are null; this means a dummy activity, i.e. a project activity without duration and resource.
- The critical factor (theta) of the project activity must be unitary for cases where the modelled value is less than or equal to double of the estimated value, and the critical factor (theta) must be greater than one for cases where the modelled value is greater than double what was estimated value.

Table 2 shows the mathematical formulations for one project activity of the Coupled Estimate Technique.

4. Implementation of the direct kinematic model with CET

To demonstrate how the direct kinematic model is implemented, this section presents a didactic example with scheduling of duration in days, one type of resource (financial in dollar) and the critical factor unitary for all activities present in the critical path, according to the four sequential tasks. The main objective is to obtain the modelled schedule or project duration modelled (pdm) and modelled budget or project resource modelled (prm) with the effects of uncertainties of the estimates in the planned schedule or project duration estimated (pde) and planned budget or project resource estimated (pre) of the project. Assuming a critical path with the activities in the project precedence diagram which have the attributes:

- A0: Project beginning
- A1: Duration estimation equal to 30 days with uncertainty equal to 45°; resource estimation equal to $100 with uncertainty equal to 45°

- A2: Duration estimation equal to 20 days with uncertainty equal to 45°; resource estimation equal to \$50 with uncertainty equal to 45°
- A3: End of project

As task 1 for implementation of the direct kinematic model with CET, the planned schedule and budget without the uncertainties must be determined; thus, the planned schedule or project duration estimated (pde) is equal to 50 days, and the planned budget or project resource estimated (pre) is equal to \$150.

In task 2, the project precedence diagram is modelled with IDEF0 from the attributes of the project activities. **Figure 7** presents the IDEF0 with:

- Activities estimates (te1, re1; te2, re2)
- Activity uncertainties ($\alpha 1$, $\beta 1$, $\theta 1$; $\alpha 2$, $\beta 2$, $\theta 2$)
- Activity variables to be modelled (tm1, rm1; tm2, rm2)
- Project duration estimated (pde) and project resource estimated (pre)
- Project duration modelled (pdm) and project resource modelled (prm) to be modelled

During the third task, the modelled variables are calculated according to the kinematic equations of the CET presented in **Table 2** and **Figure 4**. From the estimates and uncertainties of the activities, the modelled variables of project activities are presented in **Table 3**.

And finally, the fourth task describes that the project duration modelled (pdm) is equal to the sum of the modelled durations, and project resource modelled (prm) is equal to the sum of the modelled resources. **Table 4** presents the project scheduling modelled according to the direct kinematic model with CET.

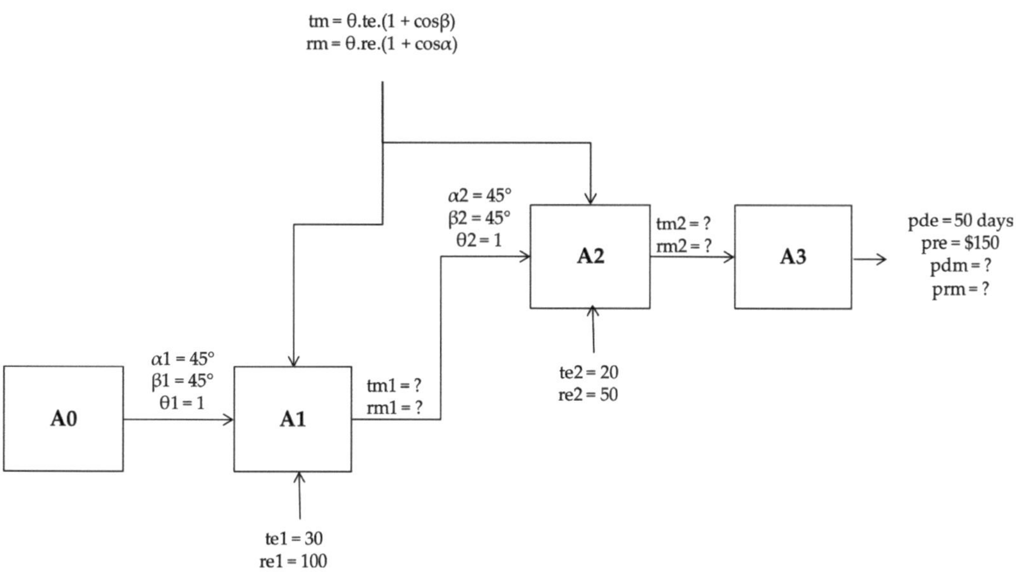

Figure 7. Project precedence diagram with IDEF0 for direct kinematic model with CET.

Activities	Estimates	Uncertainties	Modelled variables
A1	te1 = 30 days	$\alpha 1 = 45°$	51 days
	re1 = \$100	$\beta 1 = 45°$	\$170
A2	te2 = 20 days	$\alpha 2 = 45°$	34 days
	re2 = \$50	$\beta 2 = 45°$	\$85

Table 3. Modelled variables of the project activities.

Project	Estimates	Modelled	Effects
Duration	pde = 50 days	pdm = 85 days	35 days besides of planned
Resource	pre = \$150	prm = \$255	\$85 besides of planned

Table 4. Project scheduling modelled according to the direct kinematic model with CET.

Analyzing the information in **Table 4**, due to the uncertainties in the estimation of the project duration, the schedule should be increased in 35 days that represent 70% besides of the project duration estimated. In the same manner, due to the uncertainties in the estimation of the project resource, the budget should be increased in \$85 that represents 70% besides of the project resource estimated.

5. Implementation of the inverse kinematic model with CET

To demonstrate how the inverse kinematic model is implemented, this section presents a didactic example with scheduling of duration in days, one type of resource (financial in dollar) and the critical factor unitary for all activities present in the critical path, according to the four sequential tasks. The main objective is to obtain the modelled uncertainties in the estimation of the project durations and resources from estimated and performed values of the project activities. Assuming a critical path with the activities in the project precedence diagram which have the attributes:

- A0: Project beginning
- A1: Duration estimation equal to 30 days and duration performed equal to 51 days; resource estimation equal to \$100 and resource performed equal to \$170
- A2: Duration estimation equal to 20 days and duration performed equal to 34 days; resource estimation equal to \$50 and resource performed equal to \$85
- A3: End of project

As task 1 for implementation of the inverse kinematic model with CET, the planned schedule and budget as well as the performed schedule and budget with the effects of the uncertainties must be determined; thus, the planned schedule or project duration estimated (pde) is equal to 50 days, the planned budget or project resource estimated (pre) is equal to \$150, the performed schedule or project duration performed (pdp) is equal to 85 days and the performed budget or project resource performed (prp) is equal to \$255.

In task 2, the project precedence diagram is modelled with IDEF0 from the attributes of the project activities. **Figure 8** presents the IDEF0 with:

- Activity estimates (te1, re1; te2, re2)
- Activity uncertainties (α1, β1, θ1; α2, β2, θ2) to be modelled
- Activity variables, which for inverse kinematic model, the performed duration and resource (tp1, rp1; tp2, rp2)
- Project duration estimated (pde) and project resource estimated (pre)
- Project duration performed (pdp) and project resource performed (pre)

For the third task, the kinematic equations of the CET presented in **Table 2** must be tailored where the modelled variables are substituted by the performed variables according to **Figure 5**. From the estimated and performed values, the modelled uncertainties of project activities are presented in **Table 5**.

In task 4, the uncertainty of the project must be analyzed. There are some ways to determine the project uncertainty; frequently, this choice is realized by the project team taking account the nature and challenges of the project. The following are some suggestions:

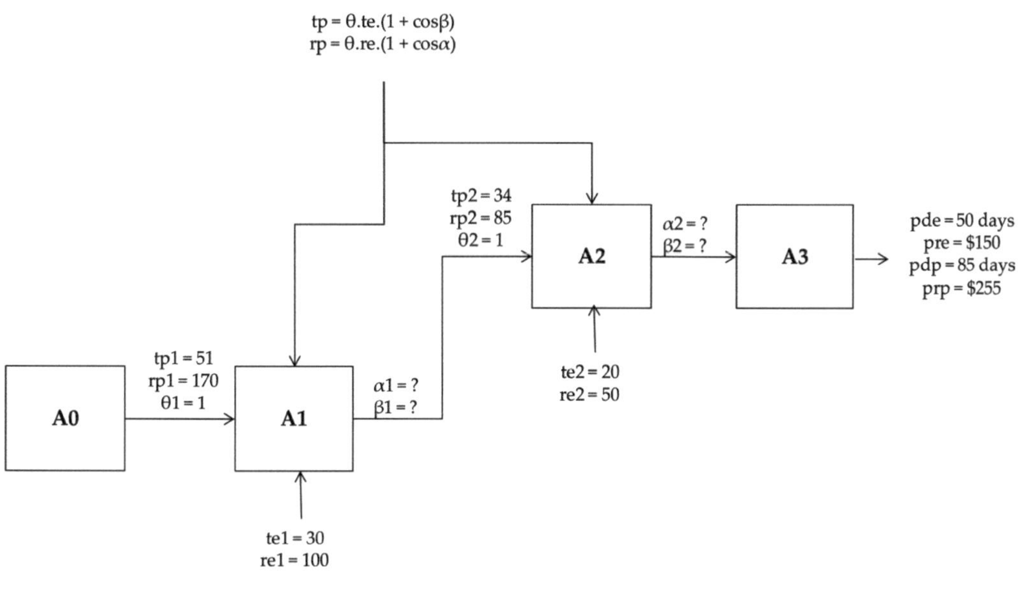

Figure 8. Project precedence diagram with IDEF0 for inverse kinematic model with CET.

Activities	Estimate	Performed	Modelled uncertainties
A1	te1 = 30 days	51 days	$\alpha 1 = 45°$
	re1 = \$100	\$170	$\beta 1 = 45°$
A2	te2 = 20 days	34 days	$\alpha 2 = 45°$
	re2 = \$50	\$85	$\beta 2 = 45°$

Table 5. Modelled uncertainties of the project activities.

- The project uncertainty may be determined by the average between the uncertainties of the activities, where for the case studied in this section, the project uncertainty for the duration and resource are equal to 45°.
- The project uncertainty may be determined by the greater or lower uncertainty of the activities.
- The project uncertainty may be determined by the sum with the weight of uncertainties of the activities.

6. Comments and conclusions

This chapter represented one interesting example of cross-fertilization between different areas, where the manipulator kinematic concepts (direct and inverse) were tailored and implemented in order to provide an innovative solution for project scheduling with constrained resources under uncertainties. Basically, the homogeneous transformation matrices (4×4) were used to model the schedule (time) and budget (resource) of the projects taking account the uncertainties in the estimates.

In the perspective of the project management, the direct kinematic model may be used to the project scheduling in order to model the schedule and budget when the duration and resource estimations are not known completely; therefore, there are degrees of uncertainties related to duration and resource estimation. And, the inverse kinematic model may be used to the implemented projects in order to model or quantify the degrees of uncertainties related to duration and resource estimation. These modelled uncertainties of the implemented project may be used during the project scheduling of similar projects.

The modelled outcomes provide information that can enhance the processes for schedule and budget of the project management, helping to achieve project scheduling of a project. They also provide information about project risk management processes, helping to identify, analyze and respond to uncertainties that are not definitely known but pose the potential for favourable or, more likely, adverse consequences on project objectives (duration and resources).

However, the model represents a conjecture of a phenomenon and, therefore, an approximation of the behaviour of that phenomenon. Thus, the modelled outcomes of the kinematic model for project scheduling with constrained resources under uncertainties must be critically analyzed by a project team.

Author details

Giuliani Paulineli Garbi[1*] and Francisco José Grandinetti[2]

*Address all correspondence to: giuliani.garbi@gmail.com

1 Brazilian Institute of Space Research, São José dos Campos, Brazil

2 Taubate University, Taubate, Brazil

References

[1] Besikci U, Bilge U, Ulusoy G. Multi-mode resource constrained multi-project scheduling and resource portfolio problem. European Journal of Operational Research. 2015;**240**:22-31

[2] Maravas A, Pantouvakis J. Project cash flow analysis in the presence of uncertainty in activity duration and cost. International Journal of Project Management. 2012;**30**:374-384

[3] Herroelen W, Reyck B, Demeulemeester E. Resource-constrained project scheduling: A survey of recent developments. Computers & Operations Research. 1998;**25**:279-302

[4] Peteghem VV, Vanhoucke M. An experimental investigation of metaheuristics for the multi-mode resource-constrained project scheduling problem on new dataset instances. European Journal of Operational Research. 2014;**235**:62-72

[5] Yang Q, Lu T, Yao T, Zhang B. The impact of uncertainty and ambiguity related to iteration and overlapping on schedule of product development projects. International Journal of Project Management. 2014;**32**:827-837

[6] Balouka N, Cohen I, Shtub A. Extending the multimode resource-constrained project scheduling problem by including value considerations. IEEE Transactions on Engineering Management. 2016;**63**(1):4-15

[7] Xu J, Zheng H, Zeng Z, Wu S, Shen M. Discrete time–cost–environment trade-off problem for large-scale construction systems with multiple modes under fuzzy uncertainty and its application to Jinping-II Hydroelectric Project. International Journal of Project Management. 2012;**30**:950-966

[8] Scholten L, Schuwirth N, Reichert P, Lienert J. Tackling uncertainty in multi-criteria decision analysis—An application to water supply infrastructure planning. European Journal of Operational Research. 2015;**242**:243-260

[9] Ward S, Chapman C. Transforming project risk management into project uncertainty management. International Journal of Project Management. 2003;**21**:97-105

[10] Hans EW, Herroelen W, Leus R, Wullink GA. Hierarchical approach to multi-project planning under uncertainly. The International Journal of Management Science. 2007;**35**:563-577

[11] Klerides E, Hadjiconstantinou E. A decomposition-based stochastic programming approach for the project scheduling problem under time/cost trade-off settings and uncertain durations. Computers and Operations Research. 2010;**37**:2131-2140

[12] Vonder SV, Demeulemeester E, Herroelen W, Leus R. The trade-off between stability and makespan in resource-constrained project scheduling. International Journal of Production Research. 2006;**44**:215-236

[13] Fu KS, Gonzalez RC, Lee CSG. Robotics: Control, Sensing, Vision and Intelligence. New York: McGraw-Hill Book Company; 1987

[14] Lipschutz S. Linear Algebra. Schaum Outline, McGraw-Hill. 4th ed2011

[15] Richard P. Robot Manipulators—Mathematics, Programming and Control. Cambridge, MA: MIT Press; 1982

[16] Goldman R. Design of an Interactive Manipulator Programming Environment. Ann Arbor, MI: UMI Research Press; 1985

[17] Craig JJ. Introduction to Robotics—Mechanical and Control. 3rd ed., Hardcover ed2004

Chapter 8

WMR Kinematic Control Using Underactuated Mechanisms for Goal Direction and Evasion

Jorge U. Reyes-Muñoz, Edgar A. Martínez-García,
Ricardo Rodríguez-Jorge and Rafael Torres-Córdoba

Additional information is available at the end of the chapter

http://dx.doi.org/10.5772/intechopen.70811

Abstract

This work presents the mechanical design and the kinematic navigation control system for a tricycle-wheeled robot (one drive-steer and two lateral fixed passive) with two underactuated mechanisms: a global compass and local evasive compass. The proposed goal-reference mechanism is inspired by the ancient Chinese south-seeking chariot (c. 200–265 CE) used as a navigation compass. The passive lateral wheels transmit an absolute angle from its differential speeds to automatically steer the front wheel. An obstacle-evasive compass mechanism is commutated for steering control when detecting nearby obstacles. The absolute and local compass mechanisms commutate each other to control to the robot's steering wheel to reach a goal while avoiding collisions. A kinematic control law is described in terms of the robot's geometric constraints and is combined with a set of first-order partial derivatives that allows interaction between the global and local steering mechanisms. Animated simulations and numerical computations about the robot's mechanisms and trajectories in multi-obstacle scenarios validate the proposed kinematic control system and its feasibility.

Keywords: WMR, kinematic control, south-seeking chariot, underactuated compass, potential-field, self-steer, navigation

1. Introduction

So far today, numerous types of modern robotic platforms that perform complex tasks are composed of underactuated mechanisms that were deployed by ancient civilizations. Underactuated mechanisms prevail as the most efficient systems because they take more advantage applying the law of conservation of energy than redundantly kinematic systems. Unlike redundant systems, the underactuated systems pose a reduced number of actuators and less independent control variables and naturally take more advantage of the inertial and gravity forces. The redundant systems have a larger number of control variables than variables in the working space. As a consequence, each controlled actuator somehow has to counteract gravity and inertial forces to establish own kinematic behaviors. A major aspect in robotics engineering concerns physical modeling and control of mobility for a robot to provide autonomous navigation. Similar to any biological entity, the ability to purposely navigate is fundamental for an intelligent robot. Over 2000 years ago, the Chinese invented the south-seeking chariot (SSC), which was used to maintain an absolute orientation along very long trips of hundreds of kilometers. Thus, they basically created one of the first absolute direction compass devices that did not require any other element to function, but its inner mechanisms only. The invention's compass was used to adjust toward a desired orientation at the beginning of a trip, and it was invented nearly 800 years in advance when the magnetic compass was invented. The SSC is basically a differential gearing system with a pointing-out silhouette above (**Figure 1**). The gearing system compensates the chariot turns by gear transmission relations, keeping the statue arm pointing out always to the same direction. **Figure 1a** shows the SSC design, and **Figure 1b** depicts the SSC prototype with its gearing system made of straight wooden gears. However, this system has largely been studied in modern times by numerous authors [1–7].

Any type of navigational and path-tracking task depends on steering systems. And, in order to infer navigation references, mobile robots use a diversity of exteroceptive sensing devices, such as ultrasound sonar, infrared range detectors, cameras, GPS, and so forth. Nevertheless, a common disadvantage using these types of devices is that they have to obtain external measurements with respect to (w.r.t.) the robot's fixed Cartesian coordinate system and use them to estimate orientation through geometry models with cumulative errors.

Moreover, some types of proprioceptive sensing devices offer inner measurements that are relative to global inertial system (e.g., magnetic compass, GPS). Unlike local exteroceptive sensors, global proprioceptive sensors yield noncumulative overtime measurement errors. For instance, a magnetic compass implemented as a global orientation system measures angle w.r.t. the earth magnetic axis. And then, such measurement angle is arithmetically used to infer a global destination angle likely described in another external inertial system. However, magnetic compasses are sensitive and are affected by other nearby magnetic fields or tided to suffer errors from rotations that set the device perpendicular to the earth's magnetic axis. Therefore, other

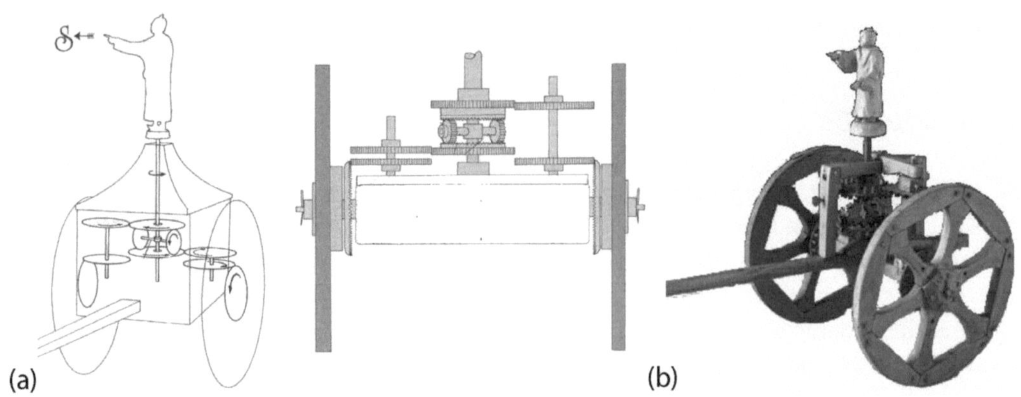

Figure 1. South-pointing chariot. (a) Modern design and (b) modern prototype.

types of sensing devices must be combined to recover from missed observations. Furthermore, the global positioning system (GPS) is another type of global orientation measurement device, which requires at least two successive position observations overtime to provide an instantaneous robot's global angle w.r.t. the earth's geographic north pole. As a major disadvantage, GPS cannot provide measurements nearly of inside buildings, forests, or in cloudy days because electromagnetic signals are blocked missing observations during arbitrary periods of time.

A diversity of works has reported navigation systems that combine numerous types of measuring orientation devices such as GPS and magnetic compasses, which provide high precision. This type of technology produces sensing measurements tided to established references (e.g., magnetic/north pole). Hence, useful global-specific references have to be inferred by different methods, which may imply in these calculations inverse/direct solutions of geometric triangulations or algebraic models [8–12].

Reliable local/global navigation for a wheeled mobile robot has fundamentals on controllability and maneuverability. And, both robot's abilities, respectively, must depend on robust driving control models and the steering kinematic designs [13, 14]. These models are the bases for planning and motion control [15, 16] in navigation, and most of them relay on different numerical mathematical solutions in robotics [17]. There are some complex works on robot's navigation seeking absolute orientation references [19, 20]. There are other navigation works with major emphasis on collision avoidance relaying on kinematic approaches [18, 21].

In this work, the mechanical design, the physical model, and a control system for a tricycle-wheeled robot with fundamentals on underactuated mechanical functions are proposed. The self-steer robot design proposed in this work has been inspired on the south-seeking chariot, in part, to take advantage of the underactuated mechanical compass with absolute direction to the robot's goal. Therefore, a specific and complex gearing mechanism system was designed to self-steer commuting between an obstacle-avoiding compass and a leading-to-the-goal compass (**Figure 2**).

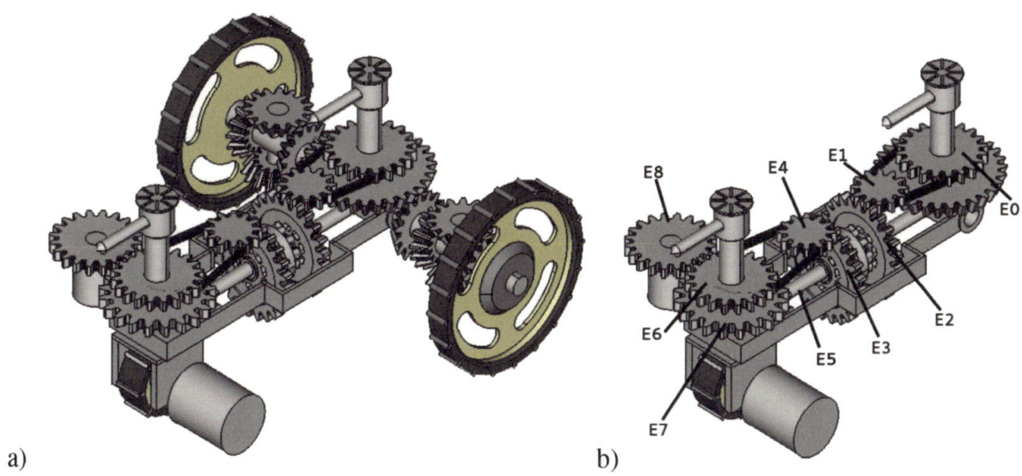

Figure 2. Proposed tricycle robot system. (a) General system view and (b) absolute/local compass mechanism.

The global goal-reference compass permanently maintains orientation information available in a direct manner for the robot without sensors, despite experimenting multiple evasive maneuvers.

A kinematic model for the local/global compass and commutation mechanisms are deduced. Moreover, a kinematic control law deduced for a three-wheeled structure with one drive-steer frontal wheel and a pair of lateral passive wheels is analyzed and disclosed. The proposed control law estimates the robot's posture and combines the interactive switching between goal-leading and obstacle avoidance navigation control. One underactuated mechanism directly leads the robot to the goal. Another underactuated mechanism leads the robot toward free-collision routes.

This manuscript presents simulation results to validate this novel approach that combines ancient underactuated automaton type, with a modern-wheeled robot focus. So far today, the authors are not conscious about other similar approaches reported in the scientific literature. This work does not pretend to introduce a comparative analysis nor efficiency with other state-of-the-art robotic trends. The authors of this research believe that the preliminary results presented in this manuscript will evolve into an efficient technological approach in the near future. Its application will establish a novel approach because it allows to directly have absolute angle observation overtime, allowing the mechanisms to lead the robot with naturally global navigation, reducing computational efforts to other algorithmic tasks, and complementing other sensing devices to improve control and perception.

Sections of this chapter have been organized in the following manner. Chapter 2 discloses the kinematic models for the absolute/local mechanisms. In Chapter 3, the commutation system kinematic and time delays are discussed. In Chapter 4, the control model and simulation results are presented. Finally, Chapter 5 discusses some conclusive remarks about this work.

2. Absolute/local compass mechanism

2.1. Global underactuated compass

The proposed system sets the robot's desired Cartesian goal as the absolute reference orientation. The absolute compass directly compensates its bearing by the differential angle provided by the two lateral wheels (**Figure 3a**). The mechanical compass directly self-steer the robot toward the global reference without any actuator (**Figure 3b**). When the robot is not leading along the global compass angle, the synchronization chain gradually reorientates the robot's steering wheel until matching the goal's absolute angle. **Figure 3c** illustrates the self-steering mechanism parts: (1) passive lateral wheels, (2) frontal driving wheel with passive steering, (3) wheel differential gearing mechanisms, (4) transmission gears coupling the absolute compass, (5) global goal-angle compass, (6) driving actuator/motor, (7) self-steer synchronization chain, and (8) lateral wheel shaft.

In **Figure 4**, a more detailed depiction of the differential mechanism is illustrated. The absolute compass (5) is composed of three differential systems (3) and (5), which transmit rotary motion from the lateral wheels (1) up to the steering wheel (2) (see **Figure 4b**). The compass (5) is

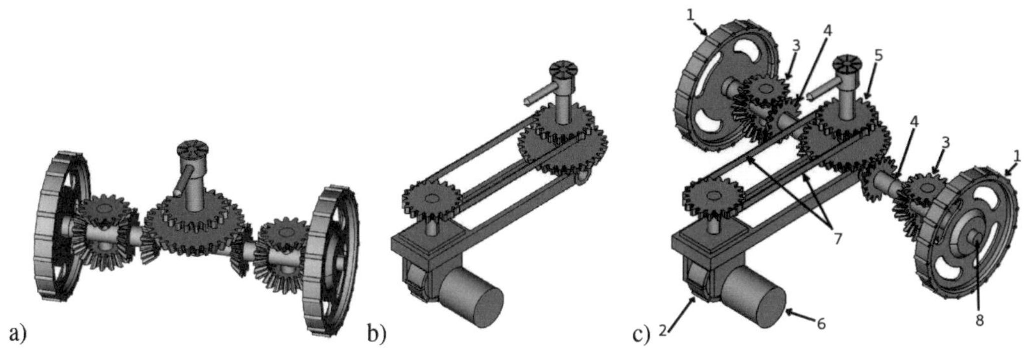

Figure 3. Absolute compass system. (a) Wheel differential mechanisms, (b) self-steer mechanism, and (c) goal-reference self-steer robot mechanism.

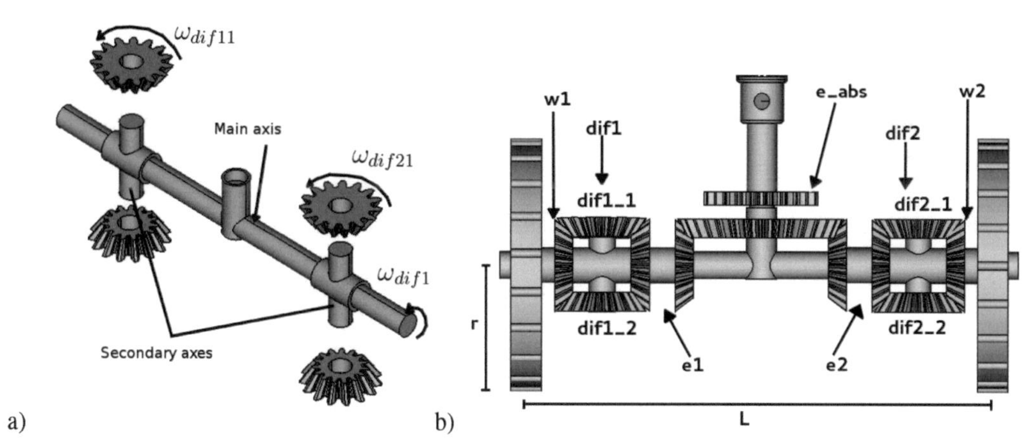

Figure 4. Differential gearing system. a) wheel main shafts, b) absolute direction differential.

compensated by the gearing relations (**Figure 4b**), and the differential rotary motion is transmitted to the gear (5), namely, e_{abs}. It is worth noting that the differential motion (3) of the lateral wheels' shaft (8) is fixed to the main axis (4). The lateral wheels' angular speeds, dubbed w_1 and w_2, are asynchronous w.r.t. the main shaft (4), similarly differential gears e_1 and e_2. The differential shafts of gears dif_1 and dif_2 (**Figure 4b**) are perpendicularly joined to the main axis (**Figure 4a**), and the rotary speed is basically the same for both differentials (4). Nevertheless, the angular velocity for gears ($dif1_1$, $dif1_2$, $dif2_1$, and $dif2_2$) might be different for each differential mechanism (3); it would depend on each wheel's (1) instantaneous velocity.

Finally, the compass, differential mechanism (5) that is composed of the gears e_1, e_2, and e_{abs} is the mechanism providing the absolute orientation reference toward the robot's goal.

Without loss of generality, it follows that the deduction of the differential angular velocity model transmitted between e_1 and e_2 is provided next. Thus, the angular velocity for ω_{e1} is an averaged differential value:

$$\omega_{e1} = -\frac{\omega_1 - \omega_2}{2} \tag{1}$$

where both lateral angular speeds have clockwise (left-sided wheel) and counterclockwise (right-sided wheel) signs, respectively:

$$\omega_{e1} = -\omega_{e2} \tag{2}$$

The differential angular speed is equivalent to the main shaft rotary speed, such that

$$\omega_{dif1} = \omega_{dif2} = \frac{\omega_1 - \omega_2}{2} \tag{3}$$

and, in general, it is assumed that the gears' (*dif*1 and *dif*2) angular velocities are averaged values:

$$\omega_{dif11} = \omega_{dif12} = \frac{\omega_1 + \omega_{e1}}{2} \tag{4}$$

as well as

$$\omega_{dif21} = \omega_{dif22} = \frac{\omega_2 + \omega_{e2}}{2}. \tag{5}$$

In such a manner, the angular velocity transmitted to the compass *e_abs* poses the following relation:

$$\omega_{e_{abs}} = \frac{r_{e_{abs}}}{r_{e_1}}\omega_{e1}, \tag{6}$$

In the proposed design, it is assumed that the gear *e_abs* doubles its radius w.r.t. gears e_1 and e_2. Therefore, by rewriting Eq. (6)

$$\omega_{e_{abs}} = \frac{1}{2}\omega_{e1}, \tag{7}$$

and by substituting in the previous expression, the differential rotary speed model ω_{e1} since ultimately depends on velocities ω_1 and ω_2:

$$\omega_{e_{abs}} = -\frac{\omega_1 - \omega_2}{4}. \tag{8}$$

Therefore, from the previous expression, let us assume that the robot moves along a straight trajectory line, then $\omega_1 = \omega_2$, and no compass lateral motion is yielded; hence, $\omega_{e_{abs}} = 0$. Following **Figure 2b** notation for the gears, the steering wheel with gear E_7 is synchronized with $\omega_{e_{abs}}$ by the mechanism system illustrated in **Figure 5a**.

A first gear connection $E0 - E1$ has the following speed relation:

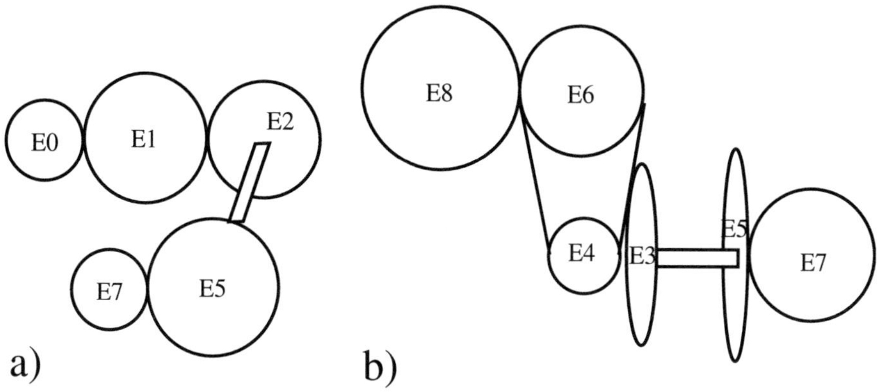

Figure 5. Steering mechanisms. (a) Goal-direction system and (b) evasive compass gearing.

$$\dot{\varphi}_1 = -\frac{r_0}{r_1}\dot{\varphi}_0; \tag{9}$$

Similarly, the gear connection $E_1 - E_2$ is algebraically simplified:

$$\dot{\varphi}_2 = -\frac{r_1}{r_2}\dot{\varphi}_1 = \frac{r_0}{r_2}\dot{\varphi}_0; \tag{10}$$

The parallel connection E_5 with E_2 has the same relation:

$$\dot{\varphi}_5 = \dot{\varphi}_2 = \frac{r_0}{r_2}\dot{\varphi}_0; \tag{11}$$

Finally, the angular velocity for E_7 w.r.t. E_0

$$\dot{\varphi}_7 = -\frac{r_5}{r_7}\dot{\varphi}_5 = -\frac{r_0 r_5}{r_2 r_7}\dot{\varphi}_0. \tag{12}$$

Thus, the relation (12) means that E_7 poses opposite rotary movement w.r.t. the absolute compass sense of rotation and half of its angular speed as well.

2.2. Local evasive compass

Another compass mechanism with local reference frame steers the robot for obstacle avoidance leading the robot along a safe instantaneous angle (**Figure 5b**). When the robot detects near obstacles, the absolute compass is suspended, and a commutator device switches to the evasive local compass mechanism, activating the gear E_8. **Figure 5b** describes the gearing mechanism that steers the robot for evasive navigation.

In the kinematic model for evasive steering through E_7, w.r.t. E_8 is deduced next. The relation between E_8 and E_6 is

$$\dot{\varphi}_6 = -\frac{r_8}{r_6}\dot{\varphi}_8; \tag{13}$$

and the angular velocity for E_4 is

$$\dot{\varphi}_4 = \frac{r_6}{r_4}\dot{\varphi}_6 = -\frac{r_8}{r_4}\dot{\varphi}_8, \tag{14}$$

For the perpendicular connection with E_3, the angular speed model is

$$\dot{\varphi}_3 = -\frac{r_4}{r_3}\dot{\varphi}_4 = \frac{r_8}{r_3}\dot{\varphi}_8; \tag{15}$$

For E_3 and E_5 connected in parallel

$$\dot{\varphi}_5 = \dot{\varphi}_3 = \frac{r_8}{r_3}\dot{\varphi}_8; \tag{16}$$

Finally, the model for E_7 is

$$\dot{\varphi}_7 = -\frac{r_5}{r_7}\dot{\varphi}_5 = -\frac{r_8 r_5}{r_2 r_7}\dot{\varphi}_8. \tag{17}$$

Let us highlight that for the evasive local compass E_8, the escaping orientation is instantaneously set up by an actuator only when any nearby obstacle is detected. And, such local orientation transmits motion to the steering gear (E_7) through the mechanism of **Figure 5b**. Thus, the local/global compass mechanisms are physical controllers that substitute models and computer algorithms.

3. Absolute/local commuting mechanism

The commutation mechanism interactively couples and uncouples either the absolute compass or the local compass (**Figure 6a**). The commuting device switches into the local compass immediately where any nearby obstacle is detected. Alternatively, it switches to the global compass retaking orientation toward the goal as soon as obstacles are no longer detected. **Figure 7a** depicts the gearing transmission that commutates the different steering mechanisms, the local compass (the front), and the global compass (the back).

According to **Figure 6b**, the commutation mechanism yields linear motion (the commutator device of **Figure 6a**); it activates the local compass by rotating a servomotor and shrinking a sliding crank link *L*. Inversely, the absolute compass is activated by turning off the servomotor, and then a spring-mass-damper system stretches the sliding crank link *L*.

The motion transmission system (**Figure 6a**) is inspired by the model of a modern vehicle's speed transmission box. For the present case, a common shaft works for two asynchronous speed gearing systems. In addition, **Figure 7a** and **7b** illustrates the motion transmission flow

http://dx.doi.org/10.5772/intechopen.70811

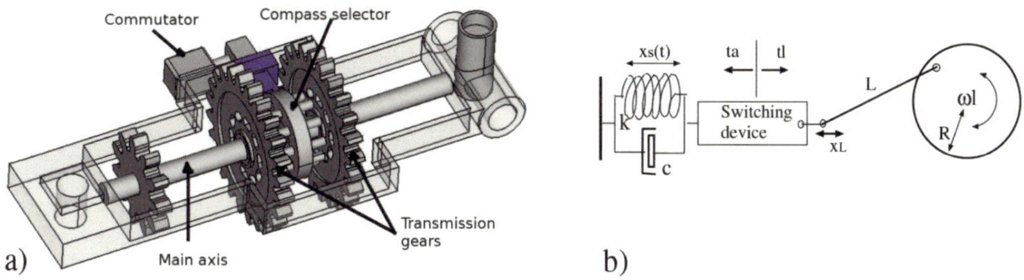

Figure 6. Commutator device. (a) Absolute/evasive motion transmission system and (b) basic commuter mechanism.

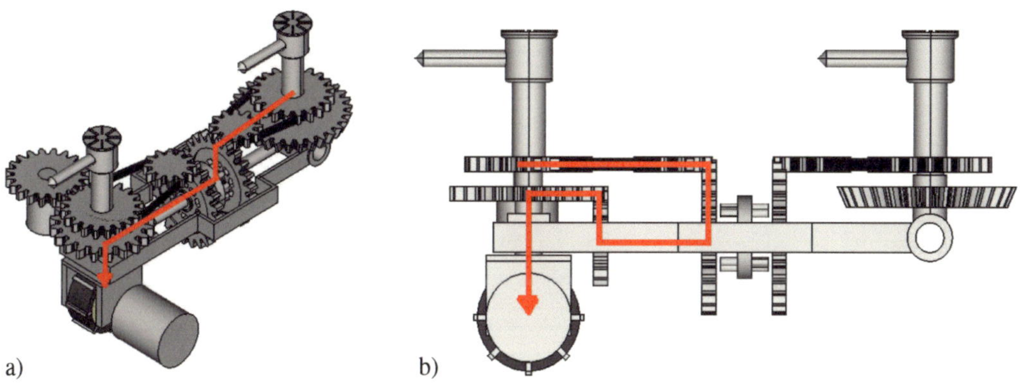

Figure 7. Motion transmission flow (red arrowed lines). (a) Goal-reference and (b) obstacle evasive.

by the red color arrowed lines. When either the local compass mechanism or the global compass mechanism is commutated, the flow transmission motion only concerns with its own gearing mechanism featuring a physical continuous controller.

The angular speed model ω_l $[s^{-1}]$ for the commutation rotary servomotor (**Figure 6b**) to switch into the local compass was obtained through fitting empirical observations with the next theoretical model:

$$\omega_l = \kappa_1 e^{\kappa_2 t}; \tag{18}$$

where κ_1 [rad/s] is an amplitude factor and κ_2 [1/s] is a slope rate factor. Subsequently, to estimate the commutation time t_l [s] required to mechanically coupling into the local compass mechanisms, the next model is deduced:

$$t_l = \frac{\ln\left(\frac{\omega_l}{\kappa_2}\right)}{\kappa_1}. \tag{19}$$

Moreover, when no obstacles are detected, the commuter device switches into its initial state by coupling the absolute compass. In this case, an underactuated system commutates the state by a spring-mass-damper mechanical system with critically dumped configuration, modeled by second-order linear differential equation. Where m [kg] is the mass pulling the spring, c [kg/s] is the damping effect constant, k [kg/s ^{2}] is the spring elasticity constant, and x, $\dot{x}$, and $\ddot{x}$ are the spring distance, velocity, and acceleration, respectively:

$$m\ddot{x} + c\dot{x} + kx = 0 \tag{20}$$

Thus, it may be solved as a first-order linear equation by temporally omitting the second-order term $m\ddot{x}$ such that

$$c\frac{\mathrm{d}x}{\mathrm{d}t} = -kx, \tag{21}$$

reorganizing and completing the integrals

$$\int_x \frac{\mathrm{d}x}{x} = \frac{-k}{c}\int_t \mathrm{d}t, \tag{22}$$

Thus, by solving the improper integrals and multiplying both sides of the equation by the Euler number e,

$$\mathrm{e}\left[\ln(x) = -\frac{k}{c}t + C_1\right] \tag{23}$$

and a solution is obtained; to simplify let us define $\lambda := -k/c$,

$$x(t) = \mathrm{e}^{\lambda t + c_1} \tag{24}$$

Assuming the integration constant $C_1 = 0$ and developing the higher order solutions

$$x(t) = \mathrm{e}^{\lambda t};\ \dot{x}(t) = \lambda \mathrm{e}^{\lambda t};\ \ddot{x}(t) = \lambda^2 \mathrm{e}^{\lambda t}. \tag{25}$$

In addition, by substituting such functions in the next expression

$$m\lambda^2 \mathrm{e}^{\lambda t} + c\lambda \mathrm{e}^{\lambda t} + k\mathrm{e}^{\lambda t} = 0, \tag{26}$$

In order to decrease the commutation time of the sliding linear mechanism, a critically damped system is assumed and modeled by $(c^2 - 4mk) = 0$, with terms $\lambda_1 = \lambda_2$:

$$\lambda_{1,2} = \frac{-c}{2m}, \tag{27}$$

It is desired to speed up as much as possible the commutation time. Then, the mechanical device linear displacement is modeled as a function of time by

$$x(t) = (a_1 + a_2)\mathrm{e}^{-ct/2m}. \tag{28}$$

And, solving for the commutation time for switching to the absolute compass t_a,

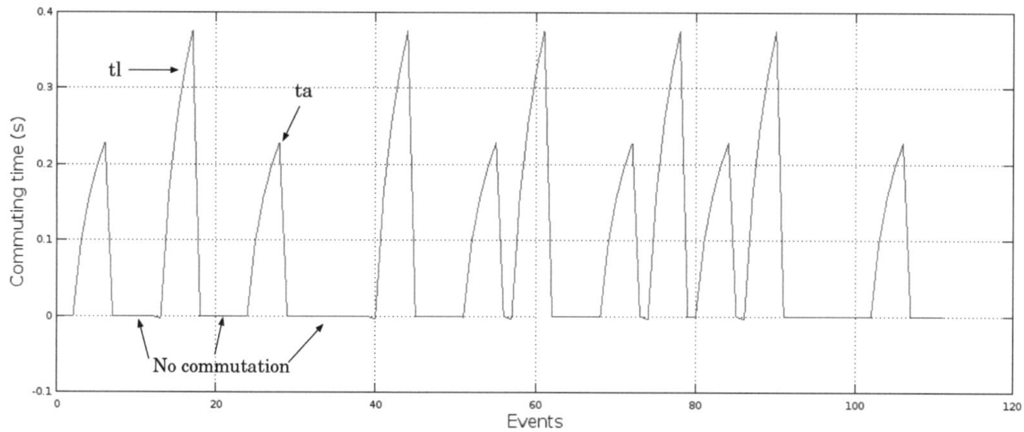

Figure 8. Commuting times: t_a (absolute mechanism) and t_l (local mechanism).

$$t_a = -\frac{2m}{c}\ln\left(\frac{x}{a_1 + a_2}\right) \tag{29}$$

Figure 8 depicts a sequence of transition times taken by the commuting mechanism. Times t_l are the times behavior of the slider crank servomotor when shrinking the link. Times t_a are the times behavior of the spring-mass-damper device when stretching the link. Such time magnitudes are reached due to mechanical motions. However, such commutation times are physically fast enough, and a robot's trajectory is not affected in the precise points when commutations occur.

4. Kinematic control law

In this section, a kinematic control law is deduced and analyzed. The proposed controller simultaneously controls driving and self-steering velocities and keeps track of the robot's posture. The robotic platform is a tricycle-type kinematic structure with two lateral passive wheels at the back (**Figure 9a**) and a central active-drive and passive self-steering wheel at the front (**Figure 9b**). Each wheel kinematic is described by three parameters, α, ℓ, and r, where α [rad] is the angle of a wheel w.r.t. the x-axis of the robot's fixed inertial system, ℓ [m] is the distance between the robot's centroid and each wheel's contact point with the ground, and r [m] is an ideal wheel's radius. In addition, the robot's kinematic structure considers two controlled kinematic variables β_t and φ_t, where β_t [rad] is the instantaneous steering angle of the front-sided wheel and φ [rad] is the instantaneous driving wheel's angle.

The kinematic parameters and variables describing the robot platform are summarized in **Table 1** according to each type of wheel (fixed passive, steerable, and drivable active).

The robot's wheel kinematic parameter is modeled by the following constraint equation, which is stated from the wheel plane,

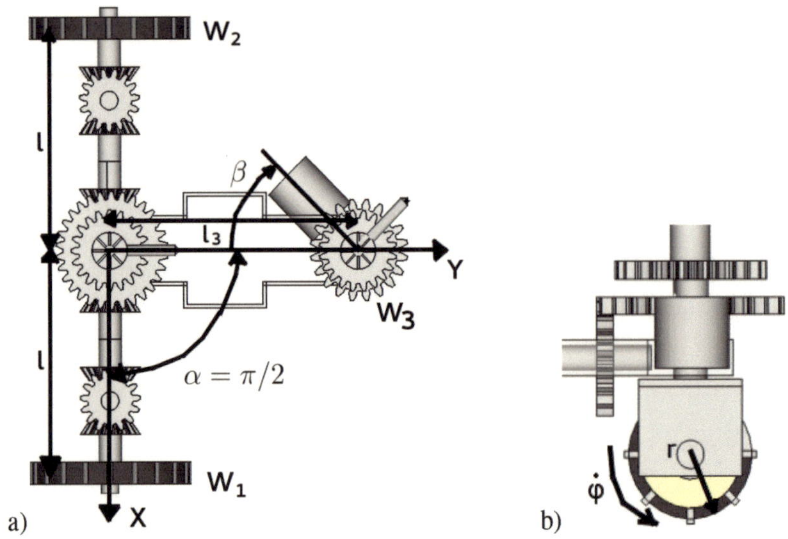

Figure 9. Robot's kinematic structure. (a) Top-view kinematic structure and (b) side-view drive-steer wheel.

Wheel type	α	β_t	ℓ
W_1 (fixed)	0	0	ℓ
W_2 (fixed)	π	0	ℓ
W_3 (drive/steer)	$\pi/2$	–	ℓ_3

Table 1. Tricycle-type robot's wheel kinematic parameters and variables.

$$\mathbf{J}_1(\beta_t)\mathbf{R}(\theta_t)\dot{\xi}_t + \mathbf{J}_2\varphi_t = 0, \tag{30}$$

Likewise, the kinematic constraint is modeled in the wheel's orthogonal plane by

$$\mathbf{C}_1(\beta_T)\mathbf{R}(\theta_t)\dot{\xi}_t = 0. \tag{31}$$

where $\mathbf{J}_1, \mathbf{C}_1 \in \mathbb{R}^{N\times 3}$ basically describes the wheel kinematic constraints. Matrix $\mathbf{J}_2$ poses the ideal wheel's radius, and $\mathbf{R}(\theta)$ is the Euler rotation matrix. Vector ξ_t describes the state space system, and its components are the robot's posture $\xi=(x, y, \theta, \beta)^{\mathrm{T}}$, as well as the front-sided wheel driving speed φ_t, and its product with $\mathbf{J}_2$ establishes a diagonal matrix of the wheels' tangential velocities.

Moreover, for the particular case of fixed wheels and central orientable wheels, the following kinematic constraints apply for individual wheels.

In the wheel's plane

$$\left[-\sin\left(\alpha+\beta_t\right) \quad \cos\left(\alpha+\beta_t\right) \quad \ell\cos\left(\beta_t\right)\right]\cdot\mathbf{R}(\theta_t)\cdot\dot{\xi}_t + r\dot{\varphi}_t = 0. \tag{32}$$

In the wheel's orthogonal plane

$$\left[\cos\left(\alpha+\beta_t\right) \quad \sin\left(\alpha+\beta_t\right) \quad \ell\sin\left(\beta_t\right)\right]\cdot\mathbf{R}(\theta_t)\dot{\xi}_t=0. \tag{33}$$

where matrix $\mathbf{J}_1$ is the fixed passive back-sided wheels matrix of kinematic constraints, which is obtained by substituting the parameters of **Table 1** in expression (32). Such that, $\mathbf{J}_f$ for the fixed wheels is

$$\mathbf{J}_f=\begin{pmatrix}-\sin(0) & \cos(0) & \ell\cos(0)\\ -\sin(\pi) & \cos(\pi) & \ell\cos(0)\end{pmatrix}, \tag{34}$$

and by simplifying

$$\mathbf{J}_f=\begin{pmatrix}0 & 1 & \ell\\ 0 & -1 & \ell\end{pmatrix}. \tag{35}$$

Taking into account the kinematic constraints that are similar to the central orientable wheel, the vector model $\mathbf{J}_o$ is obtained:

$$\mathbf{J}_o=\left(-\sin\left(\frac{\pi}{2}+\beta_t\right) \quad \cos\left(\frac{\pi}{2}+\beta_t\right) \quad \ell_3\cos\left(\beta_t\right)\right) \tag{36}$$

For the front-sided central orientable wheel, since the angle β_t is nonstationary, and in order to easy its algebraic solution the following trigonometric identities are substituted in (36) and obtain a simplified form for the vector $\mathbf{J}_o$. Thus,

$$\sin(a+b)=\sin(a)\cos(b)+\cos(a)\sin(b),$$

Likewise,

$$\cos(a+b)=\cos(a)\cos(b)-\sin(a)\sin(b);$$

and the following row vector is produced:

$$\mathbf{J}_o=\left(-\cos\left(\beta_t\right) \quad -\sin\left(\beta_t\right) \quad \ell_3\cos\left(\beta_t\right)\right). \tag{37}$$

Therefore, for a tricycle-type robotic structure as in the present context, and with parameters of **Table 1**, the matrix $\mathbf{J}_1$ is consequently defined as

$$\mathbf{J}_1=\begin{pmatrix}\mathbf{J}_f\\ \mathbf{J}_o\end{pmatrix}=\begin{pmatrix}0 & 1 & \ell\\ 0 & -1 & \ell\\ -\cos\left(\beta_t\right) & -\sin\left(\beta_t\right) & \ell_3\cos\left(\beta_t\right)\end{pmatrix}, \tag{38}$$

where $\mathbf{J}_f$ represents the fixed wheel kinematic constraints and $\mathbf{J}_o$ represents the central orientable wheel constraints. The constraints denoted in the general form (31) implicate that vector $\mathbf{R}(\theta_t)\cdot\dot{\xi}_t$ belongs to the null space $\mathbf{C}_1^*$ denoted by

$$\mathbf{C}_1^*\left(\beta_t\right)=\begin{pmatrix}\mathbf{C}_f\\ \mathbf{C}_\beta\end{pmatrix} \tag{39}$$

being $\mathbf{C}_f$ and $\mathbf{C}_\beta$ as vectors of the fixed and the central orientable wheel constraint, respectively, with notation

$$\mathbf{R}(\theta_t) \cdot \dot{\xi}_t \in \mathcal{N}\left[\mathbf{C}_1^*(\beta_t)\right]. \tag{40}$$

Hence, the robot's posture first-order derivative $\dot{\xi}(t)$ or state vector is constrained into a distribution defined as

$$\mathcal{N}\left[\mathbf{C}_1^*(\beta)\right] = \text{span}\{\text{col}\boldsymbol{\Sigma}(\beta)\}, \tag{41}$$

and the null space of any matrix $\mathbf{A}$, written as a $\mathcal{N}(\mathbf{A})$, is the set of all solutions of $\mathbf{A} \cdot \mathbf{x} = 0$ and stated in set notation by

$$\mathcal{N}(\mathbf{A}) = \{\mathbf{x} \in \mathbb{R}^x | \mathbf{A} \cdot \mathbf{x} = 0\}. \tag{42}$$

Thus, for the particular case of Eq. (41), the null space is

$$\mathcal{N}\left[\mathbf{C}_1^*(\beta_t)\right] = \{\mathbf{x} \in \mathbb{R}^3 | [C_1^*(\beta_t)]\mathbf{x} = 0\}. \tag{43}$$

A manner to solve for the null space matrix condition is to reduce it to the echelon form Ψ:

$$\mathbf{C}_1 = \begin{pmatrix} 1 & 0 & 0 \\ -1 & 0 & 0 \\ -\sin(\beta_t) & \cos(\beta_t) & \ell_3 \sin(\beta_t) \end{pmatrix}, \tag{44}$$

Thus, a vector $\mathbf{x}$ that satisfies Eq. (43) must be defined then:

$$\begin{pmatrix} 1 & 0 & 0 \\ -1 & 0 & 0 \\ -\sin(\beta_t) & \cos(\beta_t) & \ell_3 \sin(\beta_t) \end{pmatrix} \cdot \begin{pmatrix} x_1 \\ x_2 \\ x_3 \end{pmatrix} = \begin{pmatrix} 0 \\ 0 \\ 0 \end{pmatrix}. \tag{45}$$

The previous expression (45) is a set of linear systems and is rewritten as an augmented matrix $[\mathbf{C}_1 | \mathbf{0}]$:

$$[\mathbf{C}_1 | \mathbf{0}] = \begin{pmatrix} 1 & 0 & 0 & | & 0 \\ -1 & 0 & 0 & | & 0 \\ -\sin(\beta) & \cos(\beta) & \ell_3 \sin\beta & | & 0 \end{pmatrix}. \tag{46}$$

The augmented matrix (46) is algebraically solved by reducing it to echelon form, starting by making zeros the first column, but one the first element:

$$\Psi = \begin{pmatrix} 1 & 0 & 0 \\ 0 & 0 & 0 \\ -\sin(\beta_t) & \cos(\beta_t) & \ell_3 \sin(\beta_t) \end{pmatrix}, \tag{47}$$

Then, sum up $\Psi_{1,i}\Psi_{3,1}$ where $\Psi_{3,1} = \sin(\beta_t)$:

$$\Psi = \begin{pmatrix} 1 & 0 & 0 \\ 0 & 0 & 0 \\ 0 & \cos\left(\beta_t\right) & \ell_3 \sin\left(\beta_t\right) \end{pmatrix}. \tag{48}$$

Since the pivot element $\Psi_{2,2}=0$, rows 2 and 3 are exchanged:

$$\Psi = \begin{pmatrix} 1 & 0 & 0 \\ 0 & \cos\left(\beta_t\right) & \ell_3 \sin\left(\beta_t\right) \\ 0 & 0 & 0 \end{pmatrix}. \tag{49}$$

Now, $\Psi_{i,2}=\mathbf{0}$ except its pivot element $\Psi_{2,2}$, thus let us divide $\Psi_{2,i}/\cos(\beta_t)$ to set to 1 such pivoting element:

$$\Psi = \begin{pmatrix} 1 & 0 & 0 \\ 0 & 1 & \ell_3 \tan\left(\beta_t\right) \\ 0 & 0 & 0 \end{pmatrix}. \tag{50}$$

Thus, by having Ψ in the reduced form, now the system is solved as the system (45):

$$\begin{pmatrix} 1 & 0 & 0 \\ 0 & 1 & \ell_3 \tan\left(\beta_t\right) \\ 0 & 0 & 0 \end{pmatrix} \cdot \begin{pmatrix} x_1 \\ x_2 \\ x_3 \end{pmatrix} = \begin{pmatrix} 0 \\ 0 \\ 0 \end{pmatrix}. \tag{51}$$

Hence, rewriting such a solution in the matrix form

$$\mathbf{x} = \begin{pmatrix} 0 \\ -\ell_3 \tan\left(\beta\right) \\ 1 \end{pmatrix} x_3 \tag{52}$$

Being the vector $\mathbf{x}$ of this null space of matrix $\mathbf{C}_1$, such that

$$\mathcal{N}\left[\mathbf{C}_1^*\left(\beta\right)\right] = \text{span}\begin{pmatrix} 0 \\ -\ell_3 \tan\left(\beta\right) \\ 1 \end{pmatrix} x_3; \tag{53}$$

For our application purpose, $x_3=\cos\beta$, then the null space solution $\mathcal{N}\left[\mathbf{C}_1^*\left(\beta\right)\right]$ is

$$\mathcal{N}\left[\mathbf{C}_1^*\left(\beta\right)\right] = \text{span}\begin{pmatrix} 0 \\ -\ell_3 \sin\beta \\ \cos\beta \end{pmatrix}. \tag{54}$$

Therefore, Eq. (41) is rewritten because for each instant time t, a temporal variable v_t exist .

$$\dot{\xi} = \mathbf{R}^{\mathrm{T}}(\theta_t)\boldsymbol{\Sigma}(\beta_t)v_t, \tag{55}$$

and

$$\dot{\beta}_t = w_t. \tag{56}$$

where v_t is the instantaneous robot's absolute velocity and w_t is instantaneous robot's yaw rate. As a matter of fact $\boldsymbol{\Sigma}(\beta_t)$ is the null space of the kinematic constraint matrix in the wheel's orthogonal plane (Eq. (54)), defined by

$$\boldsymbol{\Sigma}(\beta_t) = \begin{pmatrix} 0 \\ -\ell_3 \sin(\beta_t) \\ \cos(\beta) \end{pmatrix}. \tag{57}$$

This state control law is expressed in a more compact form:

$$\dot{\mathbf{z}}_t = \mathbf{B}(z)\mathbf{u}_t; \tag{58}$$

where

$$\mathbf{z} = \begin{pmatrix} \xi_t \\ \beta_t \end{pmatrix}; \ \mathbf{B}(z_t) = \begin{pmatrix} \mathbf{R}^{\mathrm{T}}(\theta_t)\boldsymbol{\Sigma}(\beta_t) & 0 \\ 0 & 1 \end{pmatrix}; \ \mathbf{u} = \begin{pmatrix} v \\ w \end{pmatrix}. \tag{59}$$

Thus, by substituting expression (57) in the model (58), the kinematic control law is rewritten:

$$\begin{pmatrix} \dot{x} \\ \dot{y} \\ \dot{\theta} \\ \dot{\beta} \end{pmatrix} = \begin{pmatrix} \ell_3 \sin\theta \ \sin\beta & 0 \\ -\ell_3 \cos\theta \ \sin\beta & 0 \\ \cos\beta & 0 \\ 0 & 1 \end{pmatrix} \cdot \begin{pmatrix} v \\ w \end{pmatrix}. \tag{60}$$

The robot's posture is essentially modeled by Eq. (60) and validates by the Cartesian trajectory depicted in **Figure 10a**. For this case, v was set at constant value overtime, and w gradually increased according to the range $\{0, \ldots, \pi/4\}$. Then, the robot turns into its original angle and then continues straight.

Figure 10b and **10c** shows β and θ behaviors w.r.t. orientation error θ_E, when $\theta_E = 0$ the robot is bearing toward the desired goal. Actually, β depicts how the steering wheel performs from its starting angle up to the final angle. Because of the absence of obstacles, the steering wheel's angle experiences no commutation toward the local mechanism. However, the absolute compass controls the wheel from the beginning and gradually steers until orientation is aligned to the global destination. Its behavior along the navigation task is illustrated in plot 10b.

Moreover, in order to instantaneously estimate the deviation angle θ_E, let us define the global destination θ_G [rad]. And, such desired orientation is perturbed by escaping angle θ_R [rad] due to relative obstacle locations (**Figure 11a**). Likewise, θ_R is established every time the local compass mechanism is switched to. Obstacles are detected within a territorial radius d_l, or ignored when the robot moves farther away from the scope d_l (**Figure 11b**).

http://dx.doi.org/10.5772/intechopen.70811

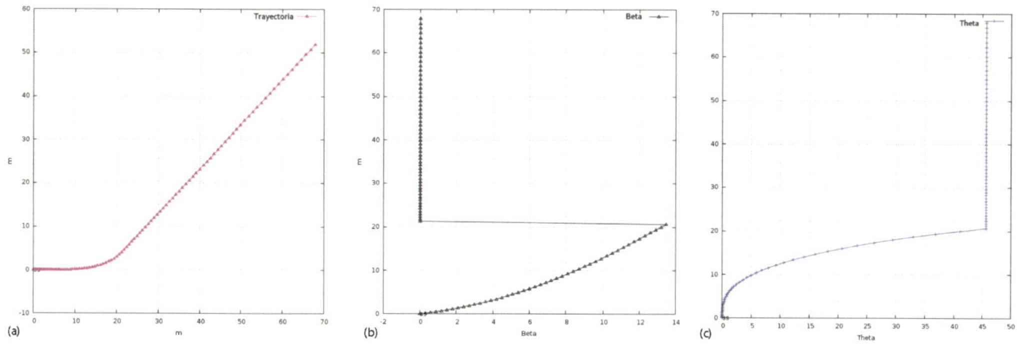

Figure 10. Control law (60) simulation results. (a) Trajectory control with v=cte and w ranging $\beta = \{0, \ldots, \pi/4\}$, (b) wheel's steer angle β behavior, and (c) robot's angle θ behavior.

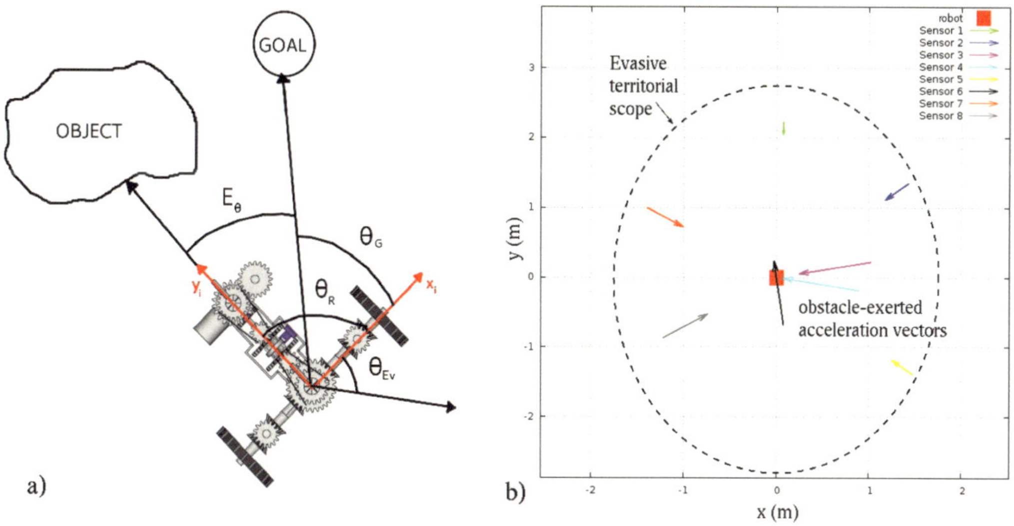

Figure 11. Evasive system. (a) θ_R and θ_G kinematics and (b) evasive territorial scope and obstacle-repulsive acceleration vectors.

For instance, sensor 1 detects a far obstacle, and its repulsive acceleration influence is low. However, sensor 6 yields a greater repulsive acceleration influence since it is nearer to the robot's location.

Thus, let us postulate:

$$\theta_E = \theta_R - \theta_G \tag{61}$$

From the control law, the state vector $\mathbf{z}$ is now estimated by directly using the instantaneous error angle θ_E:

$$\dot{\mathbf{z}}\left(\dot{x}, \dot{y}, \theta_E, \beta\right) = \mathbf{B}(z) \cdot \mathbf{u}(v, w), \tag{62}$$

and by obtaining the inverse solution for $\mathbf{u}$, now the input system involves error angle:

$$\mathbf{u}(v, w) = \mathbf{B}(z)^{-1} \cdot \dot{\mathbf{z}}(\dot{x}, \dot{y}, (\theta_R - \theta_G), \beta). \tag{63}$$

The robot's controlled orientation is fed back by θ_E and calculated by numerical successive approximations. This controller is a recursive system that involves both, the direct and the inverse solutions, to reduce the numerical error θ_E.

The instantaneous escaping angle θ_R is obtained within a local coordinate system by directional derivatives, which approach a cosine function, similarly to the work reported in Ref. [20]. A gradient operator of a cosine navigation function approximates a repulsive partial differential equation to evade any observable obstacle. The robot's repulsive acceleration vector function $\mathfrak{a}_\alpha = (a_x, a_y)^\top$ produces its magnitude effect w.r.t. the obstacle's range measurement δ_α [m]:

$$\mathfrak{a}_\alpha(x, y) = -\nabla_{x,y} k_\alpha (k_2 - r_\alpha) \cdot \cos(\phi_\alpha) \tag{64}$$

where k_2 [m] is the obstacle diametrical territory and k_α $[ms^{-2}]$ is an adjustment constant factor of the physical acceleration amplitude. Thus, the Cartesian geometric model (not measurement) r_α describes the distance between the robot $\mathbf{x}_r = (x_r, y_r)^\top$ and any obstacle $\mathbf{x}_o = (x_o, y_o)^\top$, which is geometrically modeled by

$$r_\alpha = \sqrt{x^2 + y^2} = \sqrt{(x_r - x_o)^2 + (y_r - y_o)^2} \tag{65}$$

The angle limits $\{0 \le \phi_\alpha \le \pi/2\}$ is a transformation relationship of the obstacle range δ_α, and the condition $\delta_\alpha < d_l$ limits the repulsive acceleration effects:

$$\phi_\alpha = \frac{\delta_\alpha \pi}{2 d_l}; \quad \text{where } d_l \propto \frac{\pi}{2} \quad \text{and } d_l = \delta_\alpha \Rightarrow \|\mathfrak{a}\| = 0 \tag{66}$$

Substituting the functional form of δ_α in (64) and temporally considering $k_\alpha = 1$ for analysis purpose

$$\mathfrak{a}_\alpha(x, y) = -\nabla_{x,y}\left(\left(k_2 - \sqrt{(x_r - x_o)^2 + (y_r - y_o)^2}\right) \cdot \cos(\phi_\alpha)\right) \tag{67}$$

and algebraically expanding

$$\mathfrak{a}_\alpha(x, y) = -\nabla_{x,y}\left(-k_2 \cos(\phi_\alpha) + \sqrt{(x_r - x_o)^2 + (y_r - y_o)^2} \cdot \cos(\phi_\alpha)\right) \tag{68}$$

Thus, applying the gradient operator $\nabla_{x,y}$ w.r.t. x and y components and algebraically simplifying

$$\frac{\partial \mathfrak{a}_\alpha}{\partial x} = \frac{(x_r - x_o) \cdot \cos(\phi_\alpha)}{\sqrt{(x_r - x_o)^2 + (y_r - y_o)^2}} \tag{69}$$

and

$$\frac{\partial \mathfrak{a}_\alpha}{\partial y} = \frac{(y_r - y_o) \cdot \cos(\phi_\alpha)}{\sqrt{(x_r - x_o)^2 + (y_r - y_o)^2}} \tag{70}$$

Therefore, the obstacle-repulsive directional vector expressed in terms of Cartesian components XY is

$$\mathfrak{a}_\alpha(x,y) = k_\alpha \frac{\cos(\phi_\alpha)}{\sqrt{(x_r - x_o)^2 + (y_r - y_o)^2}} \cdot \begin{pmatrix} x_r - x_o \\ y_r - y_o \end{pmatrix} \tag{71}$$

Thus, the model for multiple obstacles α produces a controlled escaping direction in robot's local coordinate framework:

$$\mathbf{a}_T = k_\alpha \sum_\alpha \frac{\cos(\phi_\alpha)}{\sqrt{(x_r - x_o)^2 + (y_r - y_o)^2}} \cdot \begin{pmatrix} x_r - x_o \\ y_r - y_o \end{pmatrix} \tag{72}$$

It follows that the evasive acceleration magnitude is defined by

$$\|\mathbf{a}_T\| = \sqrt[2]{\left(\frac{\partial \mathfrak{a}_\alpha}{\partial x}\right)^2 + \left(\frac{\partial \mathfrak{a}_\alpha}{\partial y}\right)^2}, \tag{73}$$

and the instantaneous escaping angle θ_R

$$\theta_R = \arctan\left(\frac{\left(\frac{\partial a}{\partial y}\right)}{\left(\frac{\partial a}{\partial x}\right)}\right). \tag{74}$$

Finally, the proposed kinematic controller has the following scheme described in Algorithm 1.

Algorithm 1. Local/global robot's underactuated system controller

---

$\omega_l = 0.0$

while $(\|\xi_t\| - \|(x_G, y_G)\|) > 0$ **do**

  switching to either local or global mechanism

  if $\delta_\alpha > 0$

    Commuter's motor speed $\omega_l = k_1 e^{k_2 t}$

    Servo's angle $\phi = \frac{\omega_l \kappa_1}{\ln\left(\frac{\omega_l}{\kappa_2}\right)}$

    Avoidance acceleration $\mathbf{a}_T = k_\alpha \sum_\alpha \frac{\cos(\phi_\alpha)}{r_\alpha}(\mathbf{x}_r - \mathbf{x}_\alpha)$

    Avoidance angle $\theta_R = \arctan\left(\frac{\partial a/\partial y}{\partial a/\partial x}\right)$

    $\|\mathbf{a}_T\| = \sqrt[2]{(\partial a/\partial x)^2 + (\partial a/\partial y)^2}$

  else

    Turns servo off $\omega_l = 0.0$

    Release slider crank $x_a = r\cos(\phi)\sqrt[2]{(a_1 + a_2)^2 e^{-ct/m} - (r\sin(\phi))^2} + \Delta s$

    $\theta_G = -\frac{\varphi_{w1} - \varphi_{w2}}{4}$

  $\theta_E = \theta_R - \theta_G$

  $\mathbf{u} = \left(\frac{\|\mathbf{a}_T\|}{\Delta t}, \frac{d\theta_E}{dt}\right)^T$

  $\dot{\mathbf{z}}_t = \mathbf{B}(\theta_E, \beta) \cdot \mathbf{u}_t$

  $\mathbf{u}_{t+1} = \mathbf{B}^{-1} \cdot \dot{\mathbf{z}}$

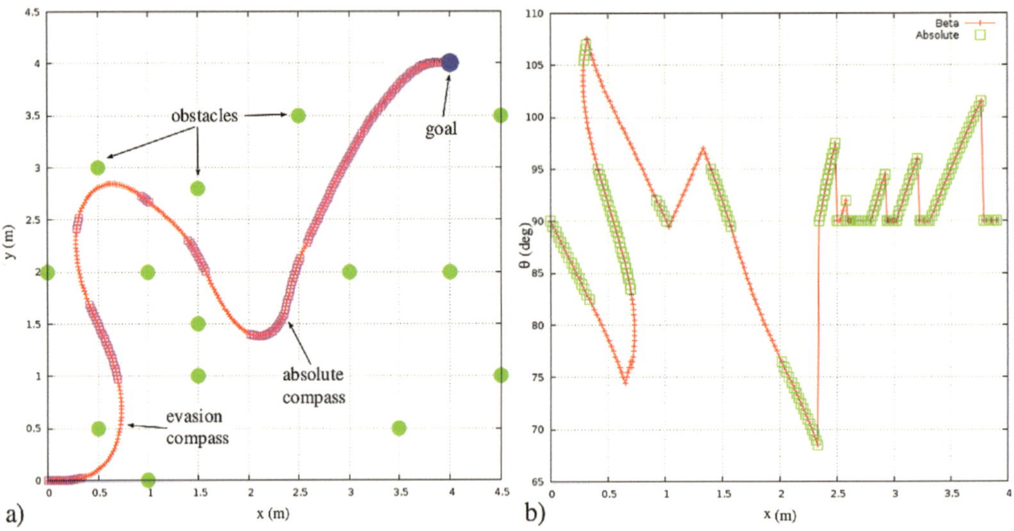

Figure 12. Controlled robot's course $v = 1.0\ rad/s$ and $d_l = 0.5\ m$. (a) Cartesian space and (b) component X versus robot's angle.

Hereafter, **Figure 12** validates our kinematic model approach by depicting how the robot's trajectory reaches the final Cartesian goal (−7, −8) in global coordinates. In addition, **Figure 12a** shows the Cartesian trajectory among multiple obstacles and the local/global commuting steering modalities. Likewise, as a manner to show validation, **Figure 12b** illustrates only the component X versus the robot's angle during the same navigational task.

5. Conclusion

The interest of this chapter was to introduce the analysis of an alternative kinematic steering controller by using underactuated mechanical compasses and to demonstrate its feasibility, controllability approach, and natural efficiency. Despite the complexity in its implementation, a compass mechanism is proposed because it allows to directly set up the global goal as an absolute reference. Moreover, the compass mechanism itself directly steers the robot to the goal simultaneously avoiding obstacles. These functional features are important advantages for a robotic platform w.r.t. other traditional orientation systems, because the global orientation depends neither on feedback from sensor devices nor on computational complexity expenses to instantaneously estimate the global destination orientation. Unlike kinematic redundant structures, the proposed approach limits electric energy use for one driving actuator, not for steering actuators. Steering controllers no longer spend algorithmic computational resources, reserving such resources either for other robotic algorithmic tasks or for increasing additional hardware devices. Mechanical controllers are slow if compared with software algorithms, even though mechanisms are fast enough w.r.t. the available robot's mobility speeds. In this regard, a critical issue is the commutation linear actuator, which is passive/active based on a slider crank combined with a spring-mass-damper system. The commutator device was fast enough, proving that sophisticated active linear actuators were not needed. It was found out that commutation times between local/global compasses did not negatively affect the system performance or the trajectory stability. Local/global compasses commuting activity warranted the robot to reach the global destination while avoiding collisions. The controlled trajectories yielded were concatenations of inter-switching segments with no discontinuities found. We concluded that from the implementation perspective, the proposed approach is neither necessarily better nor worst in effectiveness than a traditional redundantly actuated and multisensor approach. As a matter of fact, a traditional approach is easier to physically implement. Nevertheless, a redundant traditional approach is in disadvantage, if driving wheels grow in number, not to mention incrementing steering actuators. As a consequence, the more redundant is a system, the more discretized are the controlled motions and trajectories lose stability, which is an inherent difference from the naturally continuous motions produced by underactuated systems. A compact linear kinematic control law to switch inter-compass usage was deduced, with direct and inverse solutions. It elegantly combined the underactuated mechanism control with a very short algorithm to detect obstacles and to estimate the instantaneous escaping orientation. It was found that the robot's trajectory continuity may be altered, if kinematic evasion parameters are readjusted consequently changing the mechanism commutation response. Through simulation results, the inter-mechanism interactions and functions were validated. For the specific case of the proposed kinematic structure, the local/global compass showed as much efficiency as any similar redundant system. The future work will focus on fault recovery from slips, sliding, or collision dynamics that get the compass orientation uncalibrated. In addition, not only to further orientation analysis but also positioning sensors to supplement and improve this underactuated approach.

Author details

Jorge U. Reyes-Muñoz, Edgar A. Martínez-García*, Ricardo Rodríguez-Jorge and Rafael Torres-Córdoba

*Address all correspondence to: edmartin@uacj.mx

Laboratorio de Robótica, Institute of Engineering and Technology, Universidad Autónoma de Ciudad Juárez, Mexico

References

[1] Li SH. The South-Pointing Carriage and the Mariner's Compass. Taipei: Yee Wen Pub. Co.; 1959

[2] Lu ZM. An analysis of the ancient Chinese South-Pointing Chariot. Journal of Sichuan University. 1979;**2**:95-101

[3] Muneharu M, Satoshi K. Study of the mechanics of the South-Pointing Chariot (the South Pointing Chariot with the bevel gear type differential gear train). Transactions of the Japan Society of Mechanical Engineers. 1990;**56**(C):462-466

[4] Hong-Sen Y. Chap. 7. South-pointing Chariots, Reconstruction Designs of Lost Ancient Chinese Machinery, vol. 3. Dordrecht: Springer; 2007

[5] Hong-Sen Y, Chun-Wei C. A systematic approach for the structural synthesis of differential-type South Pointing chariots. JSME International Journal Series C Mechanical Systems, Machine Elements and Manufacturing. 2006;**49**((3), SI on Advanced Technology of Vibration and Sound):920-929

[6] Santander M. The Chinese South-Seeking chariot: A simple mechanical device for visualizing curvature and parallel transport. American Association of Physics Teachers. September 1992;**60**(9)

[7] Junmin W, Xiangyu Y, Wei L. Integration of hardware and software designs for object grasping and transportation by a mobile robot with navigation guidance via a unique bearing-alignment mechanism. IEEE/ASME Transactions on Mechatronics. 2016;**21**(1): 576-583

[8] Al-Faiz MZ, Mahameda GE. GPS-based navigated autonomous robot. International Journal of Emerging Trends in Engineering Research. 2015;**3**(4)

[9] Sioma A, Blok S. Finding bearing in robot navigation with the use of the Kalman filter. Solid State Phenomena. 2013;**199**:241-246

[10] Georgiou E., Dai J.S., Luck M., The KCLBOT: A double compass self-localizing maneuverable mobile robot. ASME. International Design Engineering Technical Conference and Computers and Information in Engineering Conference, Vol.3, pp. 427–435, 2011

[11] Chen w, Zhang T. An indoor mobile robot navigation technique using odometry and electronic compass. International Journal of Advanced Robotic Systems. May-Jun 2017:1-15

[12] Zhenhai H. y Shengguo H., Integrated navigation system based on differential magnetic compass and GPS, International Conference on Information Engineering and Computer Science, 2009

[13] Parhi D, Deepak B. Kinematic model of three wheeled mobile robot. Journal of Mechanical Engineering Research. 2011:307-318

[14] Campion G, Chung W, 17 C. Springer Handbook of Robotics. In: Handbook of Robotics. Springer-Verlag Berlin Heidelberg; 2008

[15] Morin P, Samson C. Chap. 34. Springer Handbook of Robotics. In: Handbook of Robotics. Springer-Verlag Berlin Heidelberg; 2008

[16] Minguez J, Lamiraux F, Lamound J. Chap. 35 Springer Handbook of Robotics. In: Handbook of Robotics. Springer-Verlag Berlin Heidelberg; 2008

[17] Martinez-Garcia EA. Numerical modelling in robotics. OmniaScience. 2015

[18] Alonsini N.I., Low cost obstacle detection system for wheeled mobile robot. UKACC International Conference on Control, pp. 529–533, 2012

[19] Martinez-Garcia EA. Robotic DCVG Planning for Searching Flaws on Buried Pipelines. Lap Lambert Academic; 2017

[20] Castro Jiménez L, Martínez-García EA. Thermal image sensing model for robotic planning and search. Sensors. 2016;**1253**:1-27

[21] Mujahed M., Fischer D. y Mertsching B., Tangential gap flow (TGF) navigation: A new reactive obstacle avoidance approach for highly cluttered environments, Robotics and Autonomous Systems, Vol.84, pp.15–30, 2016

Chapter 9

A New Methodology for Kinematic Parameter Identification in Laser Trackers

Ana Cristina Majarena, Javier Conte,
Jorge Santolaria and Raquel Acero

Additional information is available at the end of the chapter

http://dx.doi.org/10.5772/intechopen.71444

Abstract

In recent years, there has been an increasing interest in measurement systems such as laser trackers (LT) for the verification of large-scale parts in the aeronautic, spatial or naval sectors because of their advantages in terms of portability, flexibility, high speed in data acquisition, accuracy, and reliability. These systems present systematic errors caused by geometrical misalignments, environmental conditions, mechanical wear and tear and other unpredictable variables. Different standards such as the ASME B89.4.19 and the VDI 2617-10 suggest tests to calculate the geometric errors of the LT. In this work, we present an alternative calibration method based on a new errors model. The LT can be considered as an open kinematic chain, so it is possible to shape a kinematic model of the LT. Once the kinematic model has been set, the error model is defined. The model has been validated with synthetic data. Then, experimental tests based on the measurement of a mesh of reflectors placed at suitable places for different locations of the LT have been performed to ensure the reliability of the method proposed. A sensitivity analysis shows the best experimental setup to perform a calibration test. The calibration results have been validated with nominal data.

Keywords: laser tracker, modeling, kinematic parameter identification

1. Introduction

The development of more accurate large-scale measurement systems is a critical need in the verification of large parts and facilities in sectors with high-quality requirements as in naval, aeronautic, or spatial industries. In these industries, dimensional accuracy of large parts needs long-range accurate measuring devices to ensure not only the parts right dimensions but also the precise positioning of every part in large assemblies [1–4].

Their applications are very wide such as large-volume measurements [6], inspection, calibration of an industrial robot [7], reverse engineering [8], analysis of deformations [2], machine tool volumetric verification, and so on.

The laser tracker (LT) offers significant advantages such as portability, flexibility, precision, or high-speed data acquisition. However, the mechanical assembly is an important source of errors such as offsets or eccentricities, which generates errors in measurements. A disadvantage of these measurement systems is that the user cannot know whether the LT is measuring correctly. Existing standards provide tests to evaluate the performance of the LT. However, these tests require specialized equipment with a high cost. In addition, they require long time-consuming and specialized equipment.

The calibration procedure identifies the geometric parameters to improve system accuracy. However, there are few studies about the calibration of LTs.

The basis of the calibration procedure is to determine the parameters of the geometric error model measuring a set of reflectors located at fixed locations from different locations of LT. One of the advantages of this method is that specialized equipment is not required so that any user of a measurement system could perform the LT calibration. Furthermore, the time required to calibrate the TL is considerably reduced and considerably compared with the time required to carry out functional tests recommended by ASME B89.4.19 [9].

This chapter aims to present a method for performing a quick and easy calibration by measuring a mesh of reflectors to improve the accuracy of the LT. We have developed a kinematic error model and a generator of synthetic points to evaluate the procedure performance. Later, an experimental trial to identify the geometric parameters and a sensitivity analysis to determine the most appropriate instrument calibration measurement positions have been performed.

2. What a laser tracker is?

An LT is a long-range metrological device with an accuracy of some tens of micrometers. The LT is composed of an interferometer mounted on a two degrees of freedom rotatory gimbal. The laser beam reflects in a spherically mounted retroreflector (SMR), which returns the beam to the interferometer. Any displacement of the SMR is detected by position sensitivity devices, which adapt the LT position to follow the SMR position. The distance measured by the interferometer, (d), along with the readings of the encoders placed in each one of the two rotatory axes, (θ, φ), gives the spherical coordinates of the SMR centre referring to the LT reference system as shown in **Figure 1**. Knowing the spherical coordinates, the corresponding Cartesian coordinates can also be calculated based on Eqs. (1)–(3).

$$x = d \times \cos(\theta) \times \sin(\varphi) \tag{1}$$

$$y = d \times \sin(\theta) \times \sin(\varphi) \tag{2}$$

$$z = d \times \cos(\theta) \tag{3}$$

http://dx.doi.org/10.5772/intechopen.71444

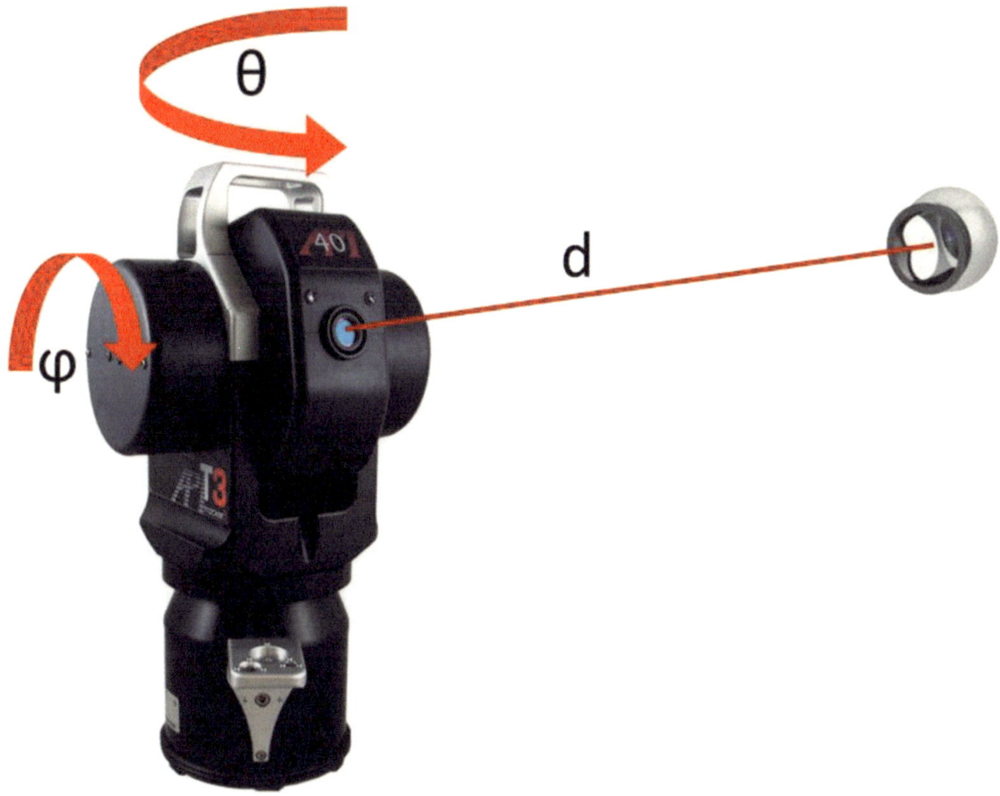

Figure 1. Laser tracker working principle.

3. Laser tracker kinematic model

The LT can be considered as an open kinematic chain with three joints: two rotary joints in the gimbal and a prismatic joint corresponding to the laser beam. The kinematic model is a mathematical expression that determines the position of the final joint of the kinematic chain (the SMR centre) with reference to the LT frame. We have used the Denavit-Hartenberg (D-H) [5] formulation to develop the kinematic model. This method defines the coordinate transformation matrices between each two consecutive reference systems j and $j-1$ as the product of the rotation and translation matrices from $j-1$ joint to j joint. The DH model needs four characteristic parameters (distances d_j, a_j, and angles θ_j, α_j) for these matrices. **Figure 2** shows the relationships between the consecutive reference systems. Knowing the kinematic parameters corresponding to every joint, the homogeneous transformation matrix from reference system j to $j-1$ is the result of the product of the four rotation and translation matrices shown in Eq. (4).

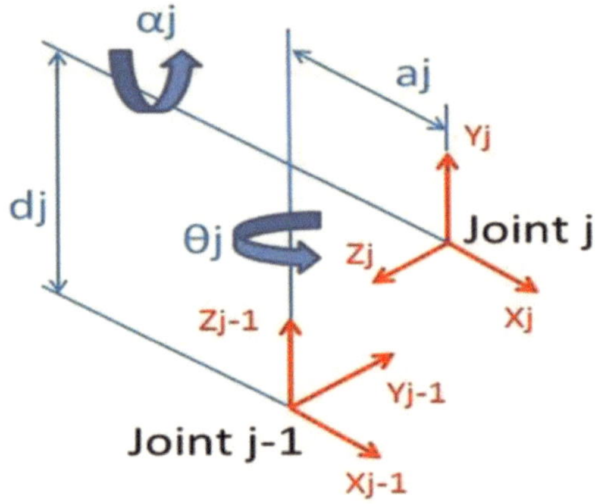

Figure 2. Denavit-Hartenberg model.

$$^{j-1}A_j = T_{z,d}R_{z,\theta}T_{x,a}R_{x,\alpha} = \begin{bmatrix} 1 & 0 & 0 & 0 \\ 0 & 1 & 0 & 0 \\ 0 & 0 & 1 & d_j \\ 0 & 0 & 0 & 1 \end{bmatrix} \times \begin{bmatrix} \cos\theta_j & -\text{sen}\,\theta_j & 0 & 0 \\ \text{sen}\,\theta_j & \cos\theta_j & 0 & 0 \\ 0 & 0 & 1 & 0 \\ 0 & 0 & 0 & 1 \end{bmatrix} \times \begin{bmatrix} 1 & 0 & 0 & a_j \\ 0 & 1 & 0 & 0 \\ 0 & 0 & 1 & 0 \\ 0 & 0 & 0 & 1 \end{bmatrix}$$
$$\times \begin{bmatrix} 1 & 0 & 0 & 0 \\ 0 & \cos\alpha_j & -\text{sen}\,\alpha_j & 0 \\ 0 & \text{sen}\,\alpha_j & \cos\alpha_j & 0 \\ 0 & 0 & 0 & 1 \end{bmatrix} = \begin{bmatrix} \cos\theta_j & -\cos\alpha_j\,\text{sen}\,\theta_j & \text{sen}\,\alpha_j \text{sen}\,\theta_j & a_j\cos\theta_j \\ \text{sen}\,\theta_j & \cos\alpha_j\cos\theta_j & -\text{sen}\,\alpha_j\cos\theta_j & a_j\,\text{sen}\,\theta_j \\ 0 & \text{sen}\,\alpha_j & \cos\alpha_j & d_j \\ 0 & 0 & 0 & 1 \end{bmatrix} \tag{4}$$

Being $T_{m,n}$ and $R_{m,n}$, the homogeneous translation matrices corresponding to translation (T) or rotation (R) n along axis m.

Following the DH model, the LT kinematic model has been determined as shown in **Figure 3**. The position of the reflector referring to the LT reference system is defined by Eq. (5).

$$^0T_3 = {}^0A_1{}^1A_2{}^2A_3 \tag{5}$$

This chapter is based on the LT API Tracker 3. **Table 1** shows the kinematic parameters corresponding to this LT model.

i	α_i (°)	a_i (mm)	d_i (mm)	θ_i (°)
1	−90	0	0	$\theta - 90$
2	90	0	0	$\varphi - 90$
3	0	0	d	−90

Table 1. Laser tracker kinematic parameters.

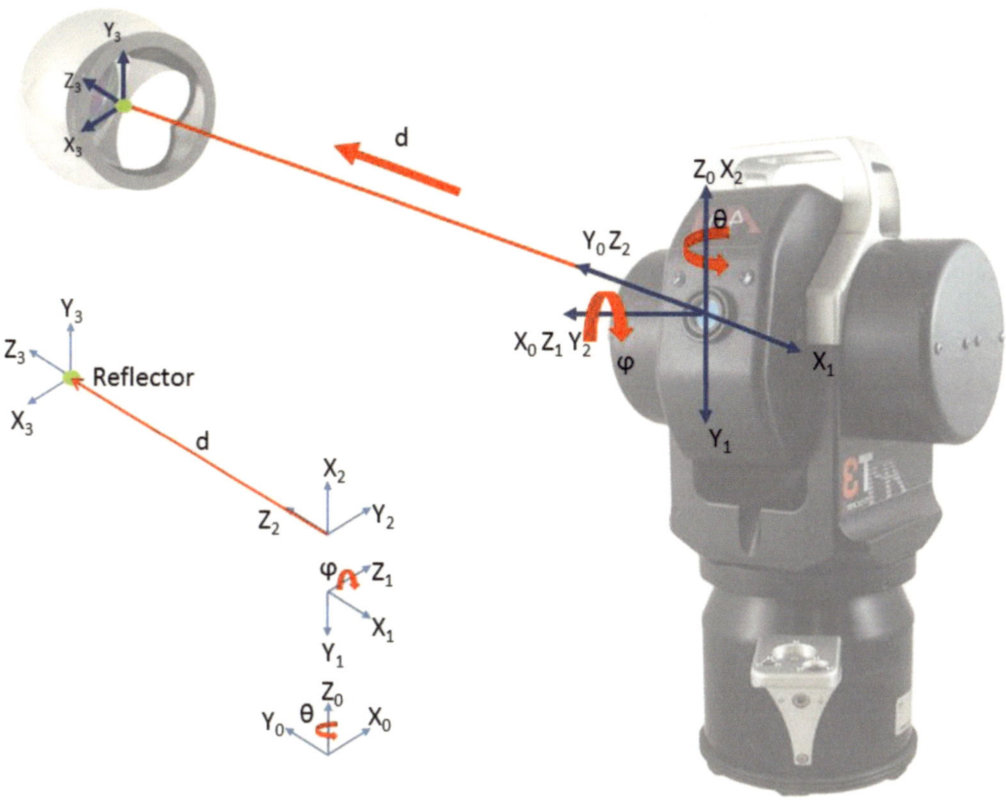

Figure 3. Laser tracker kinematic model.

4. Laser tracker error model

Relative positions between two consecutive joints are defined by the DH model. These are the nominal positions, but they are conditioned by the LT errors. This means that reference frames

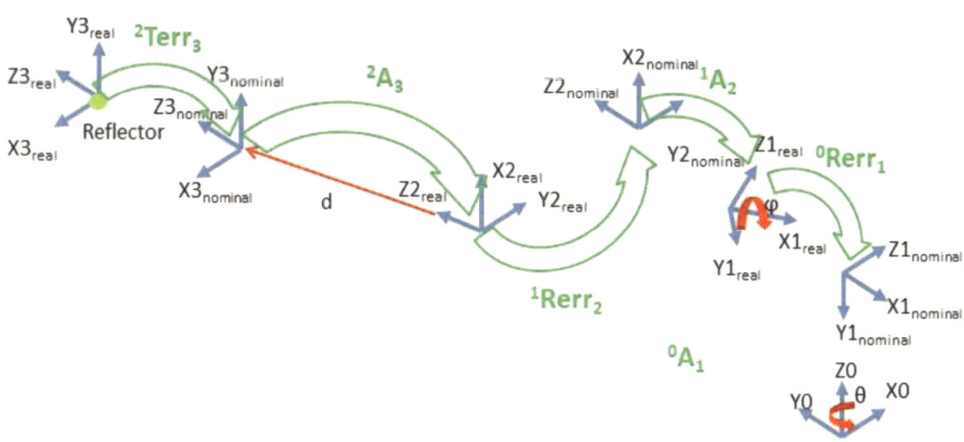

Figure 4. Error model.

are not at its expected position as shown in **Figure 4**. An error model that fits the geometry of the joints is necessary to modify the kinematic model.

In this chapter, the error model shown in **Figure 5** has been used. This model is based on six degrees of freedom error for each joint. Mathematically, the error is stated as a transformation matrix between the nominal joint frame and the real frame. The matrices are different for rotary joints (Eq. (6)) and linear joints (Eq. (7)). For each error matrix, a new set of six error parameters is defined (δx, δy, δz, εx, εy, εz). The calibration procedure will calculate the optimum parameter values to minimize the LT error.

$$R_{err} = \begin{bmatrix} \cos\varepsilon_Y \cdot \cos\theta_Z & -\cos\varepsilon_Y \cdot \sin\theta_Z & \sin\varepsilon_Y & \delta_X \\ \cos\varepsilon_X \cdot \sin\theta_Z + \sin\varepsilon_X \cdot \sin\varepsilon_Y \cdot \cos\theta_Z & \cos\varepsilon_X \cdot \cos\theta_Z - \sin\varepsilon_X \cdot \sin\varepsilon_Y \cdot \sin\theta_Z & -\sin\varepsilon_X \cdot \cos\varepsilon_Y & \delta_Y \\ \sin\varepsilon_X \cdot \sin\theta_Z - \cos\varepsilon_X \cdot \sin\varepsilon_Y \cdot \cos\theta_Z & \sin\varepsilon_X \cdot \cos\theta_Z + \cos\varepsilon_X \cdot \sin\varepsilon_Y \cdot \sin\theta_Z & \cos\varepsilon_X \cdot \cos\varepsilon_Y & \delta_Z \\ 0 & 0 & 0 & 1 \end{bmatrix} \tag{6}$$

$$T_{err} = \begin{bmatrix} 1 & -\varepsilon_\psi & \varepsilon_\theta & \varepsilon_x \\ \varepsilon_\psi & 1 & -\varepsilon_\phi & \varepsilon_y \\ -\varepsilon_\theta & \varepsilon_\phi & 1 & \varepsilon_z \\ 0 & 0 & 0 & 1 \end{bmatrix} \tag{7}$$

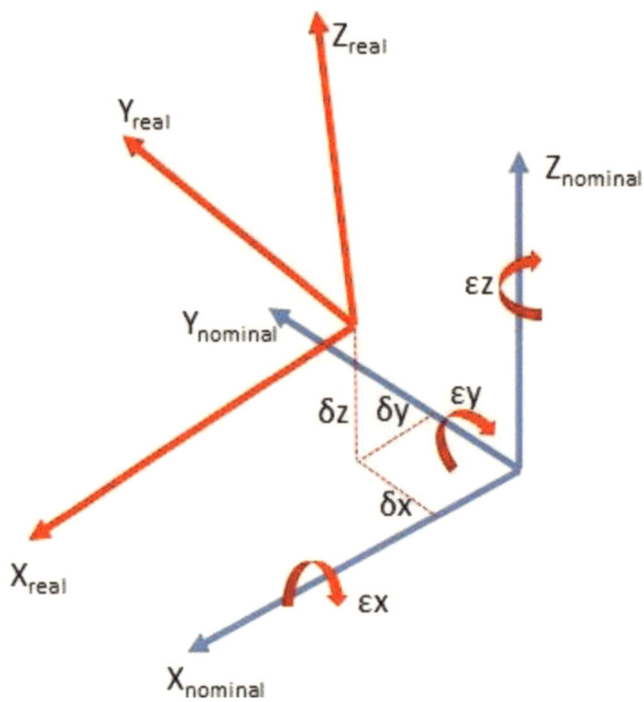

Figure 5. Error parameters.

Including the error matrices in Eqs. (5) and (8), which describes the kinematic model with the error model, is obtained.

$$ {}^{0}T_3 = {}^{0}A_1^{0}Rerr_1{}^{1}A_2^{1}Rerr_2{}^{2}A_3{}^{2}Terr_3 \tag{8} $$

By checking the behavior of the error model, it has been proven that errors depend on the joint position (the rotation angle in rotary joints and distance of the interferometer). This means that error parameters must have a formulation depending on the joint position. For the error parameters corresponding to the linear error matrix, we have used a polynomial function (see Eq. (9)) and for the error parameters corresponding to the rotary error matrices, a Fourier shape function is more convenient because of its periodic behavior (see Eq. (10)).

$$ \phi = A\phi \cdot sen\left(\frac{2\pi}{T\phi} \cdot \theta_z + \varphi\phi\right) \tag{9} $$

$$ \phi = \phi_1 + \phi_2 \cdot d + \phi_3 \cdot d^2 \tag{10} $$

For $\phi = \delta x, \delta y, \delta z, \varepsilon x, \varepsilon y, \varepsilon z$.

5. Model validation

The kinematic error model must be validated. The validation has been performed first with synthetic values and then with real values.

5.1. Synthetic data validation

A parametric generator of meshes of reflectors with known errors has been programmed. This algorithm generates a mesh of synthetic reflector coordinates with nominal position values. Then a set of error parameters is introduced, and the theoretical measurements of an LT, affected by the error parameters introduced, are calculated.

These measurements are introduced in the calibration procedure to calculate the error parameters with the optimization of the function described in Eq. (11) and correct the measurements.

Finally, the initial measurements and the corrected ones are compared with the nominal reflector positions according to Eq. (12) to calculate the calibration accuracy improvement achieved.

$$ \phi = \sum_{i=1}^{n}\left((x_i - x_{nomi})^2 + \left(y_i - y_{nomi}\right)^2 + (z_i - z_{nomi})^2\right) \tag{11} $$

$$ err_i = (x_i - x_{nomi})^2 + \left(y_i - y_{nomi}\right)^2 + (z_i - z_{nomi})^2 \tag{12} $$

The error parameter set is shown in **Table 2**.

Three meshes of synthetic reflectors have been generated:

Flat YZ plane mesh 15 × 14 = 210 reflectors (**Figure 6**).

X = 5.000 mm constant.

Y = 10.000 ÷ 10.000 Δ 1.420 mm.

Z = 1.500 ÷ 5.000 Δ 500 mm.

Cubic XYZ mesh 6 × 6 × 6 = 216 reflectors (**Figure 7**).

X – 10,000 ÷ 10,000 Δ 4,000 mm.

Y – 10,000 ÷ 10,000 Δ 4,000 mm.

Z – 10,000 ÷ 10,000 Δ 4,000 mm.

Spherical HVR mesh 12 × 8 × 5 = 480 reflectors (**Figure 8**).

H 0° ÷ 360° Δ33°.

V 77° ÷ – 60° Δ20°.

R 1,000 ÷ 15,000 Δ 3,500 mm.

–	δX (μm)	δY (μm)	δZ (μm)	εX (μrad)	εY (μrad)	εZ (μrad)
θ 0Rerr1	10	10	10	10	10	10
φ 1Rerr2	10	10	10	10	10	10
d 2Terr3	10	10	10	10	10	10

Table 2. Synthetic error parameters.

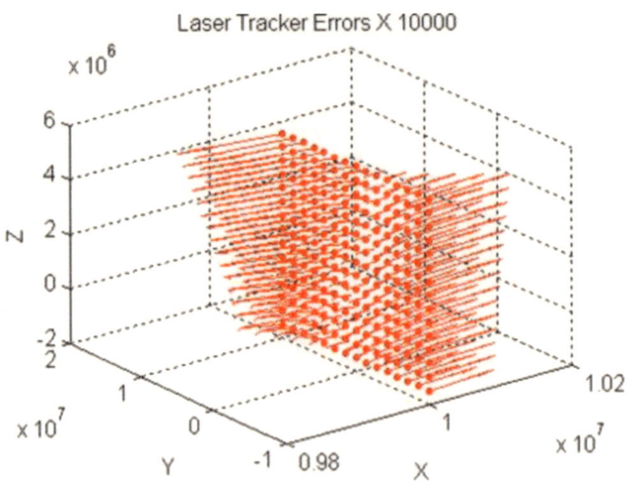

Figure 6. Errors in a plane mesh.

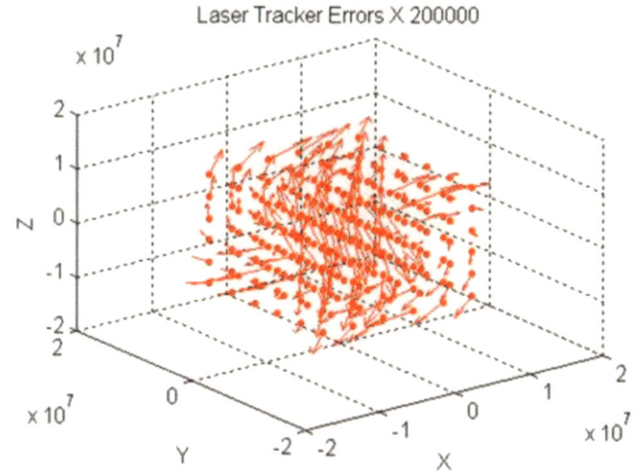

Figure 7. Errors in a cubic mesh.

The calibration procedure calculates the parameter errors and corrects the generated measurements. As the error parameters are calculated through a mathematical optimization, the result do not gives exactly the nominal parameters but provides a set of parameters that minimizes the LT error. In fact, the calibration reduces the LT error more than 98%. For example, **Table 3** shows the calculated error parameters corresponding to the spherical mesh.

5.2. Real data validation

The validation with synthetic data shows that the programmed algorithms are working properly but, as the errors have been generated with the error model purposed, it was expected to obtain a

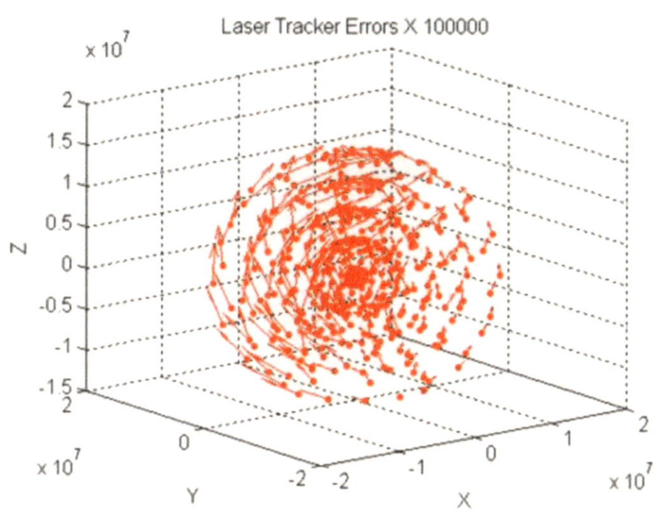

Figure 8. Errors in a spherical mesh.

–	δ_X (μm)	δ_Y (μm)	δ_Z (μm)	ε_X (μrad)	ε_Y (μrad)	ε_Z (μrad)
θ	10.001	9.977	0.000	9.994	9.994	−5.101
φ	10.045	0.000	9.969	9.986	−5.137	9.416
d	10.011	9.969	10.018	9.966	10.574	0.000

Table 3. Calculated error parameters in the spherical mesh.

good calibration result. It is necessary to check the calibration behavior with real data. To do it, an experiment has been performed. A set of 17 reflectors has been placed over the table of a coordinates measuring machine (CMM). Then the positions of these reflectors have been measured with the CMM and the LT from five different positions as shown in **Figure 9**.

The estimation of the error parameters have been performed on the basis that the distances between every pair of reflectors must be the same regardless the LT position from which they have been measured. Eq. (13) compares the distances measured by the CMM and the LT.

$$\Phi = \sum_{i=1}^{C_{n,r}} \sum_{K=1}^{LT} (d_{mik} - d_{iCMM}) \tag{13}$$

Being d_{mik}, the distance measured between the i-esim pair of reflectors from the k-esim LT position and d_{iCMM} the same distance measured by the CMM. $C_{n,r}$ is the number of possible pairs of reflectors. In this case i=$C_{2,17}$ = 136 pairs of reflectors. **Figure 10** and **Table 4** show the initial and residual errors of the reflectors mesh.

In the real calibration object of this work, we measure a mesh of reflectors out of the metrological laboratory, and there will be no nominal data to calculate the error parameters. To simulate the calibration procedure behavior and its requirements under real conditions, we have used the CMM measurements with a new optimization function. This function is equivalent to Eq. (11)

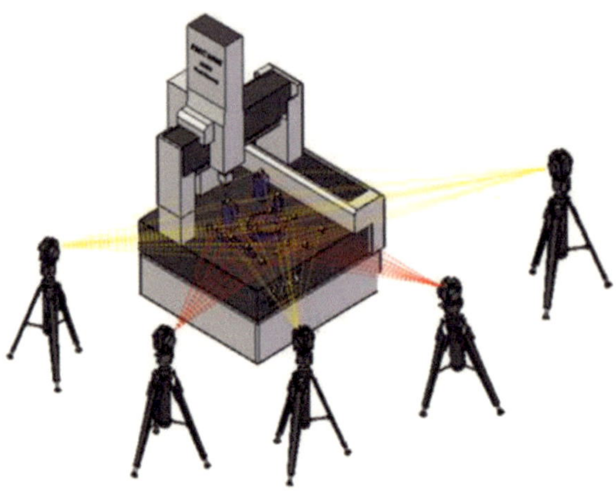

Figure 9. Real data validation experiment.

http://dx.doi.org/10.5772/intechopen.71444

but instead of comparing CMM and LT distances, we compare distances from every pair of LT positions as shown in Eq. (14).

$$f = \sum_{i=1}^{m-1} \sum_{j=i+1}^{m} \sum_{k=1}^{n-1} \sum_{l=k+1}^{n} \left| d_{kl}^{i} - d_{kl}^{j} \right| \tag{14}$$

With this optimization criterion, we found out that calibration result increased the LT error. That is due to the fact that the mathematical optimization matches the distances but to a value different from the nominal. This means that it is necessary to introduce a calibrated distances gauge in the reflectors mesh to determine its behavior. After several simulations, a gauge of four reflectors gives the best results, and the objective function is as shown in Eq. (15)

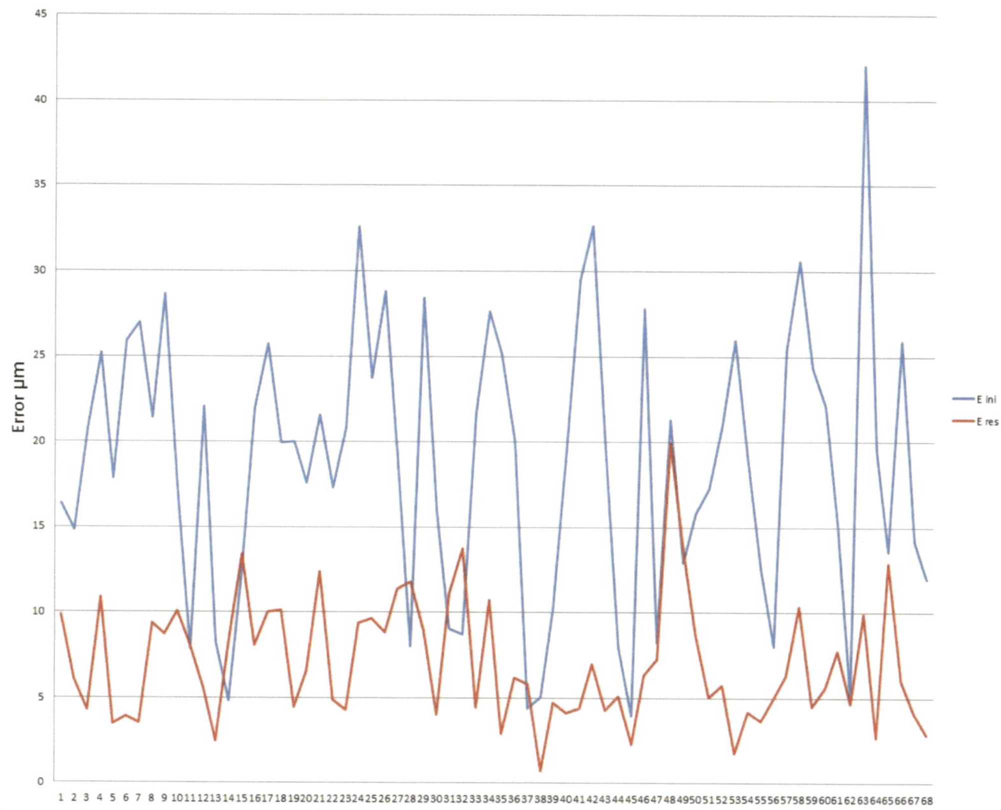

Figure 10. CMM reflectors mesh calibration.

	E_ini (µm)	E_res (µm)
Maximum	42.057	19.935
Average	18.785	7.045

Table 4. Residual errors in the CMM mesh calibration.

$$f = \sum_{i=1}^{m-1}\sum_{j=i+1}^{m}\sum_{k=1}^{n-1}\sum_{l=k+1}^{n}\left|d_{kl}^{i} - d_{kl}^{j}\right| + \sum_{i=1}^{m}\sum_{k=1}^{3}\sum_{l=k+1}^{4}\left|d_{kl}^{i} - d_{kl}^{CMM}\right| \quad (15)$$

To evaluate the calibration results, we have established two different criteria: (1) a distances criterion that evaluates the differences of distances between every pair of reflectors measured from every pair of positions of LT according to Eq. (16) and (2) a coordinates criterion that evaluates the position error for every reflector measured from every LT position (see Eq. (17)). This second criterion requires first to transform all measurements to the same reference system.

$$err = \frac{\sum_{m=1}^{4}\sum_{n=2}^{5}\sum_{i=1}^{16}\sum_{j=2}^{17}\left(\sqrt{\left(x_i^m - x_j^m\right)^2 + \left(y_i^m - y_j^m\right)^2 + \left(z_i^m - z_j^m\right)^2} - \sqrt{\left(x_i^n - x_j^n\right)^2 + \left(y_i^n - y_j^n\right)^2 + \left(z_i^n - z_j^n\right)^2}\right)}{C_{5,2}C_{17,2}} \quad (16)$$

$$err = \frac{\sum_{m=1}^{4}\sum_{n=2}^{5}\sum_{i=1}^{17}\sqrt{\left(x_{iSRO}^{m} - x_{iSRO}^{n}\right)^2 + \left(y_{iSRO}^{m} - y_{iSRO}^{n}\right)^2 + \left(z_{iSRO}^{m} - z_{iSRO}^{n}\right)^2}}{17 \times C_{5,2}} \quad (17)$$

The results of the calibration following both criteria are shown in **Tables 5** and **6**.

	SMR distance error (μm)						
	LT1	**LT2**	**LT2**	**LT4**	**LT5**	**LT1–5**	**Improvement**
Average	14.31	7.79	9.16	7.42	7.07	9.15	62.67%
Maximum	45.73	25.06	30.76	36.83	28.06	45.73	–

Table 5. Distances criteria evaluation.

	SMR coordinates error (μm)						
	LT1	**LT2**	**LT2**	**LT4**	**LT5**	**LT1–5**	**Improvement**
Average	20.26	10.25	11.78	9.80	10.28	12.47	41.79%
Maximum	33.07	17.14	22.20	22.53	23.10	33.07	–

Table 6. Coordinates criteria evaluation.

6. Sensitivity analysis

In order to know the SMR positions more appropriate to perform the calibration, a sensitivity analysis has been performed. In this analysis, the influence of every error parameter in the global LT measuring error has been analyzed. Using the synthetic data generator, many synthetic measurement meshes as error parameters that have been considered in the error model have been generated. Thus, 18 meshes are necessary. All of them have been generated with the same nominal coordinates as the spherical mesh generated in the synthetic data

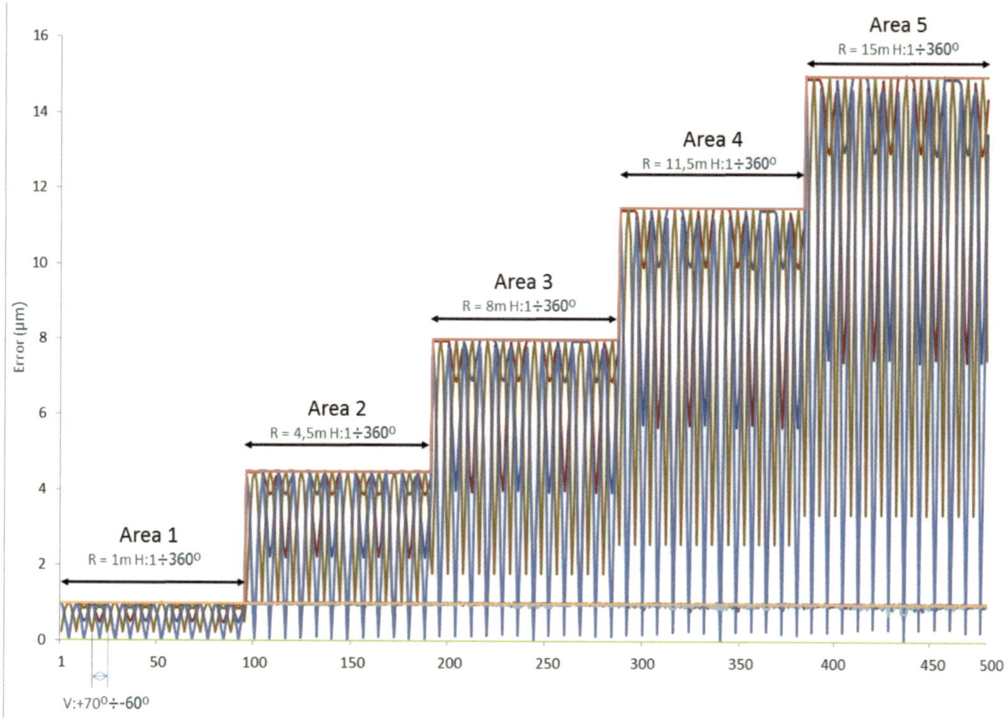

Figure 11. Sensitivity analysis.

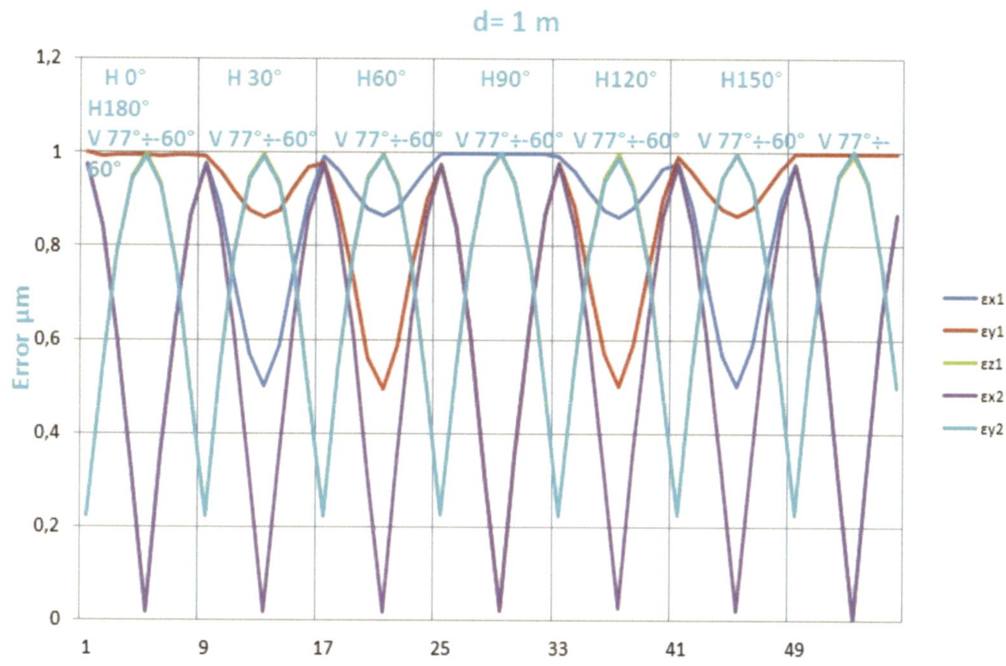

Figure 12. Sensitivity analysis of variable errors.

validation in chapter 5.1 and every mesh is affected by a single error parameter with a value of 1μm for linear error parameters or 1μrad for angular error parameters.

Figure 11 shows the error produced by every error parameter through the spherical mesh.

As a result, it is deduced that all distance error parameters (δ) produce constant errors. Errors due to parameters εx_1, εy_1 and εz_1 depend on θ, φ and d, errors due to parameters εx_2 and εy_2 depend on φ and d and errors due to parameters εz_2, εx_3 and εy_3 depend only on d. Finally, parameter εz_3 produces no errors. Taking a closer look at the parameters that produce variable errors, we can see in **Figure 12** that the maximum and minimum error values correspond to maximum, minimum and zero values of φ and also on θ every $\pi/4$.

7. Experimental setup

The sensitivity analysis proves that extreme and zero values of tilt angle are the best. It also proves that pan angle ranges must be at least 90°. According to these requirements, 24 reflectors have been placed in a corner of the laboratory. Eight of them spread on the floor (minimum tilt angle), a second set of eight reflectors on the wall at the LT height (zero tilt angle) and the last eight reflectors in the upper part of the floor (maximum tilt angle). LT has been placed at five different positions covering always a pan angle of 90° as shown in **Figures 13** and **14**.

8. Calibration results

The calibration has been performed following the model in Eq. (15) and evaluated according to Eqs. (16) and (17) in the same way as it has been done with the CMM measurements. A gauge of four reflectors has been also included in the mesh of reflectors, and this gauge has been measured in the CMM to know the real distances among its reflectors.

Figures 15 and **16** show the calibration result. In **Table 7**, the numerical values of the calibration in function of the evaluation method can be appreciated.

9. Calibration verification

Calibration results show an LT accuracy improvement according to both criteria used, but as we do not have the SMRs real positions, it is not possible to ensure without any doubt that the calibration procedure increases the LT accuracy.

A new verification is therefore necessary to assess the calibration procedure behavior. In Section 5, a set of SMRs has been measured with the LT from five different positions. These SMRs were placed on a CMM table and measured also with the CMM. These accurate measurements can be used as nominal data to check whether the calibration has improved the LT accuracy or not. The error parameters obtained in the calibration procedure have been applied

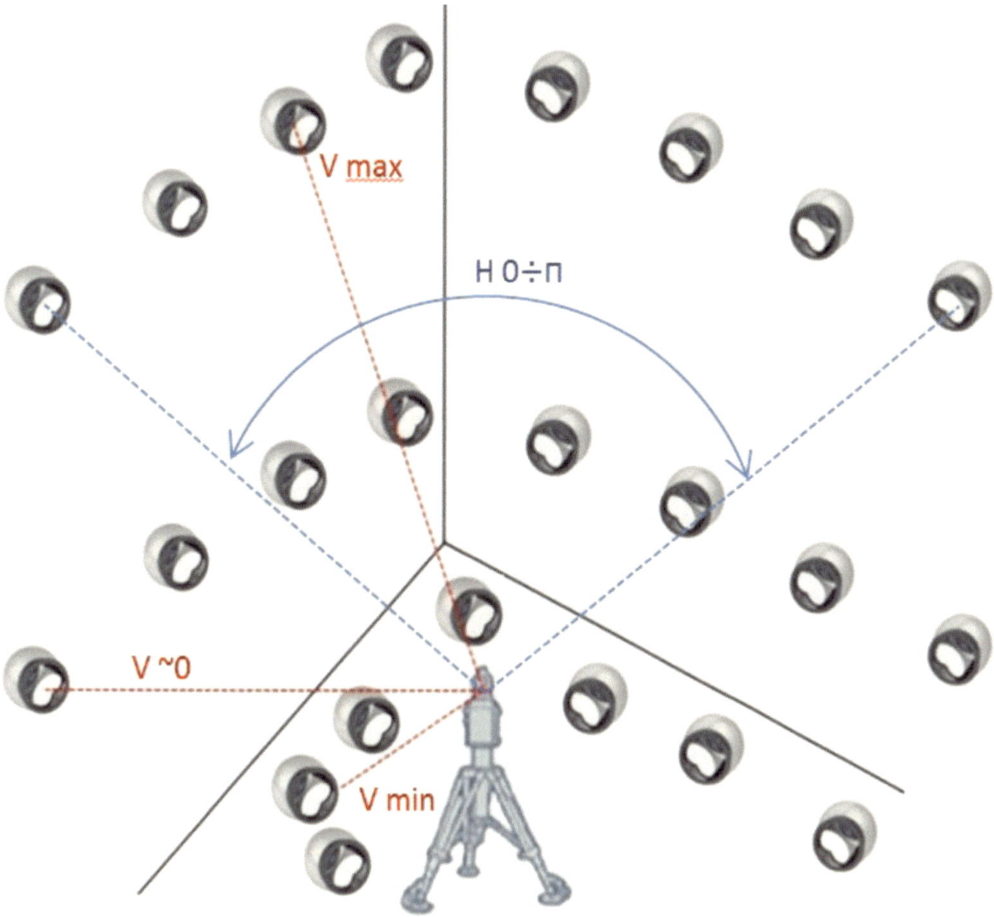

Figure 13. Calibration experimental setup.

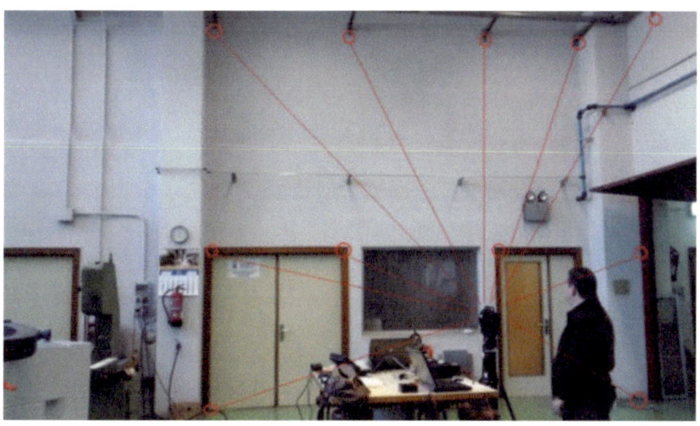

Figure 14. View of the experimental calibration process.

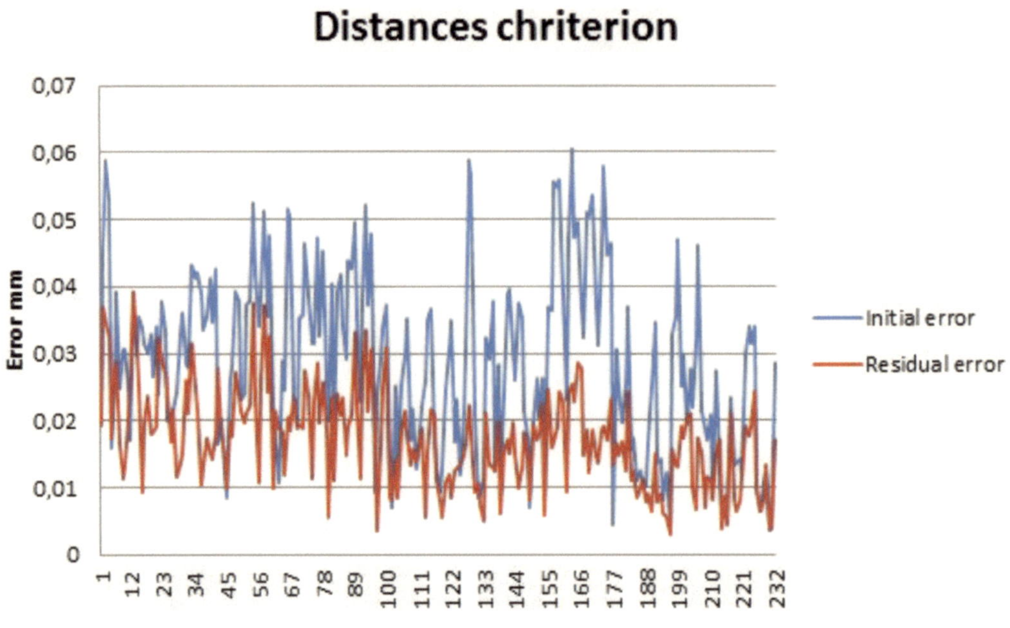

Figure 15. Calibration results distances criteria.

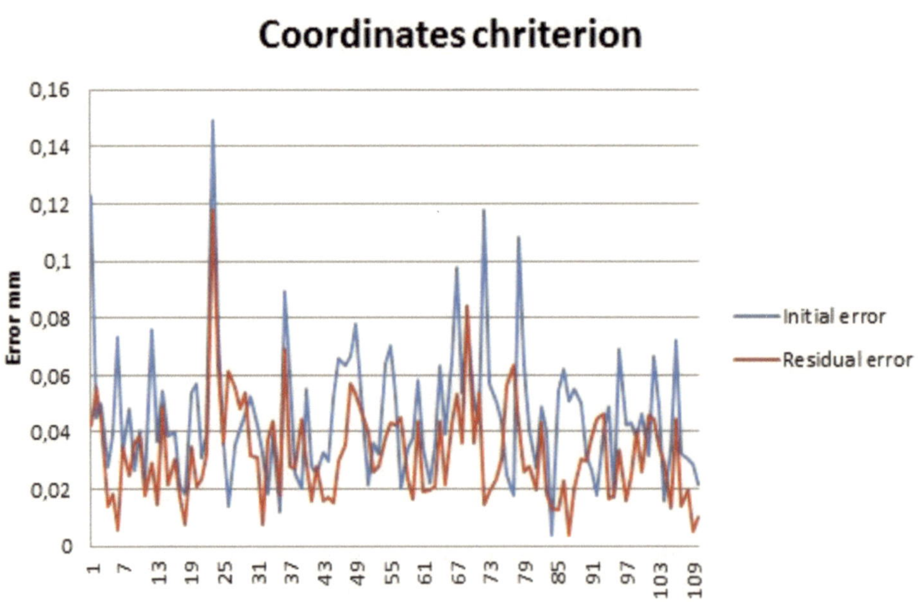

Figure 16. Calibration results coordinated criteria.

to the measurements made in the CMM table. The corrected values can be then compared to the CMM nominal measurements and can be seen in **Figures 17** and **18** and its values in **Table 8**.

http://dx.doi.org/10.5772/intechopen.71444

		Criteria	
		Coordinates	Distances
Initial error (μm)	Maximum	149	60
	Average	46	29
Residual error (μm)	Maximum	118	39
	Average	39	17
Improvement	%	14.47	40.10

Table 7. Calibration results.

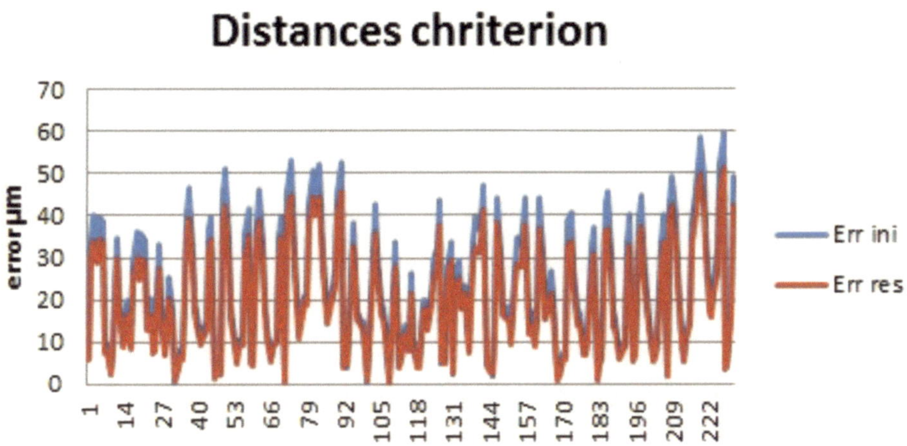

Figure 17. Calibration verification distances criteria.

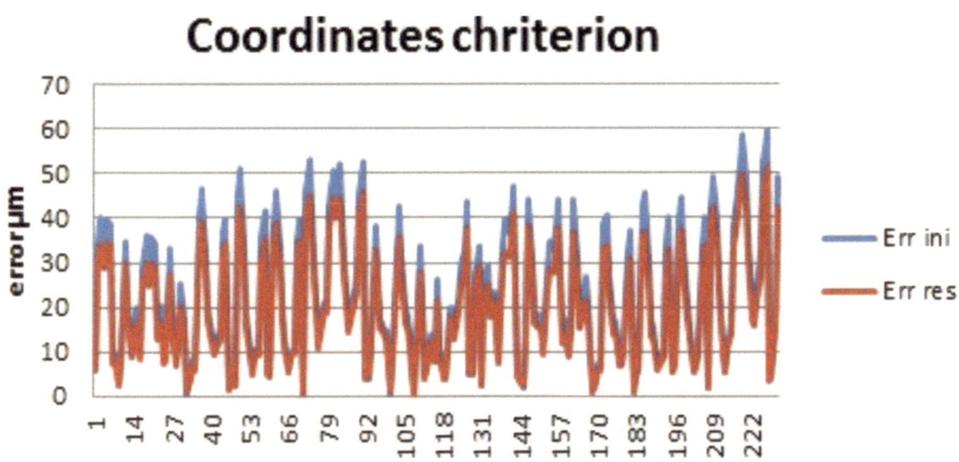

Figure 18. Calibration verification coordinates criteria.

		Criteria	
		Coordinates	Coordinates
Initial error (μm)	Maximum	54	84
	Average	21	24
Residual error (μm)	Maximum	41	53
	Average	17	18
Improvement	%	17.98	25.53

Table 8. Calibration verification results.

10. Contribution of the SMR incidence angle in the measurement uncertainty

In some of the measurements made, the position of the SMR could not be reached by the LT beam. The SMR maximum viewing angle is within ±30°, and they were placed facing a theoretical point in the middle of the LT selected positions. However, as the SMR positions and orientations are fixed along all the measurements, and they are manually placed, there is the possibility that some of them could not be visible from all the LT positions because the incidence angle was out of the SMR viewing range.

The incidence angle of the laser beam in the SMR has an important influence in the measurement accuracy, and an experiment has been performed to measure the contribution of the SMR incidence angle in the measurement uncertainty.

An SMR with its magnetic holder has been placed on a rotary worktable. The SMR has been centred; so its centre will be in the worktable rotation axis. The SMR was centred using the

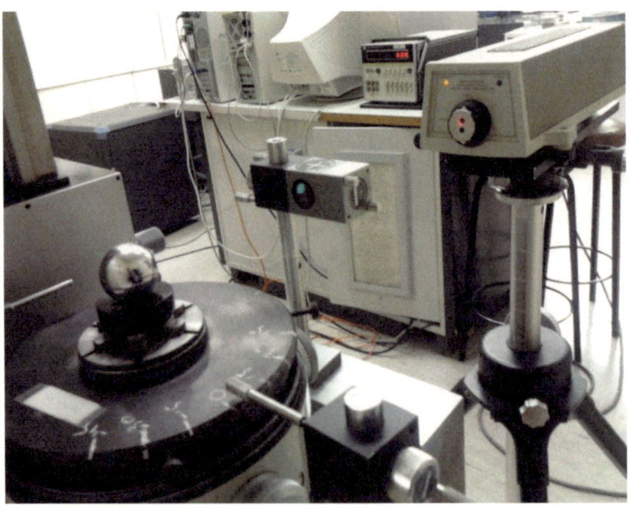

Figure 19. SMR incidence angle error measurement.

	θ (deg)									
θ (deg)		−30	−22.5	−15	−7.5	0	7.5	15	22.5	30
	−30	13.4	13	12.4	11.5	10	8.3	6.5	4.3	2.6
	−22.5	8.1	7.9	7.6	6.9	6.2	4.3	3.2	1.6	0
	−15	5	4.8	4.5	3.8	3	2.1	0.8	−0.5	−1.6
	−7.5	2.4	2.2	2	1.5	0.9	−0.2	−1.1	−1.8	−2.8
	0	1.2	1	0.8	0.4	0	−0.7	−1.5	−2.4	−3.4
	7.5	0.8	0.6	0.4	0	−0.5	−1.1	−1.8	−2.6	−3.6
	15	0.3	0	−0.2	−0.5	−1	−1.7	−2.4	−3	−3.9
	22.5	0	−0.2	−0.5	−0.8	−1.3	−2	−2.7	−3.3	−4.1
	30	−0.2	−0.5	−0.8	−1	−1.6	−2.4	−3.1	−3.6	−4.4

Table 9. Influence of the laser incidence angle in the SMR.

rotary worktable pencil and its centring accuracy has been measured to be in the range of ±0.1 μm.

The SMR has been initially located with its incidence angles equal to 0° facing to an interferometer laser beam. The interferometer measurement has been then reset to make this position as the zero length measurement.

The SMR has been then rotated on its horizontal and vertical angles within its incident available range of ±30° in 7.5° steps as shown in **Figure 19**.

Data measured by the interferometer are shown in **Table 9**. An important dependence on the angle variation can be seen, showing the influence of the vertex position error, that is, the distance between the optical centre of the CCR and the SMR sphere.

11. Conclusions

A new LT kinematic calibration has been presented and verified by comparing calibration results with nominal data measured with a CMM. The novelty of the method is that a final calibration of the LT can be made by the LT user at place just before measuring with the LT under real working conditions. This can greatly help LT measurement process by assuring a correct calibration at the moment of measuring. The only devices needed for the calibration is a calibrated gauge and a set of reflectors to be located at the measuring place.

The kinematic error model has been defined. This model has also been validated with synthetic and nominal data. The study of the influence of every error parameter in the global error of the LT has shown the best configuration for the experimental setup.

The calibration procedure has been performed with a previously calibrated LT, and the calibration has been able to improve the factory calibration of the LT.

The influence of the laser incidence angle in the measurement uncertainty shows an important contribution to the measurement errors.

The kinematic calibration model developed offers important advantages compared to the conventional methods. Existing standards require strict temperature conditions, and a large number of measurements are needed to perform the calibration. The purposed method can be used in two ways; first, the distance error calculated for every pair of reflectors measured from different LT locations gives a dimensional value of the LT accuracy, which will help the user to know whether a calibration of the device is necessary or not. In other way, if the calibration is necessary, it can be performed by the final user between the programmed calibrations without the need of a metrological laboratory. It can also be used to develop new calibration standards or complete the existing ones.

12. Future work

It is possible to find two constructive LT models from different manufacturers. The purposed method is valid for the LT having the laser source in the rotating head. The other model is characterized by having the laser source in the fixed basis of the LT. This means that they need a rotating mirror attached to the standing axis to reflect the laser beam from the source to the SMR. The calibration procedure followed in the present work can also be applied to this second LT constructive model adapting the kinematic model to the LT geometry and the laser beam path.

Along with the development of this kinematic model, further tests are convenient to study the behavior of the calibration method under different conditions such as measurement range, temperature, number and distribution of reflectors.

Author details

Ana Cristina Majarena[1]*, Javier Conte[1], Jorge Santolaria[1] and Raquel Acero[2]

*Address all correspondence to: majarena@unizar.es

1 Department of Design and Manufacturing Engineering, EINA, University of Zaragoza, Zaragoza, Spain

2 University Defense Centre, Zaragoza, Spain

References

[1] Wang Z, Mastrogiacomo L, Franceschini F, Maropoulos P. Experimental comparison of dynamic tracking performance of iGPS and laser tracker. International Journal of Advanced Manufacturing Technology. 2011;**56**:205-213. DOI: 10.1007/s00170-011-3166-0

[2] Burge JH, Su P, Zhao C, Zobrist T. Use of a commercial laser tracker for optical alignment. Optical System Alignment and Tolerancing. Proc of SPIE. 2007;**6676**(66760E):1-12. DOI: 10.1117/12.736705

[3] Huo D, Maropoulos PG, Cheng CH. The framework of the virtual laser tracker–a systematic approach to the assessment of error sources and uncertainty in laser tracker. Measurement. 2010:507-523. DOI: 10.01007/978-3-642-10430-5_39

[4] Nubiola A, Bonev IA. Absolute calibration of an ABB IRB 1600 robot using a laser tracker. Robotics and Computer-Integrated Manufacturing. 2013;**29**(1):236-245. DOI: 10.1015/2012.06.004

[5] Denavit J, Hartenberg RS. A kinematic notation for lowerpair mechanisms based on matrices. Trans. ASME Journal of Applied Mechanics. 1955;**22**:215-221

[6] Baatz R, Bogena H, Franssen HH, Huisman J, Qu W, Montzka C, Vereecken H. Calibration of a catchment scale cosmic-ray probe network: A comparison of three parameterization methods. Journal of Hydrology. 2014;**516**:231-244. DOI: 10.1016/2014.02.026

[7] Barazzetti L, Giussani A, Roncoroni F, Previtali M. Monitoring structure movement with laser tracking technology: Videometrics, Rare Imaging, and Applications XII; and Automated Visual Inspection. Proc. SPIE 8791. 2013:879106. DOI: 10.1117/12.2019997

[8] Bargigli L, Gallegati M, Riccetti L, Russo A. Network analysis and calibration of the "leveraged network-based financial accelerator". Journal of Economic Behavior and Organization. 2014;**99**:109-125. DOI: 10.1016/2013.12.018

[9] ASME B89.4.19-2006 Standard. Performance Evaluation of Laser-Based Spherical Coordinate Measurement Systems www.asme.org

Chapter 10

Optimization of Single-Sided Lapping Kinematics Based on Statistical Analysis of Abrasive Particles Trajectories

Adam Barylski and Norbert Piotrowski

Additional information is available at the end of the chapter

http://dx.doi.org/10.5772/intechopen.71415

Abstract

The chapter presents the influence of selected kinematic parameters on the geometrical results of the single-sided lapping process. The optimization of these parameters is aimed at improving the quality and flatness of the machined surfaces. The uniformity of tool wear was assumed as main optimization criterion. Lapping plate wear model was created and in detail was analyzed. A Matlab program was designed to simulate the abrasive particles trajectories and to count their distribution. In addition, the influence of additional guiding movements of the conditioning ring has been verified and the idea of a flexible single-sided lapping system assisted with a robot, which ensures the optimal constant wear over the diameter was presented.

Keywords: abrasive machining, single-sided lapping, material removal rate, particle trajectories, kinematics optimization

1. Introduction

Nowadays, products have to be manufactured with both high quality and efficiency. Finishing technologies that allow to achieve high quality surfaces with low roughness, very high accuracy of shape and dimensions are becoming increasingly important in the modern world. These results can be obtained in lapping process with the use of relatively simple means of productions. It is one of the oldest machining processes and a number of precision manufacturing applications still use the lapping process as a critical technology to achieve flatness tolerance and surface quality specification. This technology is used for machining metals and their alloys, glasses, natural materials such as marble, granite and basalt, materials used in semiconductor technology, as well as carbon, graphite, diamond and ceramics, have found use in many engineering applications [1, 2]. Typical parts processed by lapping are: pneumatic and hydraulic parts (valve plates, slipper plates, seals, cylinder parts, and castings), pump parts (seal faces, rotating valves, and body castings), transmission equipment (spacers, gears, shims, and clutch plates), inspection equipment (test blocks, micrometer anvils, optical flats, and surface plates), cutting tools (tool tips and slitter blades), automotive and aerospace parts (lock plates, gyro components, and seals) [2, 3]. All of the abrasive machining are complicated and random processes with the large amount of influencing parameters and outcomes. As a result of numerous variables affecting the process quality, the main outcomes of lapping are stock removal, roughness and flatness. In order to ensure the highest quality and accuracy on worked surfaces, researches should focus on improving lapping process by studying significant process variables.

Generally, all lapping processes can be subdivided according to the active surface of the lapping tool. The lapping process where the tool axis and the workpiece surface are parallel to each other is known as peripheral lapping, and side lapping applies when the tool axis and the workpiece surface are perpendicular to each other. More specifically, the most used lapping processes can be classified as the following: according to the generated surfaces, process kinematics and tool profiles [4]. The most well-known systems are single-sided and double-sided lapping machines. Double-side surface lapping is considered as the most accurate method in terms of parallelism and uniformity of size as two parallel even surfaces are simultaneously lapped during this process. These kinds of machine tools have a planetary system [5]. In case of flat lapping, the standard systems with conditioning rings are mostly used.

Due to the wear, the active surface of the lapping plate has some shape errors of convexity or concavity [3, 6]. This has a major impact on the shape accuracy of the workpieces. If the lapping plate is out of flat, it should be re-conditioned. Kinematic method of the correction of the tool shape errors can be applied. Workpieces' shape, size as well the lapping kinematics determine the contact between workpieces and the tool. It was observed that a trajectories distribution of abrasive particles on the lapping plate varies when the kinematic conditions are changed, e.g., by placing the workpieces at different radii, setting different rotational velocities or by introducing additional movements of conditioning ring. In this chapter, the unconventional kinematic systems were described. Carried out simulations have shown that the speed ratio k_1 and the period ratio k_2, which represent the relationships among the three basic motions of unconventional lapping systems, are major factors affecting trajectory distribution. The optimization of these parameters was conducted. The uniformity of tool wear was assumed as main optimization criterion.

2. Main factors of lapping process

The lapping system consists of several elements: lapping plate, abrasives, lapped workpieces, kinematics and machine. They influence the lapping process which determines the product quality, tool wear and efficiency of the process [1, 4]. The input factors of lapping process are of two categories: controllable and non-controllable (noise) factors. The first category includes machining parameters, its working pressures and speeds, abrasive type, characteristics of the work equipment, tool, machine, duration of machining, etc. The uncontrolled input variables includes, inter alia, environmental temperature, slurry distribution, vibrations occurring in the system, internal stress, etc. An overview of main factors of lapping process is presented in **Figure 1**.

First of the most important input parameters, which influence the surface formation and material removal in lapping process is the lapping plate, its geometry, hardness, etc. Generally, it could be established that workpieces are machined to a mirror image of the tool with respect to the flatness. During lapping process, abrasives play an important role on material removal based on size, shape, and hardness. It has been shown that material removal rate increases with increasing abrasive size up to a critical size at which it begins to stabilize and even decrease [7, 8]. Moreover, multiple studies have also shown that surface roughness improves with decreasing grain size. The properties of the abrasive vehicle, which is the medium which transports the abrasive across the workpiece, are important to how effective the abrasives are in material removal. In addition, the flatness, geometric and arrangement of workpieces have also effect on lapping results. The material hardness of workpieces has a crucial influence on a removal mechanism in abrasive processes, as well on efficient of lapping [9]. Surface formation of fragile workpieces, e.g., ceramics, occurs as a result of cracking propagation or by plastic deformation, when depth of cut is smaller than critical [10].

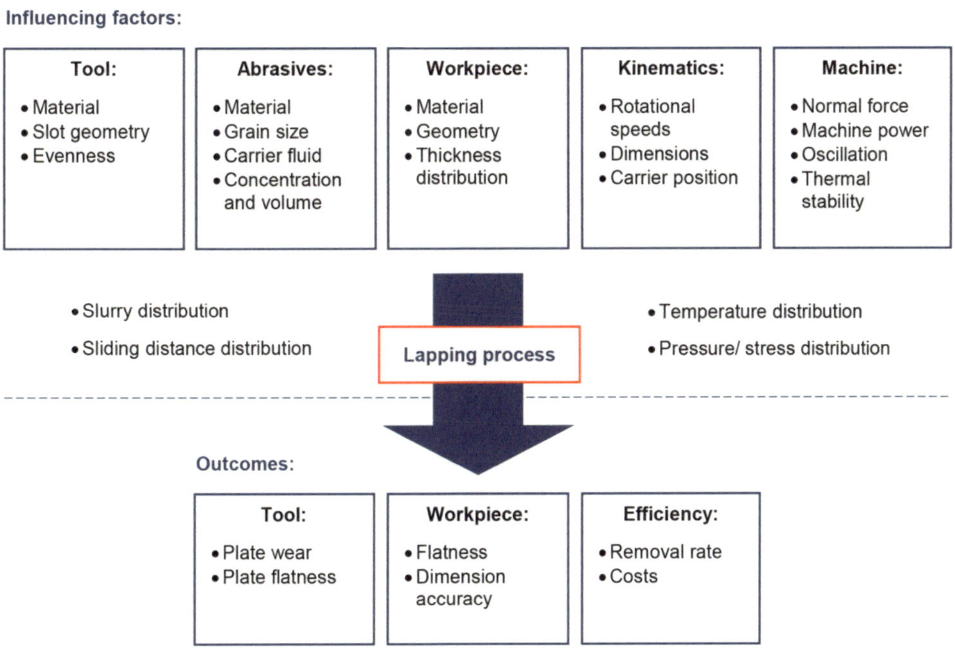

Figure 1. Main factors of lapping process.

There has been a great amount of research conducted on the kinematics of lapping [2, 3, 6, 11, 12]. The lapping pressure and lapping speed can be seen as the main variables in the lapping process. They are closely interdependent with kinematic system and machine parameters.

3. Abrasive wear in single-sided lapping process

One of the most important mechanisms of the lapping process is abrasive wear. The process describes the separation of material from surfaces in relative motions caused by protrusions and particles between two surfaces or fixed in one of them. In lapping process, abrasive grains are guided across the surface to be lapped and backed up by a lapping plate or conditioning rings. It is crucial to minimize friction and wear of the abrasive and to maximize abrasive wear of the workpiece [2, 3].

Abrasive wear is commonly classified into two types: two- and three-body abrasion. In two-body abrasion, particles are rigidly attached to the second body. When abrasive particles are loose and free to move, then we deal with three-body abrasion. Therefore, in a two-body abrasion, the abrasive grains can cut deeply into the workpiece due to the sliding motions and in the three-body abrasion, the particles spend part of time in cutting and part of time in rolling [3].

Also in the case of lapping process, several abrasives move into active positions and other grains leave the working gap. Only a specified part of all particles is able to enter the working gap with height H, which is appropriate to the lapping pressure F. **Figure 2** presents abrasive wear in single-sided lapping process and different types of lapping grains being in the working gap between tool and workpieces [3]. As can be seen in **Figure 2**, abrasives play a crucial

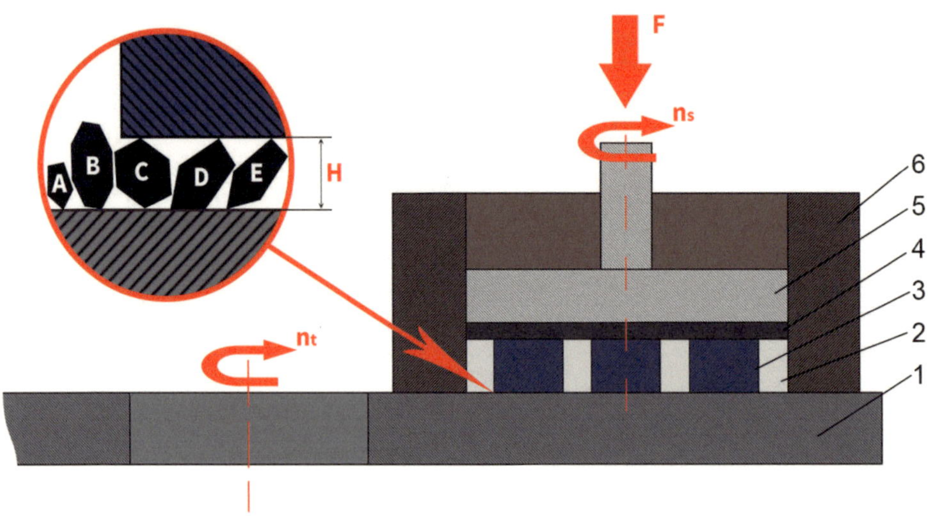

Figure 2. Single-sided lapping process: (1) lapping plate, (2) separator, (3) workpiece, (4) felt pad, (5) weight system, (6) conditioning ring.

role on material removal based on size, shape, and hardness. For example, it can be noticed that grain A is too small and grain B is too big. It means that only some of the abrasives are active and they roll like grain C or slide like grain D. The example of passive grain is grain E, which does not stick to the affecting partner.

Moreover, regular sliding grains in and out of the working gap cause the changes in the initial structure of the lapping plate. First, strongest loaded grains disintegrate into smaller and their particles are able to take part in the lapping process depending on the size and shape. Hence, the number of bigger grains is reduced, while the number of smaller grains is increased and the structure of the lapping tool is changed. The changing trajectories and velocity of abrasives cause some grains crash with each other or with grain fragments. It can be established that after crashes, grains are accumulating and can disintegrate at any time. The chain between a workpiece and lapping plate is fulfilled by active grains, which under their impact and edge transfer the normal forces into the surface of the affecting partner. Acting forces are proportional to the material volume. To reach a stationary working gap height, the amount of all normal forces transferred by active grains must be equal to the pressing force F [3].

4. Kinematic analysis of standard single-side lapping system

4.1. Single-sided lapping machine overview

Nowadays, there are many manufacturers that offer surface planarization technology with lapping kinematics. The most well-known are Peter Wolters, Lapmaster, Stähli, Engis, Hamai, Kemet, Mitsunaga and LamPlan. After a careful analysis of numerous researches and offers of many lapping machines producers, it has been emerged that most of lapping machines have standard kinematic system. Single-sided lapping machines differ in tool size, diameter of lapping plate, size and number of conditioning rings, type of pressure system (weighting or pneumatic system). More complex machines are equipped with forced drive option of conditioning rings. Such a system maintains a constant speed of workpieces.

Conventional single-sided lapping machine is shown in **Figure 3**. The key component of the machine is the tool, i.e., the annular-shaped lapping plate (1), on which the workpieces (3) are applied to. One machine usually has three to four conditioning rings (5). However, laboratory lapping machines that have one ring are also popular. The lapping plate (1) rotates with angular velocity ω_t and it drives conditioning rings, where separators (4) are placed allowing additional move of workpieces (3). Due to the frictional force, conditioning rings (5) rotate on the working surface of lapping plate (1) with angular velocity ω_s. This force depends on a radial position, velocity of conditioning rings, and friction conditions. The radial position (R) of conditioning rings can be controlled with roller forks (2). During the lapping process, a certain load is provided through felt pad (6) by weight disk (7) or pneumatic system. In this way, the parts are pressed against a film of abrasive slurry that is continuously supplied to the rotating lapping plate [2].

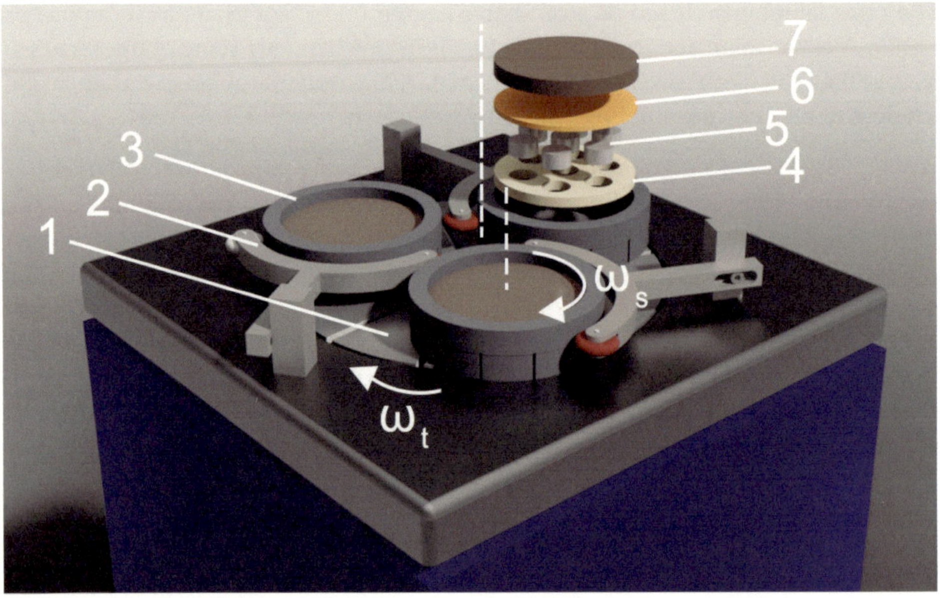

Figure 3. Single-sided lapping machine: (1) lapping plate, (2) roller forks, (3) conditioning ring, (4) separator, (5) workpieces, (6) pad, (7) weight.

4.2. Kinematical fundamentals

Due to the fact that the kinematics of the lapping is affected by a number of factors related to the influence and properties of the workpiece—abrasive slurry—lapping plate system, in the following model considerations, it is assumed that the angular velocities of conditioning rings, separator and workpieces are identical. Moreover, conditioning rings role is to even the lapping plate (**Figure 4**).

The input parameters for analysis are: angular velocity of lapping plate ω_t and of conditioning rings ω_s, inner R_{dw} and outer R_{dz} diameter of lapping plate, radial position of conditioning ring $R_{p.}$

In order to model a lapping plate, the position of any point P belonging to a workpiece must be determined. It is possible to do this by a radius vector in two coordinate systems: absolute and relative, which is related with rotating tool. The position of any point $P(r,\varphi_p)$ belonging to a workpiece in single-sided lapping system are determined in x″y″ coordinate system, which is related to conditioning ring as:

$$x_P'' = -r \cdot \cos(\varphi_s) \tag{1}$$

$$y_P'' = -r \cdot \sin(\varphi_s) \tag{2}$$

The coordinates of point P in x′y′ coordinate system are:

$$x_P' = x_P'' \cdot \cos(\omega_s \cdot t) + y_P'' \sin(\omega_s \cdot t) + R_{Px} \tag{3}$$

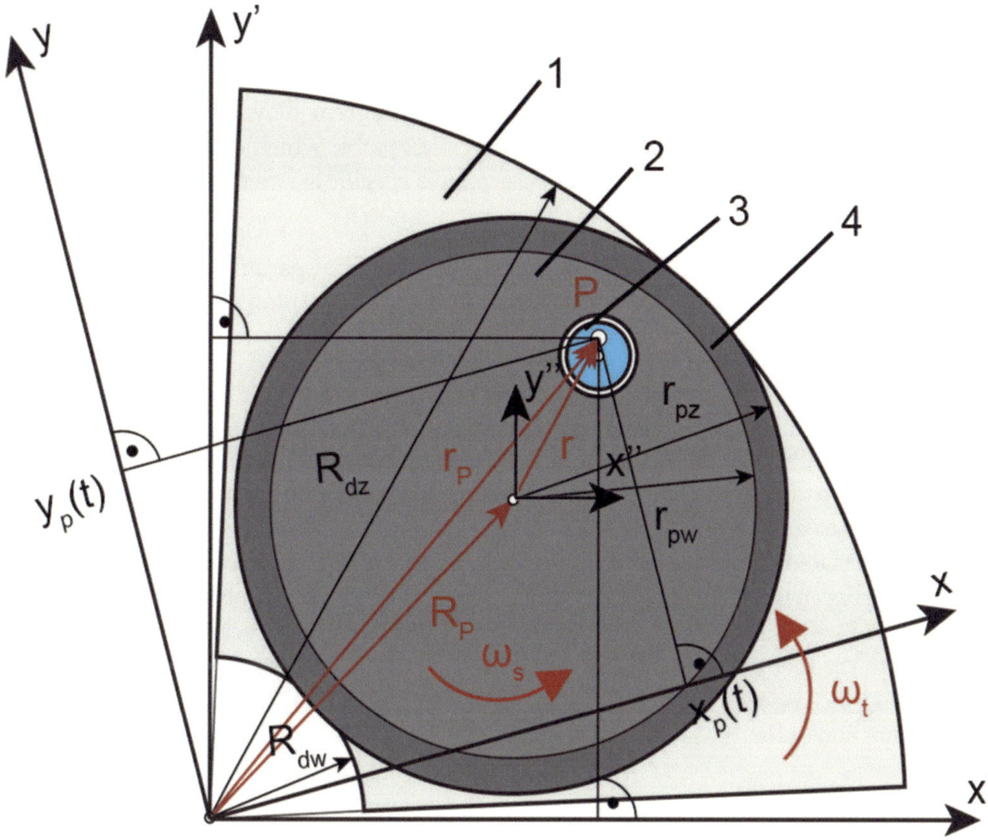

Figure 4. Kinematic diagram of single-sided lapping machine: (1) lapping plate, (2) separator, (3) workpiece, (4) conditioning ring.

$$y_P^{'} = -x_P^{''} \cdot \sin(\omega_s \cdot t) + y_P^{''} \cdot \cos(\omega_s \cdot t) + R_{Py} \tag{4}$$

After transforming it to coordinate system related to rotating lapping plate:

$$x_P = x_P^{'} \cdot \cos(\omega_t \cdot t) + y_P^{'} \cdot \sin(\omega_t \cdot t) \tag{5}$$

$$y_P = -x_P^{'} \cdot \sin(\omega_t \cdot t) + y_P^{'} \cdot \cos(\omega_t \cdot t) \tag{6}$$

$$r_P(t) = \sqrt{x_P^2 + y_P^2} \tag{7}$$

The relative velocity v of point P is defined as the derivative of the position with respect to time:

$$v = \frac{dr_P}{dt} \tag{8}$$

4.3. Single abrasive trajectories analysis

Equations of any point belonging to a workpiece position in standard single-sided lapping system were implemented in MATLAB program. This program allows to analyze single-side lapping system and can be used to mark out cycloids paths, which can be treated as areas where the lapping plate wears by the grain placed in a specific location of a conditioning ring or workpiece.

In order to analyze single abrasive trajectories, an additional parameter must be defined, which is rotational speed ratio of the conditioning ring to the lapping plate:

$$k_1 = \frac{\omega_s}{\omega_t} \quad (9)$$

Analysis of multiple simulations leads to conclusions that the k_1 parameter determines the type of cycloid path. Results are shown in **Figure 5**. It can be observed that epicycloid trajectories are marked out when the value k_1 is less than 0. When the value k_1 is close to 0, stretched epicycloid then interlaced epicycloid are received. Pericycloids appear when the value k_1 is between 0 and 1. At k_1 bigger than 1, hypocycloids can be obtained. Initially, they are stretched hypocycloids; then, they transform to interlaced hypocycloids. Attention needs to be paid to the following trajectories: cardioid ($k_1 = -1$), concentric circle ($k_1 = 0$), eccentric circle ($k_1 = 1$) and concentric ellipse ($k_1 = 2$). Moreover, the cycloid curvature radius $R_P(t)$ is a periodic function of time t and its cycle equals:

$$T = \frac{2\pi}{\omega} \quad (10)$$

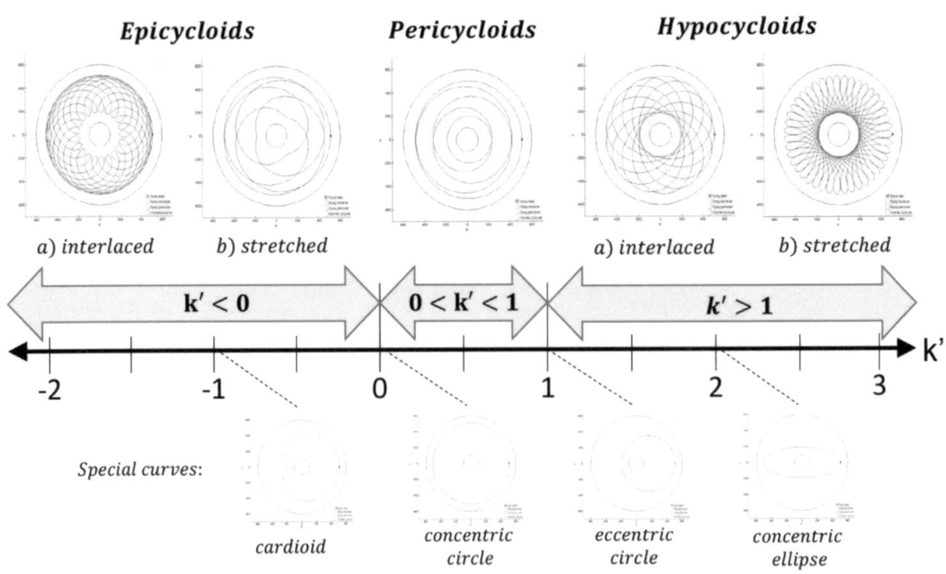

Figure 5. Path types depending on the rotational speed ratio k_1.

5. Unconventional lapping kinematics systems

Analysis of numerous researches and offers of many lapping machines producers leads to conclusion that most of lapping machines have the standard kinematic system. Unconventional lapping systems, where the conditioning ring performs an additional move such as radial, secant, swinging or off-center are presented in **Figure 6**. Simulations have shown that changing the kinematic system in single-sided lapping process causes a wide density variation of single abrasive trajectories [13].

The simulations allowed state that the most desirable system is that system, which allows to smoothly control the position of the conditioning ring on the lapping plate. It was pointed out that some systems are not very different from the standard kinematic system. Generated trajectories were almost identical. This is due to the fact that with such systems, the deflection of conditioning ring along the radius was much smaller, than in the case, for example, of the radial lapping system. In order to correct the flatness of lapping plate, the ring must be shifted toward or out of the center of the tool.

The detailed kinematic diagram of single-sided lapping system with the additional movement of conditioning ring along a chord is presented in **Figure 7**. It can be observed that in this idea, there is only one conditioning ring, which in addition to rotary motion performs a reciprocating motion between point A and B. The position of the conditioning ring is not constant as in conventional system, parameter $R_{Px}(t)$ changes the value in time. The distance from the center of the lapping plate to the chord equals R_{Py}.

The principle of kinematics calculation is identical as in case of standard single-sided lapping system. Due to additional motion of the conditioning ring, only Eq. (3) must be changed:

$$x_P' = x_P'' \cdot \cos(\omega_s \cdot t) + y_P'' \cdot \sin(\omega_s \cdot t) + R_{Px}(t) \tag{11}$$

Considering the reciprocating motion of the conditioning ring, R_{Px} is a function of time and it can be expressed as:

$$R_{Px} = R_o + \Delta R \tag{12}$$

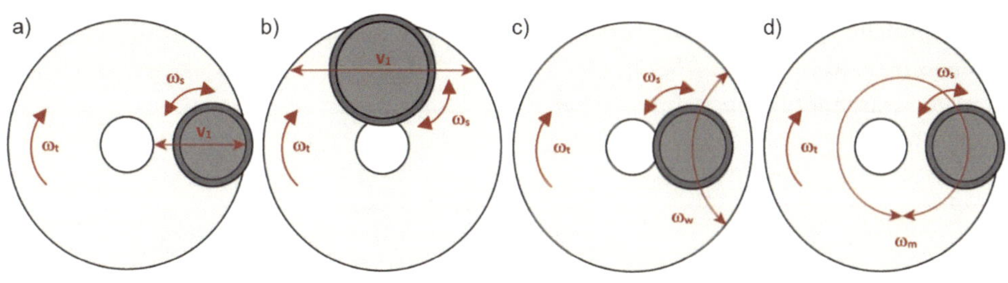

Figure 6. Examples of approach to kinematics system lapping plate—conditioning ring: (a) radial, (b) secant, (c) swinging, and (d) off-set.

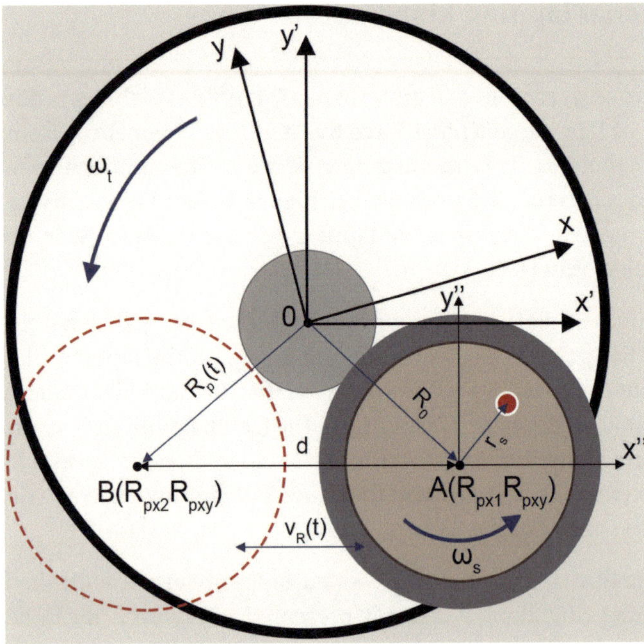

Figure 7. Diagram of single-sided lapping machine with reciprocating movement of conditioning ring.

There are many ways in positioning control applications to move from one point to another. The most common are trapezoidal and S-curve motions. In trapezoidal move profile, velocity is changed linearly and acceleration makes abrupt changes [14]. This can be quantified in an S-curve motion profile by "jerk," which is the time derivative of acceleration. S-curve motion helps to smoothen the motion and reduce the possibility of exciting a vibration or oscillation [15]. A single stroke of reciprocating motion with S-curve profile is shown in **Figure 8**. The conditioning ring moves between two points, where T_R is the reciprocating period, T_A is the acceleration time of the uniformly accelerated motion and T_D is the deceleration time of the uniformly decelerated motion. The conditioning ring has a maximum velocity v_{Rmax} at the central range of the reciprocating stroke and uniformly accelerated and decelerated motions at the two ends of the stroke.

For the simplicity, the velocity expressions of another non-dimensional parameter was introduced and defined by Eq. (12). Parameter k_2 is the ratio of the reciprocating period to the lapping plate rotary motion period:

$$k_2 = \frac{T_R}{T_T} = \frac{d \cdot \omega_t}{v_R \cdot \pi} \tag{13}$$

where d is the reciprocating stroke, T_R and T_T are the periods of the reciprocating motion of the conditioning ring and rotary motion of the lapping plate.

http://dx.doi.org/10.5772/intechopen.71415

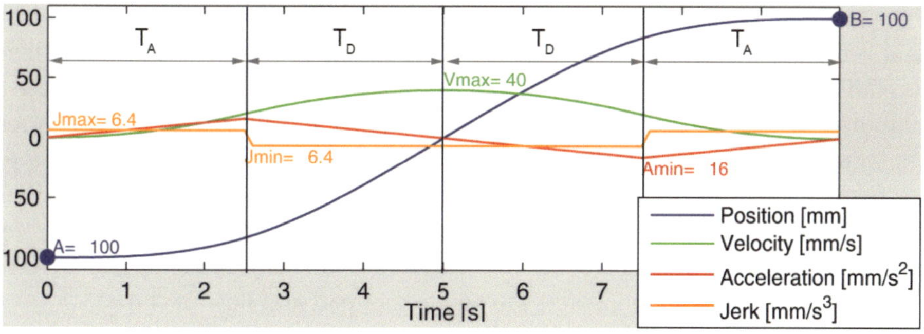

Figure 8. Single stroke of reciprocating motion with S-curve profile.

6. Kinematic optimization

Kinematic analysis showed that the basic kinematic parameters significantly affect the trajectory and velocity distributions. To improve the quality and flatness of the machined surfaces, the selection of parameter values, which are optimal was carried out. It was assumed that, the main optimization criterion is the uniformity of tool wear. In this section, the material removal rate and the lapping plate wear model are described in detail. Based on the lapping plate wear, its uniformity is calculated and the maximum value is sought.

6.1. Material removal rate

The volume of material removed by lapping process at a local position during every unit of time is an important part of study about lapping process. There are many models which describe material removal rate (MRR) during lapping, grinding or polishing. Several researchers worked on experiment and analytical models [2, 16, 17]. They show that the lapping process can be improved by optimizing both the machining efficiencies, and the consideration of the process parameter influences with surface lap height. Considerations about MRR are important because it is the only solution to provide the maintenance of reliability and lifetime of the produced workpieces.

One of the most common wear models in abrasives process is the tribological model developed by Preston, who related relative velocity and pressure to material removal rate (MRR) which is known as Preston's equation:

$$\frac{dH}{dt} = k \cdot p \cdot v \tag{14}$$

where H is a height of removed material in any point on the lapping plate, k is a constant relating to the process parameters, p is the applied pressure, and v is the relative velocity of the workpiece. The parameter k varies based on any modifications to the material removal process such as abrasive and slurry type, feed rate, and other process parameters [16].

However, some of the experiments have shown that in the case of relatively high or low velocities and pressure, linear relationship in Preston's equation does not hold true. Therefore, many modifications to the Preston's equation were proposed. Moreover, there are many formulations in literatures, which are based on the Preston's equation. Using the Preston's equation, Runnels and Eyman [18] proved that the MRR in the chemical-mechanical polishing process is related to the normal and shear stresses. Tseng and Wang [19] modified the Preston's equation to express the MRR as a function of the normal stress of the wafer, the relative velocity of the polishing and the elastic deformation of the abrasive grains. Nanz [20] provided new MRR equation considers the bending of pad and flow of slurry.

6.2. Lapping plate wear model

Intensity of lapping plate wear can be assumed as a contact intensity of the tool with the workpieces through the lapping abrasive grains. There are different methods for calculating contact intensity. This section assumes the method of calculating particles density of interpolated trajectories. Therefore, in order to simulate the trajectories and to count their distribution, Matlab program was designed.

An example of trajectories density determination steps is shown in **Figure 9**. Initially, location of random particles is generated within the conditioning ring. Then, the trajectories of the particles are calculated with a use of kinematic equations. A set of points, which are equally spaced from each other, are generated with interpolation function. The lapping plate surface is divided into small squares with the same area. Finally, a statistics function is used to count the total number of points within each square of the lapping plate surface. The contact intensity can be developed for a profile of the tool. It allows to determine if the wear causes a concavity or the convexity. The area of the lapping plate must be divided into equal rings. A measure of points in an appropriate area is determined by equation:

$$D_i = \frac{n_i}{A_i - A_{i-1}} = \frac{n_i}{(2i-1)\cdot\pi\cdot r^2} \quad (15)$$

where n is a points number in area A_i and r is rings width.

To calculate the standard deviation S_D of all the values of D_i, which is given by:

$$S_D = \sqrt{\frac{\sum(D-\bar{D})^2}{N-1}} \quad (i = 1, 2, 3 \ldots N) \quad (16)$$

where $\bar{D}$ represents the average of all the values and N is the total number of the divided rings or squares [18].

6.3. Parameters optimization

Trajectories distribution of particles in single-sided lapping process is significantly affected by rotational speed ratio of the conditioning ring to the lapping plate k_1 (Eq. (9)) in standard single-sided lapping system and in addition by period ratio of the reciprocating motion of the

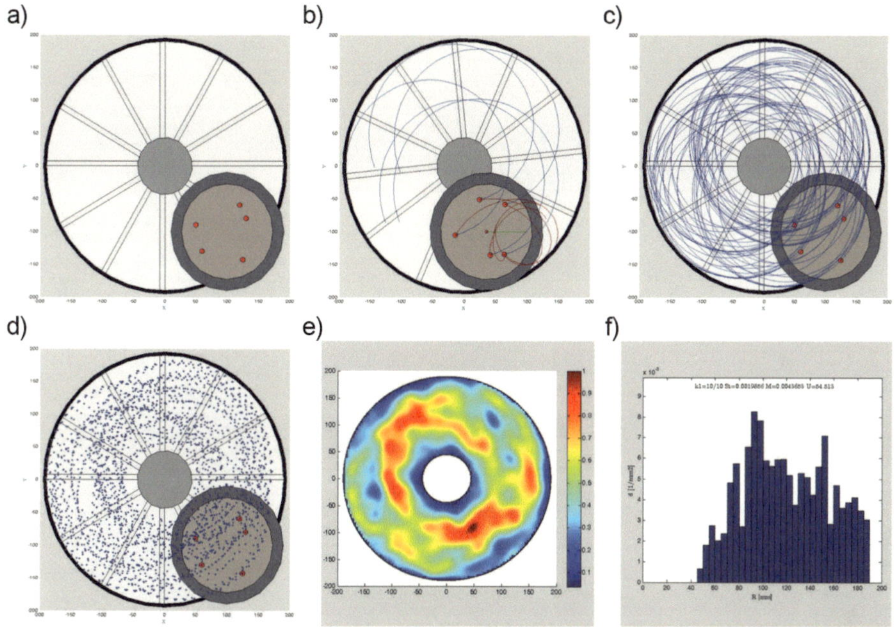

Figure 9. Trajectories density determination: (a) random generated particles, (b, c) trajectories generating, (d) trajectories interpolation, (e) density of trajectories, (f) profile density of trajectories.

conditioning ring to rotary motion of the lapping plate k_2 (Eq. (13)) in system with additional movement of conditioning ring. To obtain a better uniformity base on trajectory simulations and consequently even lapping plate wear, it is important to optimize both parameters. The uniformity of tool wear was assumed as main optimization criterion. In order to describe the evenness of the lapping wear, the trajectories distribution uniformity is defined:

$$U = \left(1 - \frac{S_D}{\overline{D}}\right) \cdot 100\% \tag{17}$$

where S_D is the standard deviation of trajectories density and $\overline{D}$ is the average value of the trajectories distribution.

It can be predicted that during the lapping process, there are more than 1 million active particles in the slurry and on the lapping plate. However, because of the calculation time, an appropriate particle number which can reflect the same regularity as the real number has to be determined. **Figure 10** shows the influence of the amount of abrasive grains to the uniformity. Results are presented for standard single-sided lapping system, when $k_1 = 0.45$ and $R_P = 125$ mm. It can be observed that for 1000 randomly distributed particles, uniformity is stable and constant.

Trajectory density uniformity for standard single-sided lapping system and for single-sided lapping system with reciprocating motion of conditioning ring is presented in **Figures 11** and **12**. Trajectories were generated during 60 s on lapping plate with internal diameter of 88 mm and

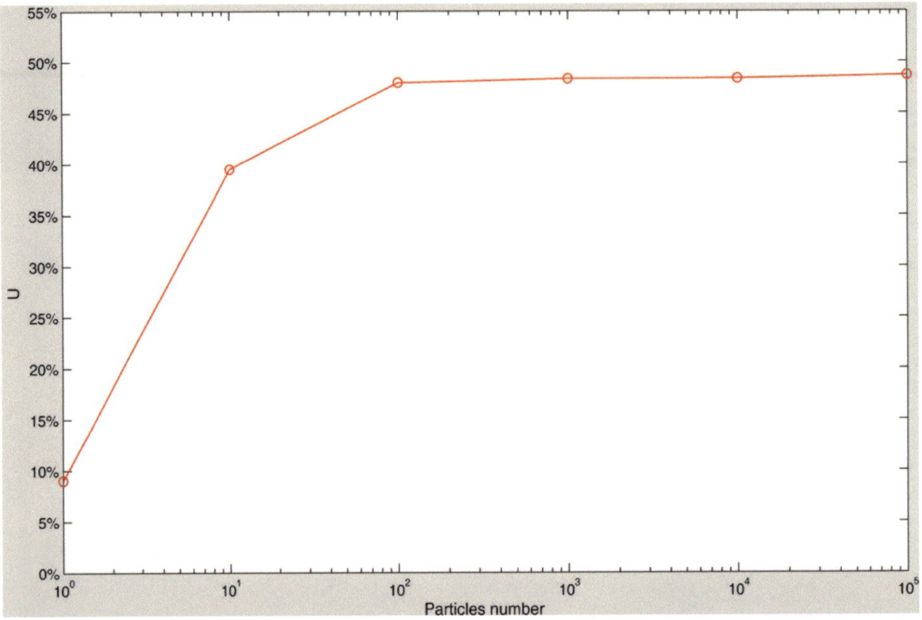

Figure 10. Uniformity of trajectories distribution produced by different number of random particles.

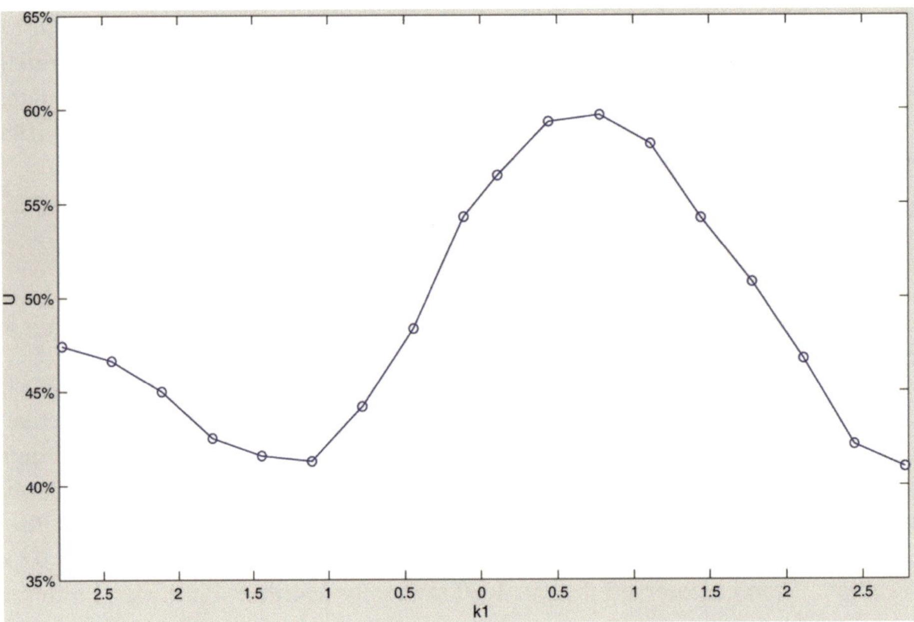

Figure 11. Trajectory density uniformity versus k_1 for standard single-sided lapping system, 1000 random particles, radial position of conditioning R = 125 mm, simulation time t = 60 s.

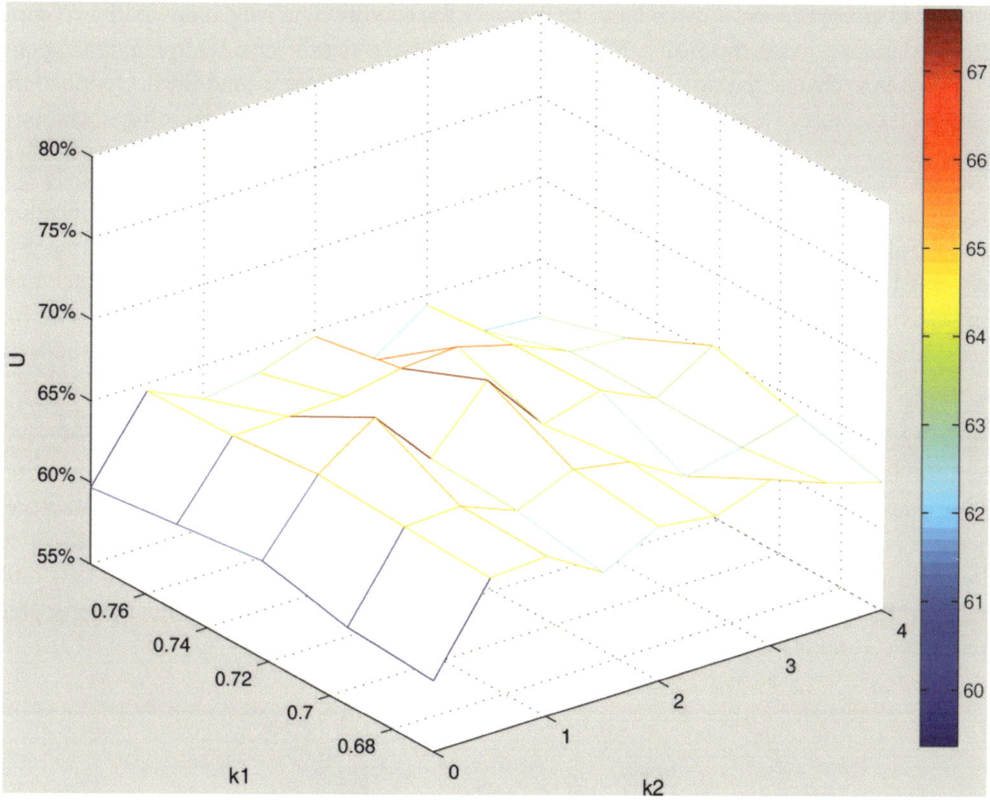

Figure 12. Trajectory density uniformity versus k_1 and k_2 for single-sided lapping system with reciprocating motion of conditioning ring, 1000 random particles, simulation time t = 60 s.

outer diameter of 350 mm. The maximum uniformity was obtained in conventional system, when k_1 = 0.6–0.9, and in non-standard lapping system, when k_1 = 0.65–0.75 and k_2 = 1–2. For single-sided lapping system with reciprocating motion of conditioning ring, the uniformity was about 10% higher.

7. Automated lapping system

The above considerations regarding complex kinematics of lapping process were the starting point to develop automated lapping system. Nowadays, machine tools are more efficient than in the past. Lapping machine manufacturers also improve their machines and supply their basic constructions with additional components. As a result of automation, some of the supporting operations can be eliminated. Lapping machines are supplied with feeding tables, loading and unloading systems of conditioning rings, which form mini-production lines. However, it was emerged that the system where the ring is led by the manipulator during the machining is novel.

In order to create universal mechanism that moves the conditioning ring at any path is complicated and in some cases impossible. Thanks to the robot that moves an effector from point to point, it is possible to change the ring trajectory at any moment and different conditions. Owing to this solution, it is possible to apply any lapping kinematics and use robot for supporting operations.

One of a number of challenges in designing automated manufacturing systems is the selection of a suitable robot. This problem has become more demanding due to increasing specifications and complexity of the robots. However, it was decided that the robot should perform material handling tasks and also lapping process. In initial selection, the articulated robots, which have four to five degrees of freedom and are powered by an electrical drive should be chosen for further evaluation. Furthermore, a continuous path or Point-to-Point control system is required.

The idea of how single-sided lapping machine and the robot working together presents **Figure 13**. The robot situated next to the lapping machine handle primarily sorted workpieces from the table to the separator, located in conditioning ring. Then robot grips the conditioning ring and shifts it onto the lapping plate, which starts to rotate. The machining is executed by the robot. It shifts the ring with workpieces in such a way to keep the flatness of the plate along the radius. After lapping process, robot shifts the conditioning ring onto collecting table and workpieces fall into the box with finished parts. Finally, the flatness of the lapping plate is controlled and is fixed in case when an error occurs.

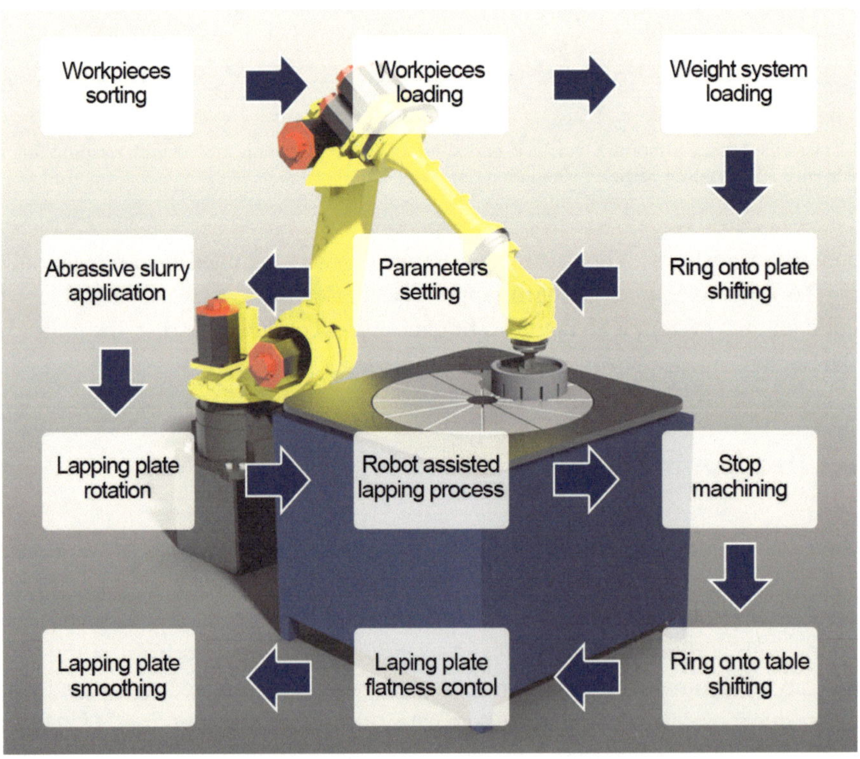

Figure 13. Idea of robotic single-sided lapping system.

http://dx.doi.org/10.5772/intechopen.71415

8. Conclusions

The aim of this chapter was to present the influence of selected factors on the geometrical results of the single-sided lapping process. One of the most crucial factors is kinematics of the executive system. Since flatness of an active surface of the lapping plate has an essential influence on the shape accuracy of lapped surfaces, the key is to maintain the flatness of the tool. It can be established that lapping plate flatness deviation can be caused by its uneven wear.

A desired distribution of the contact density, which determines a wear and allows to correct the flatness error of the lapping plate, can be obtained by choosing appropriate kinematic parameters. Therefore, the kinematic model of single-sided system was in detail analyzed and modeled in Matlab program. Based on the simulations, it was observed that a trajectories distribution of abrasive particles on the lapping plate varies when the kinematic conditions are changed, e.g., by placing the workpieces at different radii, setting different rotational velocities or by introducing additional movements of conditioning ring. Hence, the influence of additional guiding movements of the conditioning ring has been verified. Major factors affecting trajectory distribution are the speed ratio k_1 and the period ratio k_2, which represent the relationships among the three basic motions of unconventional lapping systems. The optimization of these parameters was aimed at improving the quality and flatness of the machined surfaces. Main optimization criterion was the uniformity of tool wear. The maximum uniformity was obtained in conventional system, when $k_1 = 0.6–0.9$, and in non-standard lapping system, when $k_1 = 0.65–0.75$ and $k_2 = 1–2$. For single-sided lapping system with reciprocating motion of conditioning ring, the uniformity was about 10% higher than in conventional system.

Nomenclature

ω_t	angular velocity of the lapping plate
ω_s	angular velocity of the conditioning rings
R_{dw}	inner diameter of lapping plate
R_{dz}	outer diameter of lapping plate
R_p	radial position of conditioning ring
r_P, r_S	radial position of examined points
t	machining time
H	wear removal rate
k	Preston's coefficient
p	force per unit area
v	lapping relative velocity

D_i	trajectories density
d	distance of reciprocating motion
v_R	linear velocity of reciprocating motion
T_R	reciprocating period
T_A	acceleration time
T_D	deceleration time
J_A	acceleration jerk
J_D	deceleration jerk
k_1	rotational speed ratio of the conditioning ring to the lapping plate
k_2	period ratio of the reciprocating motion of the conditioning ring to rotary motion of the lapping plate
S_D	standard deviation
U	uniformity of lapping plate wear

Author details

Adam Barylski and Norbert Piotrowski*

*Address all correspondence to: np.piotrowski@gmail.com

Faculty of Mechanical Engineering, Gdansk University of Technology, Gdansk, Poland

References

[1] Klocke F. Manufacturing Processes 2: Grinding, Honing, Lapping. Berlin, Heidelberg: Springer-Verlag; 2009

[2] Barylski A. Obróbka powierzchni płaskich na docierarkach. Gdansk: Wydawnictwo Politechniki Gdańskiej; 2013

[3] Uhlmann E, Ardelt T, Spur G. Influence of kinematics on the face grinding process on lapping machines. CIRP Annals - Manufacturing Technology. 1999;**48**(1):281-284. DOI: 10.1016/S0007-8506(07)63184-9

[4] Marinescu ID, Uhlmann E, Doi T. Handbook of Lapping and Polishing. New York, Taylor & Francis Publishing House; 2006

[5] Barylski A. Podstawy docierania jednotarczowego powierzchni płaskich. Gdańsk, Zeszyty Naukowe Politechniki Gdańskiej. Mechanika. 1992;**67**

[6] Barylski A, Piotrowski N. Koncepcje niekonwencjonalnych układów kinematycznych docierania jednotarczowego z wykorzystaniem robota. Mechanik. 2013;**78**(8-9):36-33

[7] Martin K. Neuere Erkenntnisse über den Hartmetallabtrag beim Läppen. Maschinenmarkt. 1973;**79**:103-104

[8] Ardelt T. Verfahrensvergleich Planschleifen mit Planetenkinematik – Planparallelläppen. IDR. 2001;**35**(3):214-224

[9] Neauport J, Destribats J, Maunier C, Ambard C, Cormont P, Pintault B, Rondeau O. Loose abrasive slurries for optical glass lapping. Applied Optics. 2010;**49**(30):5736-5745

[10] Molenda J, Charchalis A. Wpływ twardości materiału obrabianego na jakość powierzchni po docieraniu. Logistyka. 2015;**3**:3363-3370

[11] Ishikawa K, Ichikawa K, Suwabe H. A basic study on corrective techniques to recovering accuracy of deteriorated lapping plate. Japan Society of Precision Engineering. 1988;**3**:234-236

[12] Satake U, Enomoto T, Fujii K, Hirose K. Optimization method for double-sided polishing process based on kinematical analysis. Procedia CIRP. 2106;**41**:870-874

[13] Barylski A, Piotrowski N. New kinematic systems in single-sided lapping and their influence on lap wear. Applied Mechanics and Materials. 2016;**831**:14-24

[14] Lewin C. Mathematics of motion control profiles. Performance Motion Devices. 2013; https://pdfs.semanticscholar.org/a229/fdba63d8d68abd09f70604d56cc07ee50f7d.pdf [Accessed: 10.25.2017]

[15] Nguyen KD, Ng T-C, Chen I-M. On algorithms for planning S-curve motion profiles. International Journal of Advanced Robotic Systems. 2008;**5**(1):99-106

[16] Evans J, Paul E, Dornfeld D, Lucca D, Byrne G, Tricard M, Klocke F, Dambon O, Mullany B. Material removal mechanisms in lapping and polishing. CIRP Annals - Manufacturing Technology. 2003;**52**(2):611-633

[17] llez-Arriaga L, Téllez A. Correction of the Preston equation for low speeds. Applied Optics. 2007;**46**(9):1408-1410

[18] Runnels SR, Eyman LM. Tribology analysis of chemical-mechanical polishing. Journal of Electrochemical Society. 1994;**141**(6):1698-1701. DOI: 10.1149/1.2054985

[19] Tseng WT. Comparative study on the roles of velocity in the material removal rate during chemical mechanical polishing. Journal of the Electrochemical Society. 1999;**146**(5): 1952-1959

[20] Nanz G, Camilletti LE. Modeling of chemical-mechanical polishing: A review. IEEE Transactions on Semiconductor Manufacturing. 1995;**8**(4):382-389. DOI: 10.1109/66. 475179

Chapter 11

Optimization Approach for Inverse Kinematic Solution

Panchanand Jha and Bibhuti Bhusan Biswal

Additional information is available at the end of the chapter

http://dx.doi.org/10.5772/intechopen.71409

Abstract

Inverse kinematics of serial or parallel manipulators can be computed from given Cartesian position and orientation of end effector and reverse of this would yield forward kinematics. Which is nothing but finding out end effector coordinates and angles from given joint angles. Forward kinematics of serial manipulators gives exact solution while inverse kinematics yields number of solutions. The complexity of inverse kinematic solution arises with the increment of degrees of freedom. Therefore it would be desired to adopt optimization techniques. Although the optimization techniques gives number of solution for inverse kinematics problem but it converses the best solution for the minimum function value. The selection of suitable optimization method will provides the global optimization solution, therefore, in this paper proposes quaternion derivation for 5R manipulator inverse kinematic solution which is later compared with teachers learner based optimization (TLBO) and genetic algorithm (GA) for the optimum convergence rate of inverse kinematic solution. An investigation has been made on the accuracies of adopted techniques and total computational time for inverse kinematic evaluations. It is found that TLBO is performing better as compared GA on the basis of fitness function and quaternion algebra gives better computational cost.

Keywords: TLBO, GA, quaternion, kinematics

1. Introduction

Kinematic chain may consist of rigid/flexible links which are connected with joints or kinematics pair permitting relative motion of the connected bodies. In case of manipulator kinematics it can be categorized into forward and inverse kinematics. Forward kinematics for any serial manipulator is easy and mathematically simple to resolve but in case of inverse kinematics there is no unique solution, generally inverse kinematics gives multiple solutions. Hence, inverse kinematics solution is very much problematic and computationally expensive. For real time control of any configuration manipulator will be expensive and generally it takes long time. Forward kinematics of any manipulator can be understand with translation of position and orientation of end effector from joint space to Cartesian space and opposite of this is known as inverse kinematics. It is essential to calculate preferred joint angles so that the end effector can reach to the desired position and also for designing of the manipulator. Various industrial applications are based on inverse kinematics solutions. In real time environment it is obvious to have joint variables for fast transformation of end effector. For any configuration of industrial robot manipulator for n number of joints the forward kinematics will be given by,

$$y(t) = f(\theta(t)) \tag{1}$$

where $\theta_i = \theta(t)$, $i = 1, 2, 3, \ldots, n$ and position variables by $y_j = y(t)$, $j = 1, 2, 3, \ldots, m$.

Inverse kinematics for n number of joints can be computed as,

$$\theta(t) = f^{'}(y(t)) \tag{2}$$

Inverse kinematics solution of robot manipulators has been considered and developed different solution scheme in last recent year because of their multiple, nonlinear and uncertain solutions. There are different methodologies for solving inverse kinematics for example iterative, algebraic and geometric etc. [8] proposed inverse kinematic solution on the basis of quaternion transformation. [20–36] have proposed application of quaternion algebra for the solution of inverse kinematics problem of different configurations of robot manipulator. [35] presented a quaternion method for the demonstrating kinematics and dynamics of rigid multibody systems. [34] presented analytical solution of 5-dof manipulator considering singularity analysis. [11] presented quaternion based kinematics and dynamics solution of flexible manipulator. [14] proposed detailed derivation of inverse kinematics using exponential rotational matrices. On the other hand, after numerous surveys on conventional analytical and other Jacobian based inverse kinematics are quite complex as well as computationally exhaustive those are not exactly well suitable for the real time applications. Because of the above-mentioned reasons, various authors adopted optimization based inverse kinematic solution.

Optimization techniques are fruitful for solve inverse kinematics problem for different configurations of manipulator as well as spatial mechanisms. Conventional approaches such as Newton-Raphson can be used for nonlinear kinematic problems and predictor corrector type methods can compute differential problem of manipulator. But major drawback of these methods are Singularity or ill condition which converse to local solutions. Moreover, when initial guessing is not accurate then the method becomes unstable and does not converse to optimum solution. Therefore, recently developed metaheuristic techniques can be used to overcome the conventional optimization drawbacks. Literature survey shows the efficiency of these metaheuristic algorithms or bi-inspired optimization techniques are more convenient to achieve global optimum solutions. The major issue with these nature inspired algorithms is framing of objective function. Even these algorithms are direct search algorithms which do not require any gradient or differentiation of objective function. The comparison of the metaheuristic algorithm with heuristic algorithms is based on the convergence rate as it has been proved that the convergence of heuristic-based techniques is slower. Therefore, to adopt metaheuristic techniques such as GA, BBO, teachers learner based optimization (TLBO), ABC, ACO etc. will be suitable for enhancing the convergence rate and yielding global solution. From literature survey the teaching learning based optimization (TLBO) is similar to swarm based optimization in which the impact of learning methods from teacher to student and student to student has been highlighted. Wherein, the population or swarm is represented by group of students and these students gain knowledge from either teacher or students. If these student gain knowledge from teacher then it is called as teachers phase similarly when students learns form student then it is student phase. The output is considered as result or grades of students. Therefore, number for number of subjects resembles the variables of the function and grades or results gives fitness value, [5, 6]. There are numerous other population centered methods which have been effectively applied and shown efficiency [33]. However, all algorithms are not suitable for complex problem as proved by Wolpert and Macready. On the other hand, evolutionary strategy (ES) based methods such as GA, BBO etc. gives better results for various problems and these methods are also population based metaheuristic [16, 28]. Moreover [22] proposed inverse kinematic solution of redundant manipulator using modified genetic algorithm considering joint displacement ($\Delta\theta$) error minimization and the positional error of end effector. [32] proposed inverse kinematic solution of PUMA 560 robot using cyclic coordinate descent (CCD) and Broyden-Fletcher-Shanno (BFS) technique. [23] proposed IK solution of 4-dof PUMA manipulator using genetic algorithm. This paper uses two different objective functions which are based on end-effector displacement and joint variable rotations. [18] proposed trajectory planning of 3-dof revolute manipulator using evolutionary algorithm. [25] proposed inverse kinematics solution and trajectory planning for D-joint robot manipulator based on deterministic global optimization based method. [1] proposed inverse kinematic solution of redundant manipulator using novel developed global optimization algorithm. [4] proposed inverse kinematic solution of PUMA robot manipulator using genetic programming. In this work, mathematical modeling is evolved using genetic programming through given direct kinematic equations. [17] proposed optimization of design parameter i.e. link length using for 2-dof manipulator. [15] proposed inverse kinematic solution of 2-dof articulated robot manipulator using real coded genetic algorithm. [19] proposed inverse kinematic solution scheme of 3-dof redundant manipulator based on reach hierarchy method. [30] proposed inverse kinematic solution of 3-dof PUMA manipulator for the major displacement propose. In this work they have adopted genetic algorithm with adaptive niching and clustering. [12] proposed inverse kinematic solution of 6-dof MOTOMAN robot manipulator for positioning of the end-effector. In this work they have adopted adaptive genetic algorithm for optimum placement of the end effector. [26] proposed inverse kinematic and trajectory generation of humanoid arm manipulator using forward recursion with backward cycle computation method. [21] proposed inverse kinematic solution for 6R revolute manipulator using real time optimization algorithm. [24] proposed kinematic solution using three different methods such as bee algorithm, neural network which is later optimized by bee algorithm and evolutionary algorithm. [2] proposed kinematic solution of 3-dof serial robot manipulator using real time genetic algorithm. [13] proposed inverse kinematic solution of 6-dof robot manipulator using immune genetic algorithm. [9] proposed conventional approach i.e. penalty function based optimization method for solving IK. Even though few methods can solve hard NP problems, but it requires high-performance computing system and intricate computer programming.

On the other hand, the use of optimization algorithms is not new in the field of multi-objective and NP-hard problem to arrive at a very reasonable optimized solution, the TLBO algorithm have not been tried to solve an inverse kinematics problems and trajectory of joint variables for robot manipulator. Moreover, computational cost for yielding the inverse kinematics solution with adopted algorithms has been compared without any specialized tuning of concern parameters. Therefore, the key purpose of this work is focused on minimizing the Euclidian distance of end effector position based resolution of inverse kinematics problem with comparison of GA and TLBO obtained solution for 5R robot manipulator. The results of all algorithm are computed from inverse kinematics equations and obtained resultant error for data statistics. In other words, end-effector coordinates utilized as an input for joint angle calculations. At the end 4th order spline formula is considered for generation of end effector trajectory and analogous joint angles of robotic arm using TLBO, GA and quaternion. The sectional organization of the paper henceforth is as follows: Section 2 pertains to the mathematical modeling of the 5R robot manipulator and detail derivation of forward and inverse kinematics of 5R manipulator using quaternion algebra. In Section 3 discuss about the inverse kinematic objective function formulation for 5R manipulator. The experimental results as obtained from simulations are discussed elaborately in Section 5.

2. Quaternion vector approach for mathematical modeling

This section deals with mathematical modeling of quaternion vector algebra and application for the derivation of inverse kinematics equations. Quaternion vector methods are fruitful for both rotation and translation of a point, line, etc. with references to origin coordinate system irrespective of homogeneous transformation matrix. The Interpolation of series of rotations and translations are quite complex using Euler's angle method. In other words, the variables lies in isotropic space which is nothing but sphere surface topology and complex in nature. A brief formulation of quaternion mathematics is given in this section for assessment of references and to create background for mathematical derivation of inverse kinematic.

2.1. Rotation and translation from quaternion

The above discussions gives the importance of quaternions and the necessity of it. The quaternion rotation and translation are lies in four dimensional space therefore it is quite difficult to represent here or to imagine. **Figure 1** represents the rotation through quaternion and Eq. (3) describes rotation of a point in a space mathematically.

$$h = \cos\left(\frac{\theta}{2}\right) + i^{*}\sin\left(\frac{\theta}{2}\right) + j^{*}\sin\left(\frac{\theta}{2}\right) + k^{*}\sin\left(\frac{\theta}{2}\right) \quad (3)$$

The 4-dimensional space, imagination of fourth axis is quite complex. Therefore, in **Figure 1** a unit distance point around axis (X, Y, and Z) is given and which traces a circle. When this rotation circle is projected on a plane then the point P_1 can be seen rotated through angle θ to point P_3 which crosses the mid-point P_2. Therefore P_1 point is transforming to P_3 following by straight line makes $\cos(\theta/2)$ and $\sin(\theta/2)$.

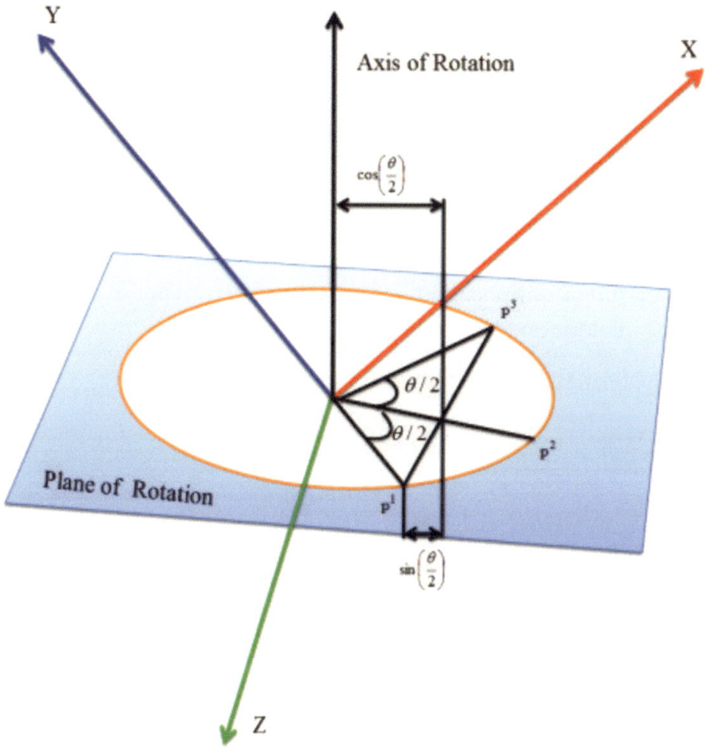

Figure 1. Rotation representation of point.

Now two quaternions can be represented on the basis of above discussed concept. If there is subsequent rotation of two quaternion h_1 and h_2 then the composite rotations $h_1{}^*h_2$ can be given by From **Figure 1**, p^1 is the point vector representing initial position and p^3 is the point vector final condition to be transformed. Therefore,

$$\begin{aligned} p^3 &= h_2{}^*\left(h_1{}^*p^{1*}h_1^{-1}\right)^*h_2^{-1} \\ &= \left(h_2{}^*h_1\right)^*p^{1*}\left(h_1^{-1*}h_2^{-1}\right) \\ &= \left(h_2{}^*h_1\right)^*p^{1*}\left(h_2^{-1}*h_1^{-1}\right) \end{aligned} \tag{4}$$

Now pure translations t_r can be done by quaternion operator that is given below,

$$t_r = h + p^1 \tag{5}$$

Quaternion transform can be given by,

$$p^2 = h^*p^{1*}h^{-1} \tag{6}$$

Therefore, an expression for the inverse of a quaternion can be given as,

$$H^{-1} = \left[< h^{-1} >, < -h^{-1} \otimes P \otimes h >\right] \tag{7}$$

where, $-h^{-1} \otimes P \otimes h = -P + [-2k(j \times (-P)) + 2j \times (j \times (-P))]$ quaternion product, which is defined in the most general form for two quaternions $h_1 = (r_1, j_1)$, and $h_2 = (r_2, j_2)$ as

$$h_1 \otimes h_2 = \left[r_1 r_2 - j_1 \bullet j_2 \;\; r_1 j_2 + r_2 j_1 + j_2 \times j_1\right] \tag{8}$$

where $j_1 \bullet j_2$ and $j_2 \times j_1$ denote dot and cross products, between the 3-dimensional vectors j_1 and j_2. Clearly, quaternion multiplication is not commutative. The set of elements {±1, ±i, ±j, ±k} is known as the quaternion group of order 8 in multiplication.

Similarly vector transformation multiplication can be given as

$$H_1 \otimes H_2 = \left(h_1, P^1\right) \otimes \left(h_2, P^2\right) = h_1{}^* h_2, \;\; h_1{}^* P^2{}^* h_1^{-1} + P^1 \tag{9}$$

where, $h_1{}^* P^2{}^* h_1^{-1} = P^2 + 2r_1\left(j_1 \times P^2\right) + 2j_1 \times \left(j_1 \times P^2\right)$

2.2. Quaternion derivation for 5R manipulator kinematics

The configuration and base coordinate frame attachment of 5R manipulator is given in **Figure 2** (a) and MATLAB plot of 5R manipulator is presented in (b). Where $\theta_1, \theta_2, \theta_3, \theta_4$ and θ_5 joint angles for are articulated arm and d_1, d_2 and d_3 are the link offset. a_1, and a_2 represents link lengths.

Now quaternion for successive transformation of each joint can be calculated from the Eq. (3) as follows,

$$H_1 = \left[< \overline{C_1} + \overline{S_1}\,\hat{k} >, < a_1 C_1 \hat{i} + a_1 S_1 \hat{j} + d_1 \hat{k} >\right] \tag{10}$$

$$H_2 = \left[< \overline{C_2} + \overline{S_2}\,\hat{j} >, < -a_2 S_2 \hat{i} - a_2 C_2 \hat{k} >\right] \tag{11}$$

$$H_3 = \left[< \overline{C_3} + \overline{S_3}\,\hat{j} >, < -d_4 S_3 \hat{i} - d_4 C_3 \hat{k} >\right] \tag{12}$$

$$H_4 = \left[< \overline{C_4} + \overline{S_4}\,\hat{i} >, < d_4 \hat{i} >\right] \tag{13}$$

$$H_5 = \left[< \overline{C_5} + \overline{S_5}\,\hat{j} >, < -d_6 S_5 \hat{i} - d_6 C_5 \hat{k} >\right] \tag{14}$$

Inverse of a dual quaternion can be calculated by Eq. (8),

$$H_1^{-1} = \left[< \overline{C_1} - \overline{S_1}\,\hat{k} >, < -a_1 \hat{i} >\right] \tag{15}$$

$$H_2^{-1} = \left[< \overline{C_2} - \overline{S_2}\,\hat{j} >, < a_2 \hat{k} >\right] \tag{16}$$

$$H_3^{-1} = \left[< \overline{C_3} - \overline{S_3}\,\hat{j} >, < d_4 \hat{k} >\right] \tag{17}$$

http://dx.doi.org/10.5772/intechopen.71409

$$H_4^{-1} = \left[< \overline{C_4} - \overline{S_4 \hat{i}} >, < -d_4 \hat{i} >\right] \tag{18}$$

$$H_5^{-1} = \left[< \overline{C_5} - \overline{S_5 \hat{j}} >, < d_6 \hat{k} >\right] \tag{19}$$

$$Q_i = H_i \otimes H_{i+1} \ldots\ldots. H_n \tag{20}$$

Where in case of 5R manipulator arm n = 5. Now calculating quaternion vector products using Eq. (20)

$$\begin{aligned} &Q_i = H_i \otimes H_{i+1} \ldots\ldots. H_n \\ &Q_5 = H_5 = \left[< \overline{C_5} - \overline{S_5 \hat{j}} >, < -d_6 S_5 \hat{i} - d_6 C_5 \hat{k} >\right] \\ &Q_4 = H_4 \otimes Q_5 = \left[< \overline{C_4} + \overline{S_4 \hat{i}} >, < d_4 \hat{i} >\right] \otimes \left[< \overline{C_5} + \overline{S_5 \hat{j}} >,\right. \\ &\qquad \left. < -d_6 S_5 \hat{i} - d_6 C_5 \hat{k} >\right] \end{aligned} \tag{21}$$

$$\begin{aligned} Q_4 = &\left[< \overline{C_4 C_5} + \overline{S_4 C_5 \hat{i}} + \overline{C_4 S_5 \hat{j}} + \overline{S_4 S_5 \hat{k}} >,\right. \\ &\left. < (d_4 - d_6 S_5)\hat{i} + d_6 C_5 S_4 \hat{j} - d_6 C_4 C_5 \hat{k} >\right] \end{aligned} \tag{22}$$

$$\begin{aligned} Q_3 = H_3 \otimes Q_4 = &\left[< \overline{C_3} + \overline{S_3 \hat{j}} >, < -d_4 S_3 \hat{i} - d_4 C_3 \hat{k} >\right] \otimes \\ &\left[< \overline{C_4 C_5} + \overline{S_4 C_5 \hat{i}} + \overline{C_4 S_5 \hat{j}} + \overline{S_4 S_5 \hat{k}} >,\right. \\ &\left. < (d_4 - d_6 S_5)\hat{i} + d_6 C_5 S_4 \hat{j} - d_6 C_4 C_5 \hat{k} >\right] \end{aligned} \tag{23}$$

$$\begin{aligned} Q_3 = &\left[< \overline{C_4 C_{3+5}} + \overline{S_4 C_{3-5}} \hat{i} + \overline{C_4 S_{3+5}} \hat{j} - \overline{S_4 S_{3+5}} \hat{k} >,\right. \\ &< (d_4 + d_4 - d_6 C_4 C_5 S_3 - d_4 C_3 - d_6 S_5 C_3 - d_4 S_3)\hat{i} \\ &\left. +(d_6 C_5 S_4 \hat{j} + (-d_4 S_3 + d_6 S_5 S_3 - d_6 C_4 C_5 C_3 - d_4 \hat{C}_3)k >\right] \end{aligned} \tag{24}$$

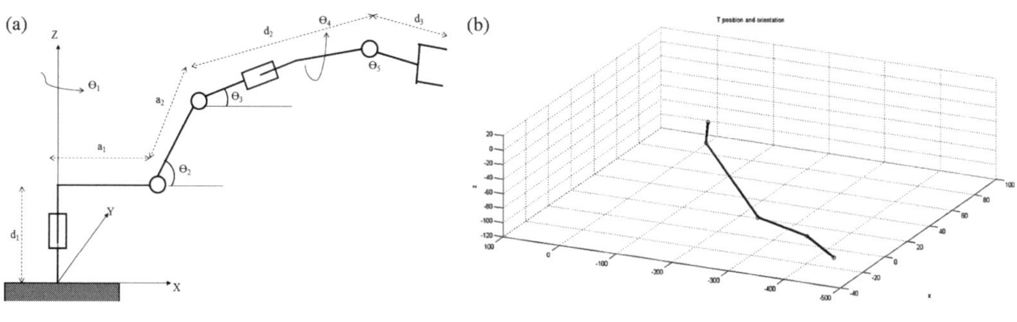

Figure 2. (a) Base frame and model of 5R manipulator; (b) configuration of 5R manipulator.

$$\begin{aligned} Q_2 = H_2 \otimes Q_3 = & \left[< \overline{C_2} + \overline{S_2}\hat{j} >, < -a_2S_2\hat{i} - a_2C_2\hat{k} > \right] \otimes \\ & \Big[< \overline{C_4C_{3+5}} + \overline{S_4C_{3-5}}\hat{i} + \overline{C_4S_{3+5}}\hat{j} - \overline{S_4S_{3+5}}\hat{k} >, \\ & < (d_4 + d_4 - d_6C_4C_5S_3 - d_4C_3 - d_6S_5C_3 - d_4S_3)\hat{i} \\ & +(d_6C_5S_4\hat{j} + (-d_4S_3 + d_6S_5S_3 - d_6C_4C_5C_3 - d_4\hat{C}_3)k > \Big] \end{aligned} \tag{25}$$

Therefore,

$$\begin{aligned} Q_2 = & \Big[< (\overline{C_2}\,\overline{C_4C_{3+5}} - \overline{S_2C_4S_{3+5}}\Big) + (\overline{C_2S_4C_{3-5}} - \overline{S_2S_4S_{3+5}})\hat{i} \\ & +(\overline{C_2C_4S_{3+5}} + \overline{S_2C_4C_{3+5}})\hat{j} + (-\overline{C_2S_4S_{3+5}} - \overline{S_2S_4S_{3-5}})\hat{k} >, \\ & < (-a_2S_2 + d_4C_2 + d_4C_2 - d_4C_{2-3} + d_6S_5C_{2+3} \\ & -d_4S_{2+3} - d_6C_4C_5S_{2+3} - d_4S_3)\hat{i} + (d_6C_5S_4\hat{j} \\ & +\Big(-a_2C_2 - d_4S_2 - d_4S_2 + d_6C_4C_5C_{2+3} + d_4S_{2-3} + d_6S_5S_{2+3} + d_4\hat{C}_{2-3})k > \Big] \end{aligned} \tag{26}$$

$$\begin{aligned} Q_1 = H_1 \otimes Q_2 = & \left[< \overline{C_1} + \overline{S_1}\hat{k} >, < a_1C_1\hat{i} + a_1S_1\hat{j} + d_1\hat{k} > \right] \otimes \\ & \Big[< (\overline{C_2C_4C_{3+5}} - \overline{S_2C_4S_{3+5}}) + (\overline{C_2S_4C_{3-5}} - \overline{S_2S_4S_{3+5}})\hat{i} \\ & +(\overline{C_2C_4S_{3+5}} + \overline{S_2C_4C_{3+5}})\hat{j} + (-\overline{C_2S_4S_{3+5}} - \overline{S_2S_4S_{3-5}})\hat{k} >, \\ & < (-a_2S_2 + d_4C_2 + d_4C_2 - d_4C_{2-3} + d_6S_5C_{2+3} - d_4S_{2+3} \\ & -d_6C_4C_5S_{2+3} - d_4S_3)\hat{i} + (d_6C_5S_4\hat{j} + \\ & (-a_2C_2 - d_4S_2 - d_4S_2 + d_6C_4C_5C_{2+3} + d_4S_{2-3} + d_6S_5S_{2+3} + d_4\hat{C}_{2-3})k > \Big] \end{aligned} \tag{27}$$

Therefore,

$$\begin{aligned} Q_1 = & \Big[< (\overline{C_1C_4C_{2+3+5}} + \overline{S_1S_4S_{2+3-5}}) + (\overline{C_1S_4C_{2+3-5}} - \overline{S_1C_4S_{2+3+5}})\hat{i} \\ & +(\overline{C_1C_4S_{2+3+5}} + \overline{S_1S_4C_{2+3-5}})\hat{j} + (\overline{S_1C_4C_{2+3+5}} - \overline{C_1S_4S_{2+3-5}})\hat{k} >, \\ & < (a_1C_1 - a_2S_2 + d_4C_2C_1 + d_4C_2C_1 - d_4C_{2-3}C_1 + d_6S_5C_{2+3}C_1 - d_4S_{2+3}C_1 \\ & + d_6C_4C_5S_{2+3}C_1 - d_6C_5S_4S_1)\hat{i} + (d_6C_5S_4 + d_6C_5S_4C_1 - a_2S_2 \\ & + a_2S_2C_1 - a_2S_2S_1 + d_4C_2S_1 + d_4C_2S_1 - d_4C_{2-3}S_1 + d_6S_5C_{2+3}S_1 \\ & - d_4S_{2+3}S_1 - d_6C_4C_5S_{2+3}S_1)\hat{j} + (d_1 - a_2C_2 - d_4S_2 - d_4S_2 \\ & + d_6C_4C_5C_{2+3} + d_4S_{2-3} + d_6S_5S_{2+3} + d_4\hat{C}_{2-3})k > \Big] \end{aligned} \tag{28}$$

$$O_{j+1} = H_j^{-1} \otimes O_j \tag{29}$$

Now calculating vector pair of quaternion using Eq. (29), to solve the inverse kinematics problem, the transformation quaternion of end effector of robot manipulator can be defined as

$$[R_{be}, T_{be}] = O_1 = \left[< w + a\hat{i} + b\hat{j} + c\hat{k} >, < X\hat{i} + Y\hat{j} + Z\hat{k} >\right] \quad (30)$$

Now using Eq. (29), O_2 will be given by,

$$\begin{aligned} &O_2 = H_1^{-1} \otimes O_1 \\ &O_2 = \left[< \overline{C_1} - \overline{S_1}\hat{k} >, < -a_1\hat{i} >\right] \otimes \\ &\left[< w + a\hat{i} + b\hat{j} + c\hat{k} >, < X\hat{i} + Y\hat{j} + Z\hat{k} >\right] \\ &O_2 = [< (o_{21}) + (o_{22})\hat{i} + (o_{23})\hat{j} + (o_{24})\hat{k} >, \\ &< (o_{25})\hat{i} + (o_{26})\hat{j} + o_{27}\hat{k} >] \end{aligned} \quad (31)$$

where, $o_{21} = w\overline{C_1} + c\overline{S_1}$, $o_{22} = a\overline{C_1} + b\overline{S_1}$, $o_{23} = b\overline{C_1} - a\overline{S_1}$, $o_{24} = c\overline{C_1} - w\overline{S_1}$, $o_{25} = X\overline{C_1} - a_1 + Y\overline{S_1}$, $o_{26} = Y\overline{C_1} - X\overline{S_1}$, $o_{27} = Z$.

Now,

$$\begin{aligned} &\mathrm{O}_3 = \mathrm{H}_2^{-1} \otimes \mathrm{O}_2 \\ &\mathrm{O}_3 = \left[< \mathrm{o}_{31} + \mathrm{o}_{32}\hat{i} + \mathrm{o}_{33}\hat{j} + \mathrm{o}_{34}\hat{k} >, < \mathrm{o}_{35}\hat{i} + \mathrm{o}_{36}\hat{j} + \mathrm{p}_z\hat{k}\right] \end{aligned} \quad (32)$$

where, $o_{31} = \overline{C_2}o_{21} + c(\overline{C_2 S_2}) - w(\overline{S_2 S_1})$, $o_{32} = \overline{C_2}o_{22} + \overline{S_2}o_{23}$, $o_{33} = \overline{C_2}o_{23} - \overline{S_2}o_{22}$, $o_{34} = \overline{C_2}o_{24} - \overline{S_2}o_{21}$, $o_{35} = -ZS_2 + XC_1C_2 + YS_1C_2 - a_1C_2$, $o_{36} = YC_1 - XS_1$, $o_{37} = a_2 - ZC_2 + XC_1S_2 + YS_1S_2 - a_1S_2$.

$$O_4 = \left[< o_{41} + o_{42}\hat{i} + o_{43}\hat{j} + o_{44}\hat{k} >, < o_{45}\hat{i} + o_{27}\hat{j} + o_{47}\hat{k}\right] \quad (33)$$

where,

$$\begin{aligned} o_{41} &= \overline{C_{2+3}}o_{21} + \overline{S_{2+3}}o_{23} \\ o_{42} &= \overline{C_{2+3}}o_{22} - \overline{S_{2+3}}o_{24} \\ o_{43} &= \overline{C_{2+3}}o_{23} - \overline{S_{2+3}}o_{21} \\ o_{44} &= \overline{C_{2+3}}o_{24} - \overline{S_{2+3}}o_{22} \end{aligned}$$

$$o_{45} = -ZS_2 - a_2S_3 - XC_1S_2S_3 - YS_1S_2S_3 + a_1S_2S_3 - ZC_2S_3 + Z - ZC_3 + XC_1C_2C_3 + XS_1C_2C_3 - a_1C_2C_3$$

$$o_{46} = YC_1 - XS_1$$

$$o_{47} = -ZS_2S_3 + a_2C_3 + XC_1C_2S_3 + YS_1C_2S_3 - a_1C_2S_3 + ZC_2C_3 + XC_1S_2C_3 + YS_1S_2C_3 - a_2S_2C_3$$

Therefore, all the joint variables can be calculated by equating quaternion vector products and quaternion vector pairs i.e. Q_1, Q_2 and Q_3 to O_1, O_2 and O_3 respectively.

$$\Rightarrow \theta_1 = a\tan\left[\frac{Y - u_y}{X - u_x} \times \frac{(a_1 + d_4C_2 + d_4C_2 - d_4C_{2-3} + d_6S_5C_{2+3} - d_4S_{2+3} + d_6C_4C_5S_{2+3})}{(-a_2S_2 + d_4C_2 + d_4C_2 - d_4C_{2-3} + d_6S_5C_{2+3} - d_4S_{2+3} - d_6C_4C_5S_{2+3})}\right] \tag{34}$$

$$\theta_2 = a\tan 2\left[\left(\frac{Z + v_x}{d_4}\right), \ \mp\sqrt{1 - \left\{\left(\frac{Z + v_x}{d_4}\right)\right\}^2}\right] \tag{35}$$

where, $d_4S_2 + d_4S_2 - v_x = Z$, $d_4S_2 + d_4S_2 = Z + v_x$, $S_2 = \left(\frac{Z+v_x}{d_4}\right)$

$$\theta_3 = a\tan 2\left[\begin{array}{l} \mathrm{m}\sqrt{1 - \left\{\left(\dfrac{-XS_2 + XC_1C_2 + YS_1C_2 - a_1C_2 - v_y}{(-d_4 - d_6S_5)}\right)\right\}^2}, \\ \left(\dfrac{-ZS_2 + XC_1C_2 + YS_1C_2 - a_1C_2 - v_y}{(-d_4 - d_6S_5)}\right), \end{array}\right] \tag{36}$$

$$\theta_4 = a\tan 2\left[\begin{array}{l} \mathrm{m}\sqrt{1 - \left\{\left(\dfrac{-ZS_2S_3 + a_2C_3 + XC_1C_2S_3 + YS_1C_2S_3 - a_1C_2S_3 + ZC_2C_3 + XC_1S_2C_3 + YS_1S_2C_3 - a_2S_2C_3}{-d_6C_5}\right)\right\}^2} \\ , \left(\dfrac{-ZS_2S_3 + a_2C_3 + XC_1C_2S_3 + YS_1C_2S_3 - a_1C_2S_3 + ZC_2C_3 + XC_1S_2C_3 + YS_1S_2C_3 - a_2S_2C_3}{-d_6C_5}\right), \end{array}\right] \tag{37}$$

$$\theta_5 = a\tan 2\left[\begin{array}{l} \left(\dfrac{-XS_2 - a_2S_3 - XC_1S_2S_3 - YS_1S_2S_3 + a_1S_2S_3 - ZC_2S_3 + Z - ZC_3 + XC_1C_2C_3 + YS_1C_2C_3 - a_1C_2C_3 - d_4}{-d_6}\right), \\ \mathrm{m}\sqrt{1 - \left\{\left(\dfrac{-ZS_2 - a_2S_3 - XC_1S_2S_3 - YS_1S_2S_3 + a_1S_2S_3 - ZC_2S_3 + Z - ZC_3 + XC_1C_2C_3 + YS_1C_2C_3 - a_1C_2C_3 - d_4}{-d_6}\right)\right\}^2} \end{array}\right] \tag{38}$$

3. Inverse kinematic solution scheme

In this section optimization algorithms are selected for computation of inverse kinematics solution of 5R manipulator. However, there are various types of optimization algorithms existed and can produce the desired IK solution, the major necessity is to achieve global optimum solution with fast convergence rate. Therefore, selection of appropriate optimization algorithm is important for fitness evaluations and GA is so far best known tool, but on the other hand TLBO has also proven its efficiency and performance. Finally selection of optimization algorithms has been made on global searching point, computational cost and quality of the result.

3.1. Optimization approach to solve inverse kinematics

Any Optimization algorithms which are capable of solving various multimodal functions can be implemented to find out the inverse kinematic solutions. The fitness function is given by the Eq. (46) fitness function F(x). Each individual represents a joint variable solution of the inverse kinematic problem for adopted population based metaheuristic algorithm. All individuals moving in D-dimensional search space and sharing the information to find out best fitness value of the function. Each individual contains set of joint angles ($\theta_1, \theta_2, \theta_3, \theta_4$ and θ_5) of 5R manipulator. The optimum set of joint angle can be find by using appropriate optimization algorithm from given desired position of end effector (X, Y, Z). In case of inverse kinematics of 5R manipulator multiple solutions exist for the single position of the end effector so it is required to find out the best set of joint angle in order to minimize whole movement of manipulator.

For the optimization of joint angle rotation of robot manipulator, one can define objective function or fitness function from joint angle rotation difference and other can be defined from end effector position displacement. These are known as joint angle error and positional function method [3, 7, 10, 29].

3.1.1. *Position based function*

The current position of the manipulator is described by (39):

$$P_c = [X_c,\ \ Y_c,\ \ Z_c] \tag{39}$$

Desired position of end effector can be denoted by (40):

$$P_d = [X_d,\ \ Y_d,\ \ Z_d] \tag{40}$$

Current position of end effector will be compared with the desired position P_d. General equation for the fitness function is given in Eq. (41) that is based on the distance norm of homogeneous Euclidian distance between the current positions to the desired position of end effector P_d evaluated by number of iterations.

$$P_{\min} = \|P_d - P_c(i)\|^2 \tag{41}$$

Current position P_c can be evaluated from Eqs. (34) through (38). Now putting the value of P_c on Eq. (42)

$$P_{\min} = \sqrt{(X_d - X_c(i))^2 + (Y_d - Y_c(i))^2 + (Z_d - Z_c(i))^2} \tag{42}$$

3.1.2. *Joint angle error*

Corresponding joint error can be given by the difference between current set of joint variables to the final required angles.

$$\theta_c = (\theta_{c1}, \theta_{c2}, \theta_{c3}, \theta_{c4}\ \theta_{c5}) \tag{43}$$

$$\theta_d = (\theta_{d1}, \theta_{d2}, \theta_{d3}, \theta_{d4}\ \theta_{d5}) \tag{44}$$

Therefore using square norm the objective function can be given as

$$\theta_{\min} = \|\theta_d - \theta_c\|^2 \tag{45}$$

Subjected to joint limits

$$\theta_1 \in [\theta_{1\min}, \ \theta_{1\max}]$$
$$\theta_2 \in [\theta_{2\min}, \ \theta_{2\max}]$$
$$\theta_3 \in [\theta_{3\min}, \ \theta_{3\max}]$$
$$\theta_4 \in [\theta_{4\min}, \ \theta_{4\max}]$$
$$\theta_5 \in [\theta_{5\min}, \ \theta_{5\max}]$$

Now overall error minimization can be given by using Eqs. (42) and (45),

$$F(x)_{\min} = \lambda\left[\sqrt{(X_d - X_c(i))^2 + (Y_d - Y_c(i))^2 + (Z_d - Z_c(i))^2}\right] + \left[\|\theta_d - \theta_c\|^2\right] \quad (46)$$

where λ is proportional weight factor for the minimization of the problem and calculation of the entire joint angles base on constraint can be achieved using fitness function (46). The performance of considered algorithm is checked with the parameters: a_1 = 60 mm, a_2 = 145 mm, d_1 = 150 mm, d_2 = 125 mm, d_3 = 130 mm. Upper and lower limit of five joint angles are: $\theta_1 = [0, 180]$; $\theta_2 = [0, 150]$; $\theta_3 = [0, 150]$, $\theta_4 = [0, 85]$ and $\theta_5 = [15, 45]$.

4. Results and discussions

TLBO and GA has been used to compute the inverse kinematics of 5-R manipulator and comparison of obtained results has been made on the basis of quality and performance. **Table 1** gives the five random position of end effector and respective inverse kinematics solutions. Current work is performed in MATLAB R2013a. The data sets are obtained from Eq. (34) through (38). The data sets are generated using quaternion vector based inverse kinematics equations as given in **Table 2**. These generated data sets are used to compare the IK solution through adopted GA and TLBO. In **Table 3**, comparative evaluations of fitness function and obtained joint variables through TLBO and GA is presented.

Positions	Joint angles				
	θ_1	θ_2	θ_3	θ_4	θ_5
P1(−76.09, 54.36, −61.94)	84.559	77.518	101.74	30.616	38.697
P2(89.69, 192.55, 90.87)	84.791	97.25	130.44	50.771	36.428
P3(−4.24, 94.08, 97.55)	18.384	78.688	35.234	77.708	34.889
P4(29.10,154.02, −31.52)	104.43	115.47	124.11	7.3372	33.774
P5(−184.33, −43.21, 8.27)	39.177	107.13	97.052	65.672	15.374

Table 1. Five different positions and joint variables.

SN	Position of joints determined through quaternion algebra							
	θ_1	θ_2	θ_3	θ_4	θ_5	X	Y	Z
1	112.5641	47.3165	8.2447	65.8373	39.8977	−186.6903	183.0670	−14.7039
2	153.1316	21.9812	126.9031	57.2629	30.4168	−92.6981	32.1423	157.3316
3	66.1779	143.2985	14.6124	73.6231	41.5228	−131.5420	−22.3866	−32.2155
4	57.6085	119.6396	104.5818	71.1946	33.8225	−10.7684	111.7435	77.4862
5	31.4308	2.9242	71.5757	63.0358	39.8749	64.7966	172.0372	151.5714
6	124.3702	116.7337	102.4999	53.1482	22.4807	−111.8590	−59.6708	60.8590
7	89.1765	13.1827	101.6747	80.3340	29.4704	−76.9533	96.2813	121.3505
8	5.6698	30.9685	57.2308	29.3079	29.6421	174.3873	107.6283	143.1839
9	131.5857	108.7086	92.8278	5.6664	36.4826	−104.6410	109.7511	40.5523
10	32.8579	102.3539	138.8770	26.4141	33.7466	146.4984	48.7416	54.4041
11	134.1878	70.4224	26.3511	82.2471	44.0566	−188.7864	15.4108	−53.9823

Table 2. Desired joint variables determined through quaternion algebra.

This work does not use special tuning of various parameters of GA and TLBO algorithm. In future research the sensitivity analysis can be performed to achieve better results. From **Table 3**, TLBO generated solutions for the position 4 is better as compared to GA in account of fitness function evaluation. There are different distance based norms, one of them is Euclidean distance

Positions	TLBO joint angles					Function value
	θ_1	θ_2	θ_3	θ_4	θ_5	
P1(−76.09, 54.36, −61.94)	86.598	72.165	72.165	40.894	30.459	0
P2(89.69, 192.55, 90.87)	83.874	69.895	69.895	39.607	30.76	0
P3(−4.24, 94.08, 97.55)	84.512	70.427	70.427	39.909	30.686	0
P4(29.10,154.02, −31.52)	85.566	71.305	71.305	40.406	30.53	0
P5(−184.33, −43.21, 8.27)	87.818	73.181	73.181	41.469	30.364	0
Positions	**GA Joint angles**					**Function value**
	θ_1	θ_2	θ_3	θ_4	θ_5	
P1(−76.09, 54.36, −61.94)	60.619	49.504	58.384	62.281	27.903	0
P2(89.69, 192.55, 90.87)	88.293	34.091	14.439	15.241	51.738	0
P3(−4.24, 94.08, 97.55)	55.004	49.274	63.942	47.842	33.633	0
P4(29.10,154.02, −31.52)	72.594	22.689	68.297	85.886	27.044	0.0137
P5(−184.33, −43.21, 8.27)	25.669	70.588	31.341	66.807	52.884	0

Table 3. TLBO results for joint variable and function value.

norm and which is used here for minimum distance between the end effector positions. If the distance between two points reached to 0 or less than 0.001 than the evolutions of fitness function can be reached best or global minimum value. It is clear that the obtained fitness value is less than the defined distance norm so adopting these algorithms are fruitful and qualitative.

Figures 3–7 signify the best fitness function value and analogous joint variables for position 1. These figures show efficiency of adopted algorithms for IK solution of 5-R manipulator. The convergence of objective function evaluation lies to zero error for GA and TLBO algorithms while for position 4, GA yields 0.013 error. It means that GA is less performing as compared to TLBO. From **Figures 8–12**, the results obtained through GA shows in terms of convergence and histogram graph and the obtained joint angles are in radian which is later converted into degree and given in **Table 3**. The GA results are obtained through MATLAB toolbox and that shows the zero convergence in single run. **Figures 8–12**, it can be seen that the generated solutions for joint angles are multiple for single position and similarly there are multiple fitness function evaluations. The best fitness function achieved here using the termination criteria and the corresponding joint variables has taken for comparison.

The proposed work is performed in dual core system with 4 GB RAM computer. It has been observe that the convergence of the solution for GA is taking less computation time as compared to TLBO and quaternion algebra. Corresponding joint angles trajectory using 4th order cubic spline is presented in **Figure 13**. Using inverse kinematic solution joint variables are used

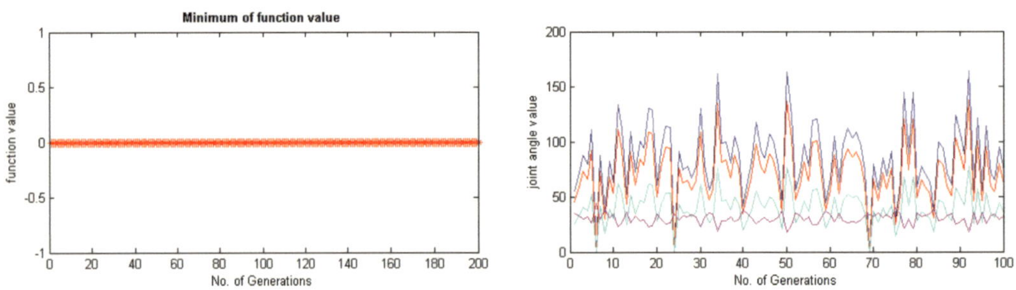

Figure 3. Joint variables and fitness function value for position P1.

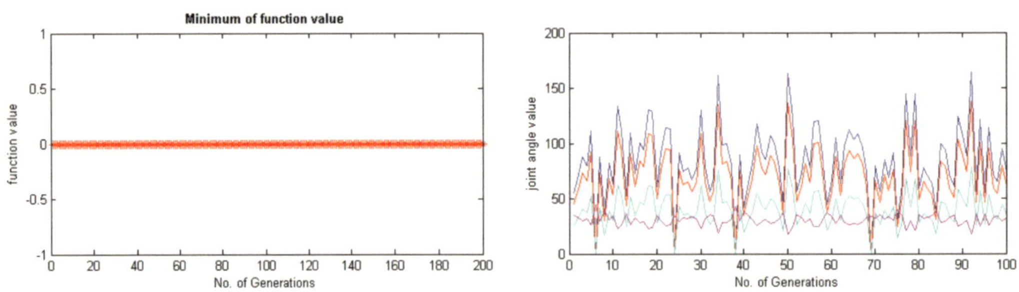

Figure 4. Joint variables and fitness function value for position P2.

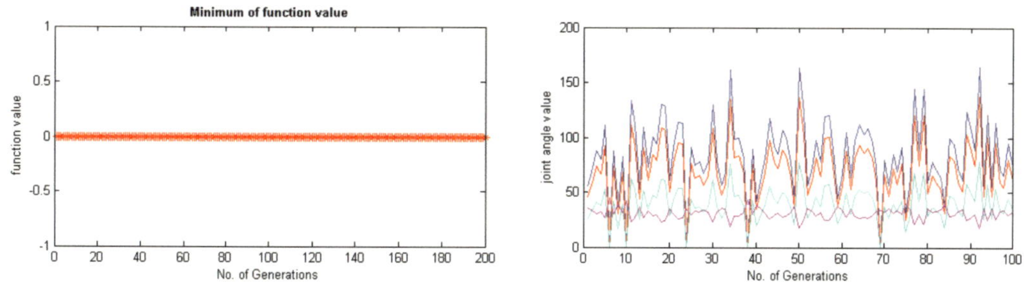

Figure 5. Joint variables and fitness function value for position P3.

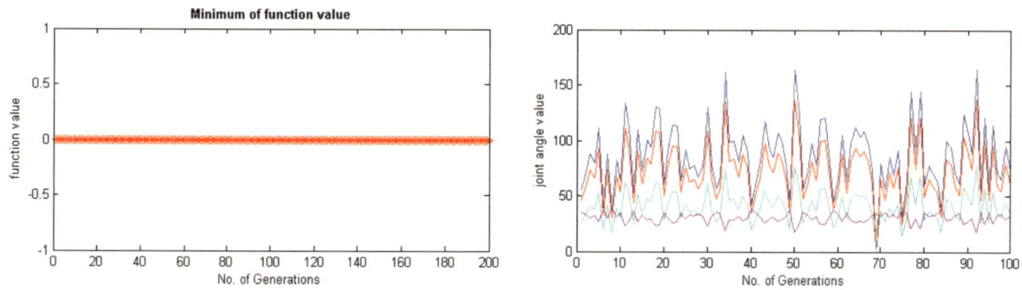

Figure 6. Joint variables and fitness function value for position P4.

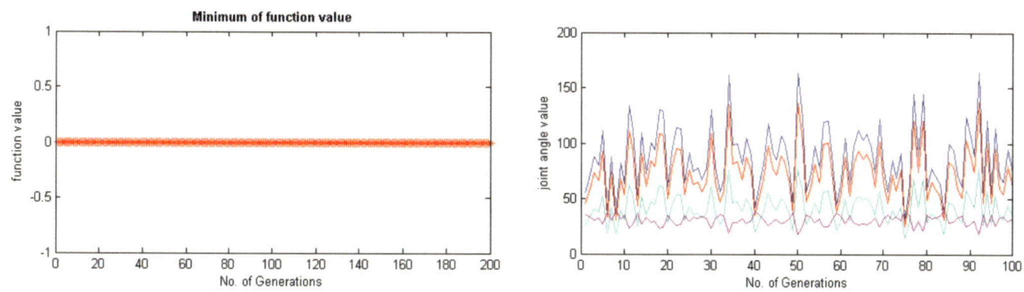

Figure 7. Joint variables and fitness function value for position P5.

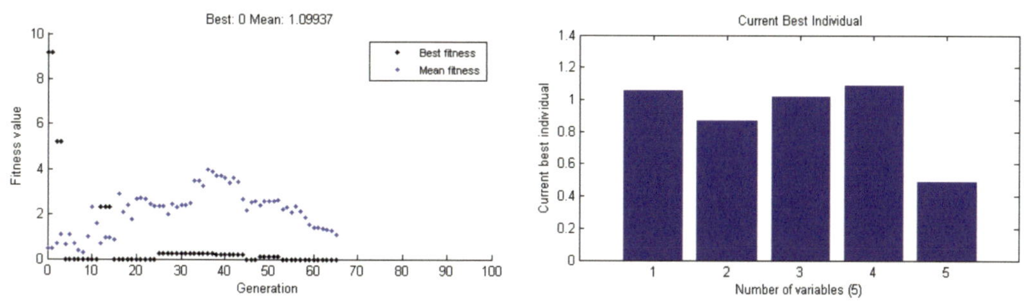

Figure 8. Joint variables and fitness function value for position P1.

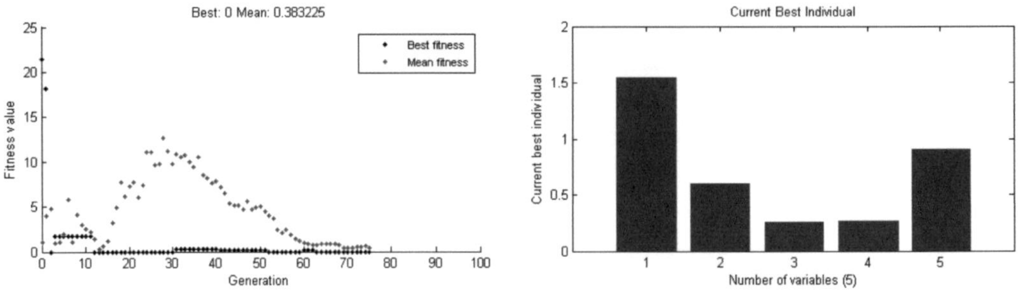

Figure 9. Joint variables and fitness function value for position P2.

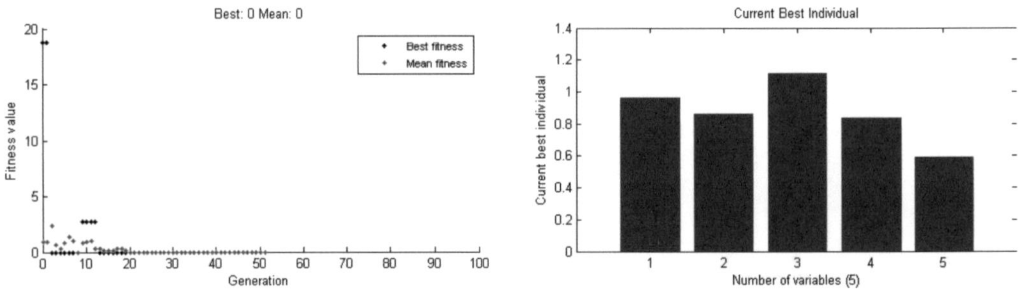

Figure 10. Joint variables and fitness function value for position P3.

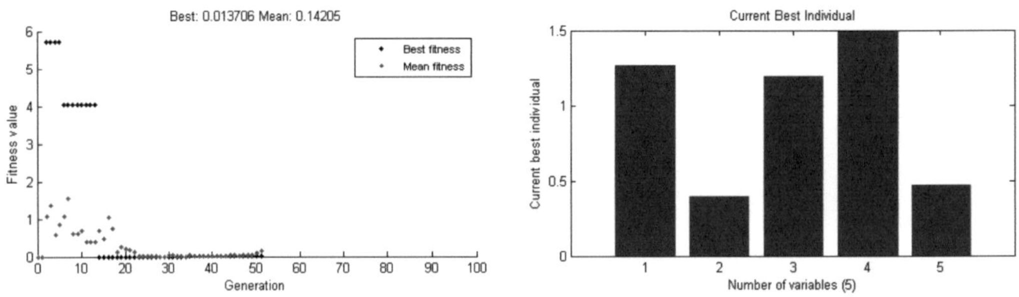

Figure 11. Joint variables and fitness function value for position P4.

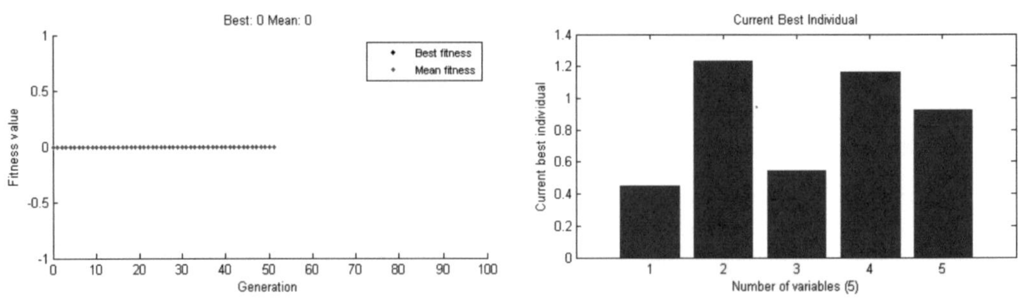

Figure 12. Joint variables and fitness function value for position P5.

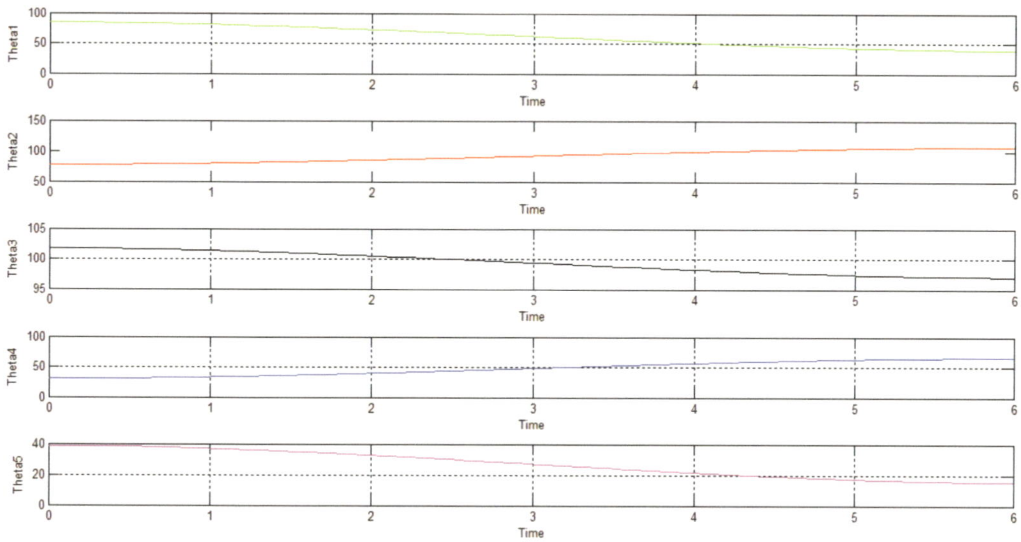

Figure 13. Trajectory of joint angle for quaternion.

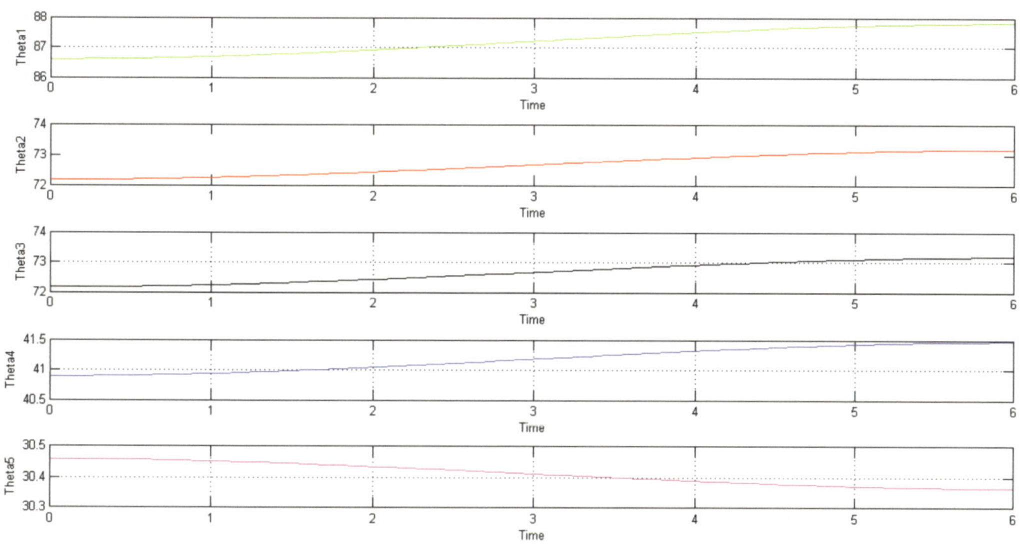

Figure 14. Trajectory of joint angle for TLBO.

to calculate the joint space trajectory for TLBO and GA as presented in **Figures 14** and **15**. Final time has been taken $t_{f=6}$ second for trajectory generation but to complete this trajectory overall computational time is 5.674 seconds. The computation time for TLBO is 15.671 seconds which is more than the GA i.e. 7.932 seconds. Therefore, on the basis of computational cost GA is performing better than TLBO while quaternion algebra taken least time (**Table 4**).

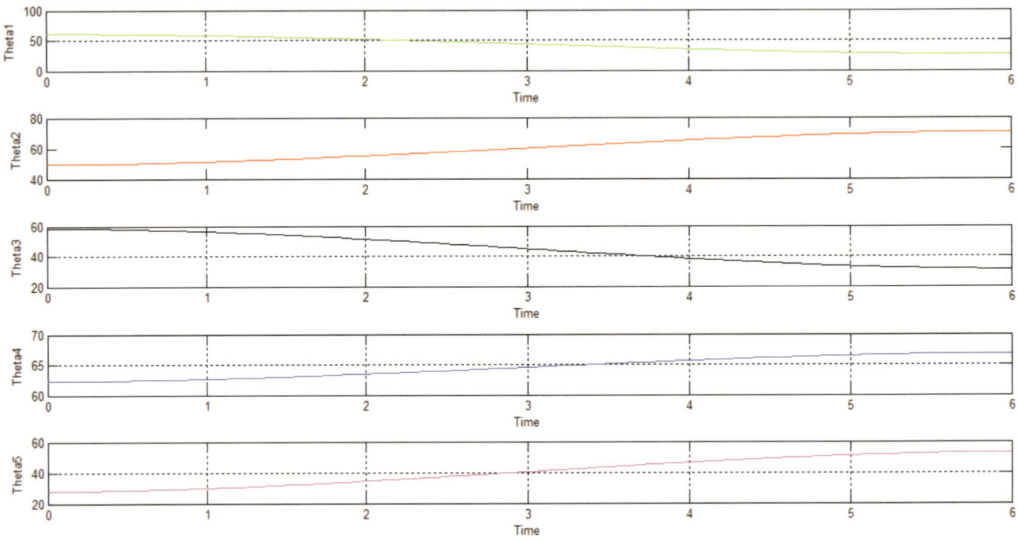

Figure 15. Trajectory of joint angle for GA.

SN	Method	Computational time
1	TLBO	15.671 s
2	GA	7.932 s
3	Quaternion	0.993 s

Table 4. Computational time for inverse kinematic evaluations.

5. Conclusions

In this paper, the work discourses the problem associated to the optimization of positional and angular error of end effector using TLBO and GA for 5R robot manipulator. Metaheuristic algorithms like PSO, GA, ABC, etc. have been used in various field of industrial robotics but the most critical issue is to solve inverse kinematic problem for any configuration of robot manipulators. Most of the optimization approach are being used for numerical solution but it has been observed that the numerical solutions does not yield solution when the manipulator is in ill-conditioned besides this it has also been observed that classical optimization methods converge in local minima. Therefore in this work global optimization method like TLBO and GA is adopted and after analyzing the results it can be concluded that adopted optimization algorithms convergence rate is higher and complexity does not increase with the manipulator configuration. Although many researchers are tried to obtain global solution but the computations cost are more in the problem henceforth overcoming the problem of computational cost with quaternion objective function.

The adopted algorithms are very much appropriate for constrained and unconstrained problems. To estimate the effectiveness of considered algorithms, comparison has been made with quaternion algebra. **Table 3** gives the comparative results of adopted algorithm and proposed quaternion solutions of 5-R manipulator. This work considered forward and inverse kinematic equations for preparing the objective function for TLBO and GA. These adopted algorithms has shown the potential of getting faster convergence and yielding global optimum solution for the stated problem. In future the tuning of various parameters of GA and TLBO can be considered so as to avoid trapping in local minimum point. Even the hybridization of these algorithms may be proposed and adopt for the IK problems.

Author details

Panchanand Jha[1]* and Bibhuti Bhusan Biswal[2]

*Address all correspondence to: jha_ip007@hotmail.com

1 Raghu Engineering College, Visakhapatnam, India

2 National Institute of Technology, Meghalaya, India

References

[1] Ahuactzin J, Gupta K. Completeness Results for a Point-to-Point Inverse Kinematics Algorithm. Detroit, MI: IEEE. Vol. 2. p. 1526-1531

[2] Albert F, Koh S, Tiong S, Chen C. Inverse Kinematic Solution in Handling 3R Manipulator via Real-Time Genetic Algorithm. IEEE. p. 1-6

[3] Ayyıldız M, Çetinkaya K. Comparison of four different heuristic optimization algorithms for the inverse kinematics solution of a real 4-DOF serial robot manipulator. Neural Computing and Applications. 2015. DOI: 10.1007/s00521-015-1898-8

[4] Chapelle F, Bidaud P. A Closed Form for Inverse Kinematics Approximation of General 6R Manipulators Using Genetic Programming. 2001;**4**:3364-3369

[5] Chen C, Her M, Hung Y, Karkoub M. Approximating a robot inverse kinematics solution using fuzzy logic tuned by genetic algorithms. The International Journal of Advanced Manufacturing Technology. 2002;**20**:375-380. DOI: 10.1007/s001700200166

[6] Dulęba I, Opałka M. A comparison of Jacobian-based methods of inverse kinematics for serial robot manipulators. International Journal of Applied Mathematics and Computer Science. 2013. DOI: 10.2478/amcs-2013-0028

[7] Eiben A, Smith J. Introduction to Evolutionary Computing. New York: Springer; 2003

[8] Funda J, Taylor R, Paul R. On homogeneous transforms, quaternions, and computational efficiency. IEEE Transactions on Robotics and Automation. 1990;**6**:382-388. DOI: 10.1109/70.56658

[9] Galicki M. Control-based solution to inverse kinematics for mobile manipulators using penalty functions. Journal of Intelligent and Robotic Systems. 2005;**42**:213-238. DOI: 10.1007/s10846-004-7196-9

[10] Gan D, Liao Q, Wei S. Dual quaternion-based inverse kinematics of the general spatial 7R mechanism. Proceedings of the Institution of Mechanical Engineers, Part C: Journal of Mechanical Engineering Science. 2008;**222**:1593-1598. DOI: 10.1243/09544062jmes1082

[11] Geradin M, Cardona A. Kinematics and dynamics of rigid and flexible mechanisms using finite elements and quaternion algebra. Computational Mechanics. 1988;**4**:115-135. DOI: 10.1007/bf00282414

[12] He G, Hongming G, Zhang G, Wu L. Using Adaptive Genetic Algorithm to the Placement of Serial Robot Manipulator. Islamabad: IEEE; 2006. p. 1-6

[13] Huang H, Xu S, Hsu H. Hybrid Taguchi DNA swarm intelligence for optimal inverse kinematics redundancy resolution of six-DOF humanoid robot arms. Mathematical Problems in Engineering. 2014;**2014**:1-9. DOI: 10.1155/2014/358269

[14] Husty M, Pfurner M, Schröcker H. A new and efficient algorithm for the inverse kinematics of a general serial 6R manipulator. Mechanism and Machine Theory. 2007;**42**:66-81. DOI: 10.1016/j.mechmachtheory.2006.02.001

[15] Kalra P, Mahapatra P, Aggarwal D. On the Solution of Multimodal Robot Inverse Kinematic Functions using Real-coded Genetic Algorithms. IEEE. 2003;**2**:1840-1845

[16] Kennedy J, Eberhart R, Shi Y. Swarm Intelligence. San Francisco: Morgan Kaufmann Publishers; 2001

[17] Khatami S, Sassani F. Isotropic Robotic Design Optimization of Manipulators Using a Genetic Algorithm Method. Vancouver: IEEE; 2002. p. 562-567

[18] Kim S, Kim J. Optimal Trajectory Planning of a Redundant Manipulator Using Evolutionary Programming. Nagoya: IEEE; 1996. p. 738-743

[19] Korein B. Techniques for generating the goal-directed motion of articulated structures. IEEE Computer Graphics and Applications. 1982;**2**:71-81. DOI: 10.1109/mcg.1982.1674498

[20] Kucuk S, Bingul Z. Inverse kinematics solutions for industrial robot manipulators with offset wrists. Applied Mathematical Modelling. 2014;**38**:1983-1999. DOI: 10.1016/j.apm.2013.10.014

[21] Liu S, Zhu S. An optimized real time algorithm for the inverse kinematics of general 6R robots. In: Control and Automation, 2007. ICCA 2007. IEEE International Conference on. IEEE, Guangzhou. 2007. pp. 2080–2084

[22] Nearchou A. Solving the inverse kinematics problem of redundant robots operating in complex environments via a modified genetic algorithm. Mechanism and Machine Theory. 1998;**33**:273-292. DOI: 10.1016/s0094-114x(97)00034-7

[23] Parker J, Khoogar A, Goldberg D 1989, Inverse kinematics of redundant robots using genetic algorithms. Proceedings of IEEE International Conference on Robotics and Automation

[24] Pham D, Castellani M, Fahmy A. Learning the Inverse Kinematics of a Robot Manipulator using the Bees Algorithm. Daejeon: IEEE; 2008. p. 493-498

[25] Piazzi A, Visioli A. A Global Optimization Approach to Trajectory Planning for Industrial Robots. Grenoble: IEEE. 1997;**3**:1553-1559

[26] Rajpar A, Zhang W, Jia D. Object Manipulation of Humanoid Robot Based on Combined Optimization Approach. Harbin: IEEE; 2007. p. 1148-1153

[27] Rao R, Savsani V, Vakharia D. Teaching–learning-based optimization: A novel method for constrained mechanical design optimization problems. Computer-Aided Design. 2011;**43**:303-315. DOI: 10.1016/j.cad.2010.12.015

[28] Rocke D, Michalewicz Z. Genetic algorithms + data structures = evolution programs. Journal of the American Statistical Association. 2000;**95**:347. DOI: 10.2307/2669583

[29] Sun L, Lee R, Lu W, Luk L. Modelling and simulation of the intervertebral movements of the lumbar spine using an inverse kinematic algorithm. Medical & Biological Engineering & Computing. 2004;**42**:740-746. DOI: 10.1007/bf02345206

[30] Tabandeh S, Clark C, Melek W 2006, A Genetic Algorithm Approach to solve for Multiple Solutions of Inverse Kinematics using Adaptive Niching and Clustering. In: IEEE, pp. 1815–1822

[31] Vrongistinos K, Wang Y, Hwang Y. Quaternion smoothing on three-dimensional kinematics data. Medicine & Science in Sports & Exercise. 2001;**33**:S84. DOI: 10.1097/00005768-200105001-00480

[32] Wang L, Chen C. A combined optimization method for solving the inverse kinematics problems of mechanical manipulators. IEEE Transactions on Robotics and Automation. 1991;**7**:489-499. DOI: 10.1109/70.86079

[33] Wolpert D, Macready W. No free lunch theorems for optimization. IEEE Transactions on Evolutionary Computation. 1997;**1**:67-82. DOI: 10.1109/4235.585893

[34] Xu D, Acosta Calderon C, Gan J. An analysis of the inverse kinematics for a 5-DOF manipulator. International Journal of Automation and Computing. 2005;**2**:114-124. DOI: 10.1007/s11633-005-0114-1

[35] Zoric N, Lazarevic M, Simonovic A. Multi-body kinematics and dynamics in terms of quaternions: Langrange formulation in covariant form – Rodriguez approach. FME Transactions. 2010;**38**:19-28

[36] Zu D. Efficient inverse kinematic solution for redundant manipulators. JME. 2005;**41**:71. DOI: 10.3901/jme.2005.06.071

Chapter 12

A Random Multi-Trajectory Generation Method for Online Emergency Threat Management (Analysis and Application in Path Planning Algorithm)

Liang Yang, Yuqing He, Jizhong Xiao, Bing Li and Zhaoming Liu

Additional information is available at the end of the chapter

http://dx.doi.org/10.5772/intechopen.71410

Abstract

This paper presents a novel randomized path planning algorithm, which is a goal and homology biased sampling based algorithm called Multiple Guiding Attraction based Random Tree, and robots can use it to tackle pop-up and moving threats under kinodynamic constraints. Our proposed method considers the kinematics and dynamics constraints, using obstacle information to perform informed sampling and redistribution around collision region toward valid routing. We pioneeringly propose a multiple path planning method using 'Extending Forbidden' algorithm, rather than using variant cost principles for online threat management. The threat management method performs online path switching between the planned multiple paths, which is proved with better time performance than conventional approaches. The proposed method has advantage in exploration in obstacle crowded environment, where narrow corridor fails using the general sampling based exploration methods. We perform detailed comparative experiments with peer approaches in cluttered environment, and point out the advantages in time and mission performance.

Keywords: multiple path planning, online emergency threat management, path switching, goal biased probability, sampling based algorithm

1. Introduction

Robot path planning have been witnessed a great achievement these years with the various application of robots [1, 2]. Problems such as path planning, motion planning, and online moving obstacle management have been widely studied toward the goal of performing autonomy. Unmanned Aerial Vehicles (UAVs), an easy access robot platform, has been increasingly applied in research and commercial areas in recent years. UAV autonomy denotes the ability of tackling with obstacle (or called no-fly zone) avoidance and trajectory planning online from a starting position to a destination while satisfying the kinematic constraints [3].

For robot path planning, emergency threat management (ETM) is one of the hardest challenges that needs to be solved, where a sudden threat may burst into view or dynamic obstacles are detected on line, especially when UAV is following the desired path. Under such conditions, UAV should consider the following attributes:

1. Time efficiency: The most important requirement for ETM algorithm is time efficiency. For general ETM, the configuration is periodically updated, such as heuristic algorithm A* [4], which it is computationally intensive if the map is represented with high resolution. In order to guarantee safety, ETM requires real-time performance.

2. Kinematic feasibility: Kinematic feasibility denotes that the output of the planner meets the kinematic constraints of the robot as well as the environment. The constraints include: (a) Path smoothness: The planner is required to output kinematic smooth path, sometimes even kinodynamically feasible as well. Thus, the path should meet the state of art tracking constraints, and enables low tracking error for UAV; (b) Minimum cost of switching: The strategy of handling the threat, especially ET, is to find the cost minimum path by generating a new path or multiple paths besides the initial one. The cost for choosing the best path should take the dynamic constraints, energy consumption and time performance into consideration.

3. Specific requirements: UAVs have already been applied to many areas, such as inspection, photography, and monitoring. They have to meet some specific requirements according to environments and system constraints. For example, best pose based illumination of tunnel inspection for crack and spalling [5], and stable tracking with obstacle avoidance as UAV photography [6] which should be able to keep stable capturing even during the flying.

Development with open robot platform [7] and field implementation [8] has witnessed the promising performance of Sampling Based (SB) methods. SB algorithms (SBA) have the advantages for planning in high dimensional space, and it is with the ability to deal with multiple classes of path or motion planning problem in both static and dynamic environment [9]. Rapidly-exploring random trees (RRTs) are single query methods which obtain Voronoi biased property and only generate homotopy paths simultaneously [12]. Although it proposes to solve the multiple degrees of freedom (DOF) operating problems in known static environments [10, 11], SBA shows great performance of dealing with any kind of path or motion planning problem in complex environments for unmanned ground robots or aerial robots.

In this paper, we introduce two biased-sampling methods, which are obstacle biased and Homologous Classes (HC) biased to perform path planning respectively. For obstacle biased path method, we have discussed in [13] with UAV demonstration. For HC classed biased approach, it aims at solving the ET problem by generating alternative paths for online dynamically switching. HC introduces an online dynamic reconfiguration approach to ensure singularity between paths, which tries to generate more paths with different obstacle reference. Thus, it can perform alternative switching online when confronted with ET. The obstacle biased planning method is called Guiding Attraction based Random Tree (GART) and HC biased is called Multi-GART (MGART). We consider the environment to be known as a priori to us, and the UAVs are with the ability to understand the clearance region. Experiments and comparative simulations are illustrated to provide the effective evaluation of the proposed methods.

2. Preliminary materials

2.1. Homology and homotopy

For path planner, the purpose is to find a feasible path p_f (cost minimum or complete) from the initial position to the goal position in the workspace $W \in R^{\{n\}}$, n denotes the dimension of space the robots locate. A general cost function can be represented as:

$$J = \int c_{energy} + c_{time} + c_{threat} dt \tag{1}$$

The c_{energy}, c_{time}, c_{threat} denote energy cost, time consumption, and threat respectively. These costs are not fixed, since the energy cost can be path length, and time consumption can change according to the velocity limitation. For cost constrained path planner, the goal is to find the asymptotic optimal rather than the completeness solution. Then, more than one path can be found during the process, and the paths can be homotopic or belong to different homotopic classes (or called homology).

It is illustrated in **Figure 1**. Given a continuous map $H: I \times I \rightarrow \Gamma$ or $H: h(s,t) = h_t(s)$, Γ denotes the topological space and $I = [0, 1]$ is the unit interval. The obstacle regions are labeled with R_{o1}, $x_{initial} = h(0, t)$ denotes start point, $x_{inter} = h(1, t)$ denotes the goal position, x_{inter} denotes an inter node for obstacle avoidance. For the continuous deformation, given $h(s, 0) = \pi_0$, $h(s, 1) = \pi_1$, the path can be continuously mapping through π_0 to π_1 with $t \in [0, 1]$. For any path deformed between, they are homotopic with π_0, π_1 if and only if they stay in the closed loop $\pi_0 \coprod -\pi_1$, where the closed loop cannot collide with any obstacle region.

Definition 1—Homotopic Paths: It denotes the equivalence class of all paths under continuous mapping $H: h(s,t) = h_t(s)$, which locates in the closed loop formed $h(s, 0) \coprod -h(s, 1)$. Any

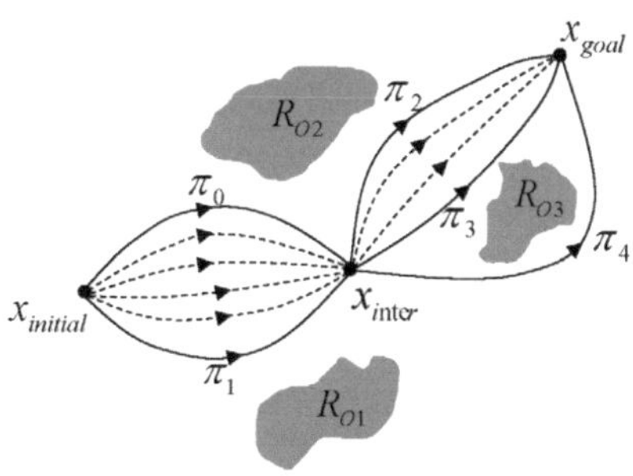

Figure 1. Homotopic and homologous classes and paths.

path in the set can be continuously deformed into any other without colliding with any obstacle in the space. For all paths in the set, they are with the same fixed end points.

We can conclude that π_2 and π_3 belong to the same homotopic class. However, we can find path π_4, which shares the same start and ending node, cannot be continuously deformed to π_3 due to the isolation of the obstacle. It means $(\pi_3 \cup -\pi_4) \cap R_{o3} \neq \Theta$. In such case, we call π_3 and π_4 are homologous, and they belong to different homotopic classes.

Definition 2—Homologous Paths: Paths, which follows the same continuous mapping $H : h(s, t) = h_t(s)$, cannot form a closed loop by $h(s, 0) \coprod -h(s, 1)$. The homologous paths belong to different homotopic classes.

2.2. Problem statement

Path planning follows a common procedure to perform trial and error process under empirical constraint to achieve completeness. The problem of path planning does not only solve a problem for exploration optimization, but also try to model the environment with a best descriptor as discussed in [13]. Let us take a look again with the problem of path planning which can be represented as:

$$H = \{h(s) | x_{\text{initial}} = h(0), x_{\text{initial}} = h(1), s \in [0, 1]\} \tag{2}$$

The path h(s) (homologous) should stay in obstacle free region R_{free}, that is, $h(s) \in R_{free}$. Usually, the path is piecewise continuously, and it can also be smoothed to obtain first order continuous thus to ensure kinematics continuous [14]. Besides the exploration to achieve completeness (in Eq. (1)), the obstacle modeling method is also important and affect the planning results.

To solve this problem, this paper proposed a multi-path online switching approach, that is, the path planner can find alternative homologous-paths. Then, this paper designs an online fast switching strategy. For multiple path planner, it aims at finding as many paths as possible,

$$H_{\text{alter}} = \cup_{i=1} h^i(x(t), u(t)) \tag{3}$$

H_{alter} denotes the set of all the alternative paths $h^i(x(t), u(t))$, x(t) denotes the state, and u(t) denotes the control. However, the mission planner cannot use all the planed paths for online switching, it should find the reasonable paths without redundancy. We propose the follow rule,

$$H_{\text{reason}} = \{h^i | h_i \underset{\neq}{H} h_j, \nabla i \neq j\} \tag{4}$$

H_{reason} denotes the paths set where any two paths are not homotopic to each other, $\underset{\neq}{H}$ denotes non-homotopy. Now, we have the paths which keep distinguishable from each other with different obstacles sequence surrounding.

http://dx.doi.org/10.5772/intechopen.71410

3. Rapidly exploring random tree path planner

In this section, we try to describe the underlying research of rapidly-exploring random tree (RRT [12], upon which we propose a novel state of art approach to facilitate the active exploration in cluttered environments). SBAs are incremental search algorithms which perform random sampling with collision checking for extension, and they were first proposed to solve high dimension planning problem. They have the merits of considering the control and kinematics constraints together, and can incrementally construct a tree from the start node (or initial state) to the goal node (or goal state) with continuously sampling and expansion.

It is shown **Figure 2**, the whole tree graph by exploration is represented as G_T, the black solid dot denotes the valid state within step accessibility under kinematics constraints, and the black solid lines connect each parent state with child state for extension. Every step, a new sample g_{sample} will be generated randomly. It should be cleared to all that the initial random sampling does not mean a fixed connection, that is, the random sampling can be a direction for extending. Then, the random sample g_{sample} tries to find the nearest state in the tree for connection under minimum Euclidean metric,

$$g_{parent} = \sup \min_{g_i \in G_T} E\left(g_i, g_{sample}\right) \tag{5}$$

Where g_i is an element of all valid states set G_T.

3.1. RRT connection with kinematic constraints

For RRT planner, given a system with state $(\dot{x_x}, \dot{x_y}, \dot{\theta})$, and a general form of system model:

$$\begin{cases} \dot{X_x} = f^x(x_x, u_x) \\ \dot{x_y} = f^y(x_x, u_y) \\ \dot{\theta} = f^\theta(\theta, u_\theta) \end{cases} \tag{6}$$

It can extend with simply random sampling with control inputs $[u_x, u_y, u_\theta]$. The random sample has to follow the kinematics constraints. Given the robot system, the differential

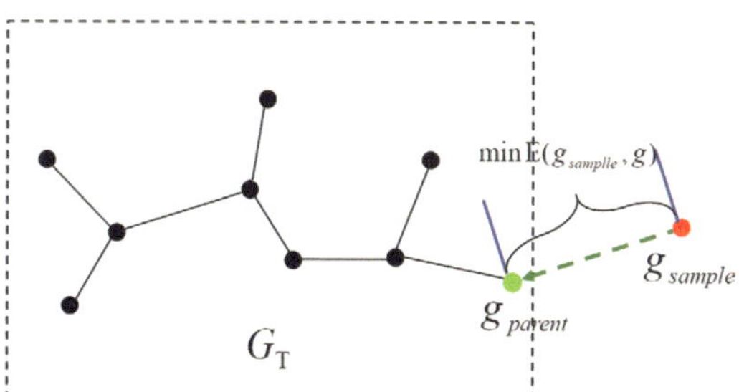

Figure 2. RRTs propagate by applying the minimal cost criterion to connect the newly sampled guard to the previous tree.

constraints can be represented as a set of implicit equations as $g(x, \dot{x}) = 0$, and it can be further represented as:

$$\dot{x} = f(x, u) \tag{7}$$

Here, x denotes the state, and $u \in U$ denotes the valid control in allowable controls set. Given the parent state $g_{parent}(t)$, the time step follows a Δt limits. Then, the control inputs vary with $u = \{u(t') \mid t \leq t' \leq t + \Delta t\}$. To compute $x(t+\Delta t)$, we can follow a typical procedure as [12]. It should be noted that the planner should extend toward the newly sampled g_{sample}. The planner first computes the possible region of reachability from current state x(t):

$$x(t + \Delta t) \in [x(t) + f(x(t), u(t) - \Delta t{\cdot}\epsilon), x(t) + f(x(t), u(t) + \Delta t{\cdot}\epsilon)] \tag{8}$$

where ϵ is the maximum first order factor of control input. RRT now picks a new state along the direction from parent to new sample, that is, $g_{new} \in [x(t) + f(x(t), u(t) - \Delta t \cdot \epsilon), x(t) + f(x(t), u(t) + \Delta t \cdot \epsilon)]$ and $g_{new} = g_{parent} + \delta(g_{sample} - g_{parent})$ with $\delta \in [0, 1]$.

3.2. Voronoi biased incremental search

Before discussing the Voronoi biased property of the SBAs, let first introduce some basic notation. Given a set of points $S = \{s_i \mid i = 1, 2, \ldots, n\}$ in a n-dimension space X. For any two distinct points s_p and s_q in set S, the dominant region (Voronoi region) of s_p over s_q is defined as the region where any inside point should be closer to s_p over s_q, that is,

$$R_{-s_p} = \left\{\chi \in \left| |s_p - \chi|^L < |s_q - \chi|^L \right.\right\} \tag{9}$$

Where χ is the dominant region corresponding to s_p, $||^L$ denotes the Lebesgue measurement. In a normal case, any point s_i has its own dominant region with,

$$R_{s_i} = \left\{\chi \in \left| |s_i - \chi|^L < |s_j - \chi|^L, \text{ for all } i \neq j \right.\right\} \tag{10}$$

Normally, random sampling of RRT follows a Monte-Carlo Method [15] to perform an uniformly sampling in a n-dimensional space under Lebesgue measurement. We can look back at the beginning of Section 3, the new sampled node tries to connect to the nearest node under Euclidean metric. We can now analyze the problem in another perspective that given g_{parent} and g_s, they connect to the same origin g_o. Then, a new sample g_{sample} is generated randomly following a Monte-Carlo process. In order to explore and reach the goal, g_{sample} tries to connect to the tree following the metric defined in Eq. (5). It means that g_{parent} and g_s can be connected for expansion under minimum distance principle, then g_{sample} has to be assigned to the dominant region which subjects to a closer point (the Voronoi region). Under this principle, g_{parent} and g_s can acquire new resource for extension with the ability to keep distinct region and extending their branches.

A typical Voronoi-biased exploration using sampling can be seen in **Figure 3**, where each branch keeps distinct with each other to form a star network like structure and it behaves the

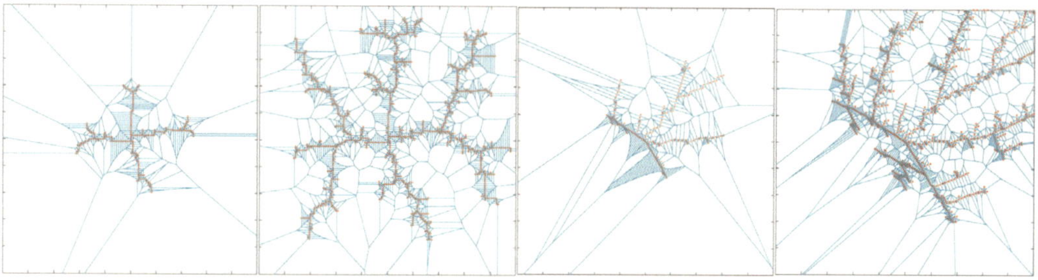

Figure 3. Results of incremental exploration of RRT and hRRT [16] after 200 and 1000 iterations, respectively.

same for heuristic informed RRT [16]. Here, unlike the dominant region of a point, RRT branch can be also treated as a distinct individual with its own Voronoi region for acquiring the extending resource.

4. Obstacle and homology biased path planner

In this Section, we propose approaches to solve two main problems, which are handling cluttered environment and online ET processing, using obstacle-biased method and homology-biased method. Collision detection during the incremental exploration is time consuming, and it follows a routine procedure to guarantee safety. It should be noted that the step validation of each new sampling state provides the directional information of obstacle distribution.

4.1. Obstacle biased path planner under kinematic constraints

SBAs mostly deploy the general idea of generating random samples for incremental exploration, and the sample locating in obstacle region will be discarded since it is time consuming and no benefits for increasing the performance of exploring. We firstly deployed a simple idea which was proved to have much higher time performance then RRT and RRT* in [17].

This paper introduces an obstacle biased algorithm, using obstacle information to help generating more samples for connection. It is shown in **Figure 4**, the newly sampled states x_1^s, x_1^s tries to connect to the nearest state in the tree. However, x_1^s leads toward the obstacle region, x_2^s locates in obstacle region. To use the obstacle information, this paper proposes an active exploring method, that is, inner propulsion and outer attraction.

For outer attraction, new sample x_1^s performs a collision checking, and find the nearest nodes ${}^{o}x_a$, ${}^{o}x_b$. We define that the further the obstacle to the sample, the more attraction it can support, that is, the attraction is proportional to the distance between obstacle and the sample using L2-norm L_2. The sample then re-allocation by add a obstacle biased attraction as:

$$\mathrm{x}_1^{s*} = \mathrm{x}_1^s + \mathrm{k}\left[\left({}^{o}x_a - \mathrm{x}_1^s\right) + \left({}^{o}x_b - \mathrm{x}_1^s\right)\right] \tag{11}$$

Where k is a constant to adjust the shifting percentage of the attraction vector.

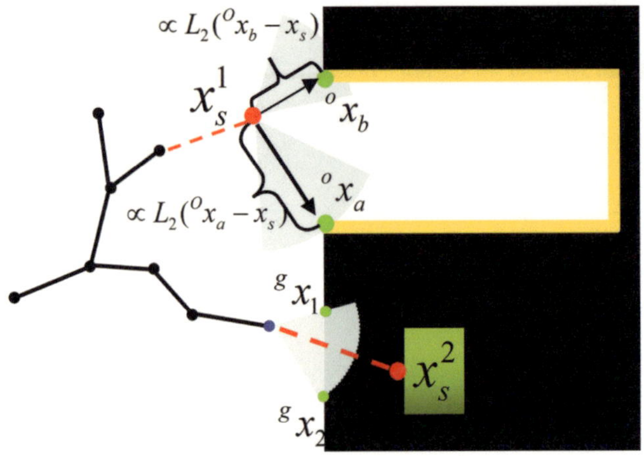

Figure 4. Obstacle biased SBA uses the obstacle location as input, with inner propulsion and outer attraction, to generate more samples for exploration. x_1^s, x_2^s are new samples, the black region denotes obstacle region.

The inner sample in collision with the obstacle is regarded to provide guiding information for the algorithm. This paper tries to find two more states ${}^g x_1, {}^g x_2$ within kinematic reachable region (discussed with Eq. (8)), it tries to find out the first two safe state with two directions which are out of obstacle region in the kinematic reachable region (the light blue fan-shaped region). Then, the two newly generated samples ${}^g x_1, {}^g x_2$ follows principle Eq. (11) to redistribute to the final position, and connect to the tree.

By using the two proposed approaches, we can generate more useful samples for extending, especially, the samples generate around the edge of the obstacles with the ability to perform more active exploration in cluttered environments. Besides, the outer attraction redistributes the samples toward the narrow corridor between the obstacle, which thus increases the probability of finding safe path through such obstacle crowed region.

4.2. Homology biased

We assume any path $h_t(s)$ generated using SBA is consisted by a set of nodes $h_t(s) = \{h_t | h_t(s'), s' \in [0, 1]\}$, as it is illustrated in **Figure 5(a)** that exploring tree is consist of the red nodes. Each red node is regarded as distinct with other nodes in the tree, with a distinct dominant region, i.e. Voronoi region. Thus, a path $h_t(s)$, which is consisted a set of states from the initial state to the goal, can be isolated with each other with a distinct region $V(b_T)$ combined by all Voronoi regions of the states.

The region dominant property differs the path with each other, where a SB tree with multiple paths (the path here may not connect to the goal, but they keep distinct with each other from the initial position) can be described by a set of branches $B_T = \{b_T(1), b_T(2), \ldots, b_T(n)\}$. For each branch, it consists of a list of states which connect with each other to form tree structure. In the tree, the end state relies on the parent state for extending as well as trajectory tracking.

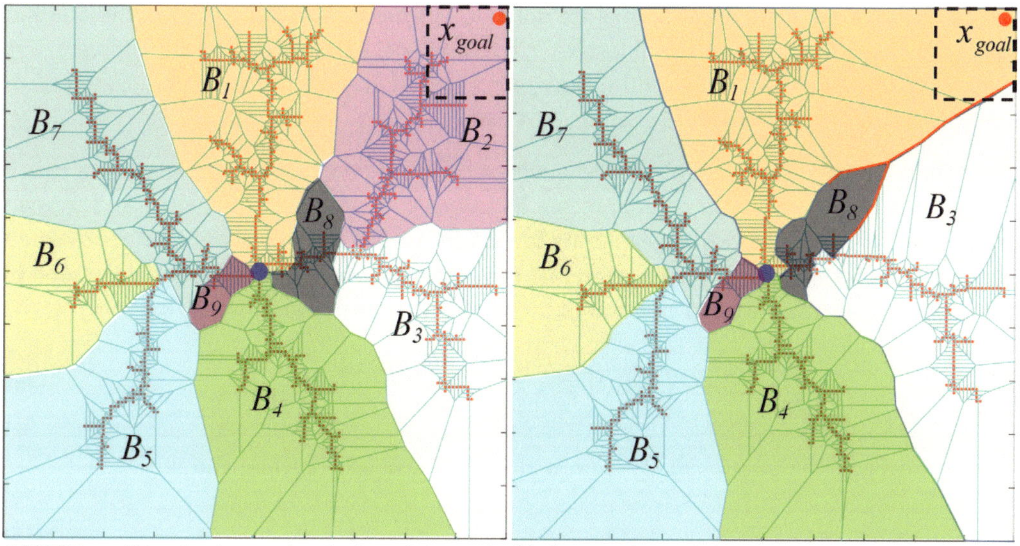

(a) RRT Voronoi biased regions, each region keep distinct with each other

(b) The Extending-Forbidden method removes branch B_2 to enable exploration of B_1, B_8, B_3 toward the goal

Figure 5. The extending-forbidden algorithm (EFA) tries to find all the states along the goal reached branch at each goal reaching checking step, such as branch B_2. Then, EFA sets the flag of the states to be inactive, switching the extending probability to the nearby branches.

4.2.1. Extending-forbidden algorithm

The path planner performs exploration following the Monte-Carlo approach. Given a configuration space C in a topological space (the rectangle region as illustrated in **Figure 5**), we denote the Lebesgue Measurement of the configuration space as $L^* \mid C|$. Then we can get the Lebesgue measurement of each branch $b_T(i)$ of the tree using the same metric. Authors in [18] proved that the dispersion of n random nodes sampled uniformly and independently in $V(b_T(i))$ is,

$$D = \sup_{b_T(i)} \left| \frac{\psi(b_t(i):n)}{n} - \frac{L^*|V(b_T(i))|}{L*|C|} \right| = o\left(\left(\frac{\log(n)}{n}\right)^{\frac{1}{d}}\right) \tag{12}$$

Where $\psi(b_t(i))$ denotes the number of samples m, $1 \le m \le n$, that lies in the sub-branch $b_T(i)$, n is the number of all the sampling, d is the dimension of the configuration space. D denotes the difference between the ration of sampling probability and ration of space Lebesgue measurement, which follows the knowledge that Monte-Carlo method performs a uniform sampling. It means the sampling probability approaches the ratio of Lebesgue measurements, that is, the exploration probability can be represented as:

$$P_{b_T(i)} = \frac{L^*|V(b_T(i))|}{L^*|C|} \tag{13}$$

However, the probability of exploring in the configuration space does not benefit the extending bias toward the goal. Let us still take a look at **Figure 5(a)**, the branch B_2 dominant the near-goal region, and other region are not able to extend toward the goal as the samples will not connect to the branches if it locates in the near-goal region. To solve this problem, this paper proposes an Extending-Forbidden Algorithm (EFA), it shifts the source for extending to other branches by forbidding the goal reached path.

Definition 3—Goal-biased Probability: Given a configuration space C, the exploring tree T and all its branches which are main branches B_T and its corresponding sub-branches. The goal-biased event denotes a branch can exploring toward the goal. If a goal region can be represented as G_r, and Voronoi($G_r 2 B_T$(i)) is the region that belongs to goal region and the Voronoi region of branch $B_T(i)$. Then, the nominal goal biased probability of branch $B_T(i)$ toward G_r is:

$$P_G^*(B_T(i)) = \frac{L^*|voronoi(G_r 2B_T(i))|}{L^*|C|} \tag{14}$$

And the real goal biased probability is normalized value of all branches, that is,

$$P_G(B_T(i)) = \frac{P_G^*(B_T(i))}{\sum_{j=1}^{n} P_G^*(B_T(j))} \tag{15}$$

Definition 4—Long Sub-branch (LSB) and Short Sub-branch (SSB): Given a tree T and all its Voronoi distinct main branches B_T. Then, we can define a length threshold δ_B. For all end vertices in each main branch, we calculate the length l_{sb} from end to the goal reach reached path (any state which firstly reached). If the length $l_{sb} > \delta_B$, then we call it Long Sub-branch (LSB). If $l_{sb} \leq \delta_B$, we call the sub-branch as Short Sub-branch (SSB).

It should be noted that the threshold is very empirical in Definition 4, and it is decided based on the configuration space and the kinematic constraints. In **Figure 6**, we set it as 15 meters, then we have SSB_1, SSB_2, SSB_3 as SSB, and LSB_1 as LSB. The reason why we have this definition is that we cannot shift all the extending resource to the neighbor branches, and we have the hypothesis that the SSB must has lower probability of finding a new path even given the resource for extending. For example, the main Voronoi dominant branches B_T keep distinct with each other, and each main branch has the probability $P_{b_T(i)}$ for exploration in the

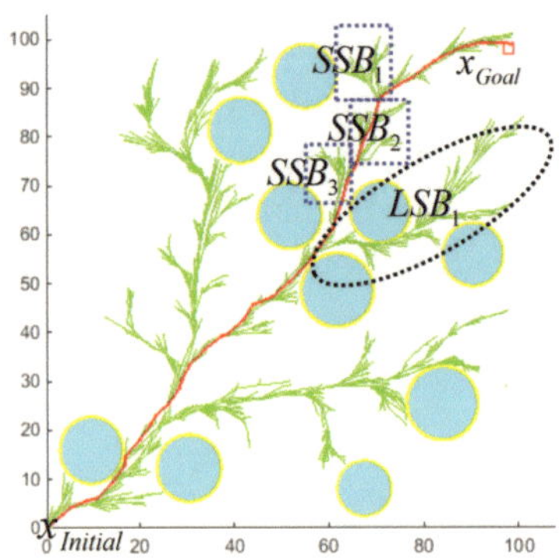

Figure 6. An intuitive example of SSB and LSB in an exploring tree, which are generated using threshold principle as defined in Definition 4.

configuration space. After the tree stops extending at a certain iteration, we can have the results as illustrated in **Figure 5(a)**. Since goal region is in Voronoi region of branch B_2, then we know that we have the goal biased probability as $P_G(B(2)) = 1$. One branch B_2 reached the goal region which is represented with dotted back rectangle, EFA searches the SSBs and LSBs based on Definition 4, and it labels states and executing forbidden. Then, we have a resource shifted Voronoi graph as illustrated in **Figure 5(b)**, where we can see that branch B_1 and B_3 obtains the Voronoi region which belongs to branch B_2. The two branches also obtain the rectangle goal region, that is, their goal biased probabilities are bigger than zero, $P_G(B(1)) > 0, P_G(B(3)) > 0$.

Since EFA can shift the goal biased by extending resource to other branches, while not all paths can obtain such resource. The following truths are hold:

1. The increasing of goal biased probability ensures the generation of a feasible path toward the goal, but not all branches with goal biased probability can reach the goal at the same time. Only one can reach the goal because of Voronoi dominant probability, thus the general SBA cannot find multiple paths.
2. The efficiency of generating multiple paths mainly depends on the environment adaptability of random exploring algorithms. For random exploring algorithm, their merits of generating multiple branches enable the generation of multiple paths.

4.2.2. Reasonable alternative reference chosen

The proposed MGART is able to perform extending-forbidden toward multiple paths, as the random exploring property guarantees completeness and diversity. However, the quality of explored paths cannot be guaranteed, particularly a large number of homotopic paths are generated. This paper proposes an approach to generate the reasonable path, and we analysis under the hypothesis that the environment is highly cluttered and it is not practical to set threshold for path planner to choose the best homological paths.

Definition 5—Reasonable Alternative Paths: Consider two homotopic paths $h(\pi_1)$ and $h(\pi_2)$ in a configuration space C. The surrounding obstacle information along each path are $\beth(h(\pi_1))$ and $\beth(h(\pi_2))$. The reasonable alternative path exists if and only the surrounding information of the two paths are not the same, such that, $\beth(h(\pi_1)) \neq \beth(h(\pi_2))$.

Given the sensing range of a robot as Υ, and the obstacles set $O = \{o_1, o_2, \ldots, o_n\}$. For path $h(\pi_1)$, it consists of a set of discrete states $X(\pi_1) = \{x(\pi_1)_1, x(\pi_1)_2, \ldots, x(\pi_1)_n\}$. In this paper, we assume that any obstacle o_i can be described with a circle or ellipse centered at o_i^c, then we can build a Delaunay Triangulation [19] connection (DTc) using the obstacle centers, the initial state, and the goal state. For DTc, it can generate a network like structure, and each two states have at most one connection. It is illustrated in **Figure 8(a)** that the green edges are the valid connections, with labels to distinct with each other. For any path, if the path intersects with an edge, the edge information should be added to the information factor, such as the solid red path intersects with edge L7, then $L7 \in \beth(h(\pi_1))$. The edge labeling method can guarantee the uniqueness toward homology, while we note that homotopic paths can also be used to perform emergency threat management. It is represented in **Figure 7(b)**, the solid red path h_s and the dotted red path h_D are homotopic to each other. Given the sensing range Υ, we have the sensing envelop which are dotted purple lines h_1^L, h_1^R for h_D and solid black lines h_2^L, h_2^R for h_s, indicating the maximum detection range for emergency threat. Then, we have the $\{o_1\} \subset \beth(h_D)$ and $\{o_1, o_2, o_3, o_4\} \subset \beth(h_S)$, thus we have h_s and h_D both regarded as reasonable alternative paths for online threat management.

The informative approach discussed thus can help to label each path in a configuration space, such as the results listed in **Table 1** of paths in **Figure 7(a)**. Then according to Definition 5, we can find the label of each path. For any several paths which have the same label, we choose the shortest path and use as the candidate for online fast switching.

4.3. Emergency threat management

The reasonable alternative path set H_{RAP} provides a network with cluttered environment adaptivity. The concept of visibility was discussed in [20], where the cycle information is used to enable fast deformation for motion planning. Visibility is defined as:

Path	Path surrounding information	
	Edge information	Obstacle information
Dotted red	L2.L3.L6.L7.L8.L19.L31	O_5, O_2
Solid red	L2.L3.L6.L7.L8.L19.L31	O_5, O_2, O_3, O_4
Dotted blue	L2.L3.L6.L14.L16.L17.L18.L31	O_5, O_2, O_3, O_4
Solid blue	L2.L3.L6.L14.L16.L20.L23.L24	O_5, O_2, O_3
Dotted pink	L2.L10.L11.L22.L27.L29.L30	O_6, O_7, O_8
Solid pink	L2.L10.L11.L22.L27.L29.L30	O_6, O_9, O_{10}, O_3

Table 1. The path information parameter is consist of two parts, which are edge information using DTc and obstacle information using sensing envelop, respectively.

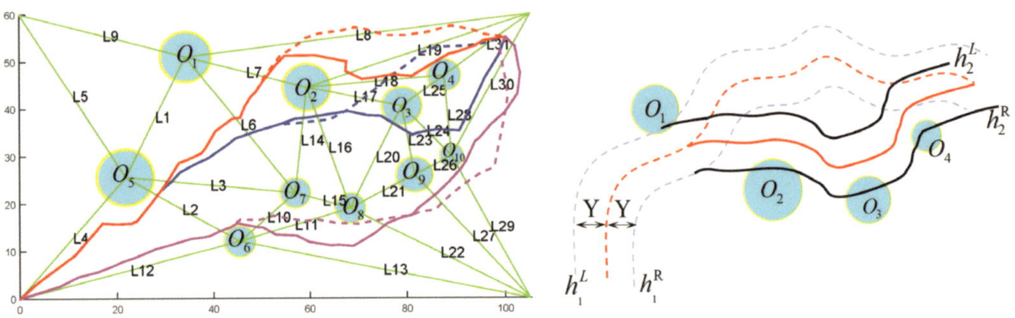

(a) An example of multiple paths generated by path planner, the green edges are generated using DT. The surounding information uses the edges for labeling.

(b) The information parameter also considers the surounding obstacles within the sensing range, thus to decide whether homotopic paths should keep.

Figure 7. An illustration of surrounding information used to find reasonable alternative path for online emergency threat management. The information parameter consists of edge information and obstacle surrounding information within sensing envelop.

http://dx.doi.org/10.5772/intechopen.71410

$$\mathrm{V}:\begin{cases} & [0,1]\times[0,1]\rightarrow\{0,1\} \\ V(x_t,x_{t'}) = & 1, if\ \mathcal{L}_{link}(x_t,x_{t'})\in C_{free}\subset C \\ V(x_t,x_{t'}) = & 0, if\ \mathcal{L}_{link}(x_t,x_{t'})\in C_{obs}\subset C \end{cases} \tag{16}$$

Where $x_{()}$ denotes a state of a path, $\mathcal{L}_{link}$ is the connection of two states, C_{free} is the free space and C_{obs} is the obstacle region. A visual illustration is provided in **Figure 8(a)**, where visibility can only in obstacle free region.

It is noted that the visibility in this paper means a possible connection to switch from one path to another for emergency threat management. For switching with visibility, given all the reasonable alternative paths H_{RAP}, the algorithm performs exploration for visibility state at each UAV state x_{UAV} among H_{RAP}. The algorithm then outputs the visible guards (states) x_{RAP} as illustrated in **Figure 8(b)**. To avoid the pop-up threat (or dynamic threat), UAV must select one entry guard from the visible guard set to reconnect to another pre-planned path to the goal. To validate the best connection, that is, the entry point and the entry connection, this paper applies the heuristic:

$$x^* = \underset{i,j}{\mathrm{argmin}}\ C_{FE}\left(x_{UAV}, H_{RAP(i)}\left(x_{RAP(j)}\right)\right) + C_{TC}\left(H_{RAP(i)}\left(x_{RAP(j)}\right)\right) \tag{17}$$

Using a simple cost based metric, where C_{FE} is the forward energy cost which is the distance and the turning cost from UAV position x_{UAV} to entry state $x_{RAP(j)}$ and the path from entry state to the goal $H_{RAP(i)}(x_{RAP(j)})$. The turning cost is the integration of heading angle difference at each state, which denotes the smoothness of the planned path. C_{TC} is the threat cost, the integration of inverse distance between state and obstacles. Using such approach, the algorithm can find five visible states $x_{RAP}=\{x_{RAP}(1), \ldots, x_{RAP}(5)\}$ as illustrated in **Figure 8(b)**. Then, it tries to find the best entry state using the minimum cost principle (Eq. (17)).

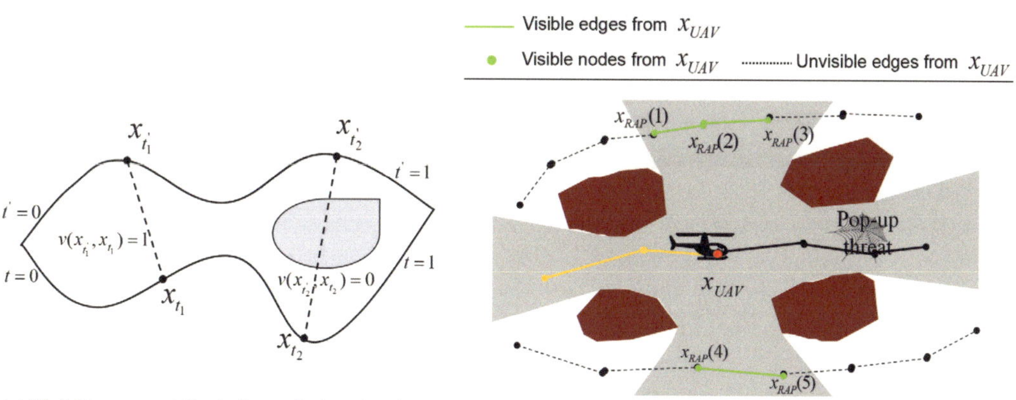

(a) Visibility concept illustration. $x()$ denotes the state of a path. Purple region denotes the obstacle.

(b) UAV is able to find visiable nodes of the multiple paths, and then pick the visible nodes out for switching. The UAV is also able to remember the taveled path as represented in yellow.

Figure 8. For online switching, UAV should follow the visible-node selection algorithm (a) to explore the possible switching route, then it switches to the cost minimum path for ETM (b).

We further consider a situation that there may have no visible states at current location. The paper proposes to use a long-term memory approach to handle this problem, that is, the travel path should be stored in memory, such as the orange edges and states illustrated in **Figure 8(b)**. In the meanwhile, the method stores the visible state along the traveled path. Then, the UAV has to fly back to find a cost minimum path toward the goal if it confronts with pop-threat and has no visible states.

5. Experiment and discussion

In this section, we highlight the performance of the obstacle biased and homology biased path planner with the ability of emergency threat management (avoiding pop-up and dynamic threat online). In the section, we will discuss the following points: (1) How the threshold of EFA affects the performance of MGART. (2) The time performance and reliability of reasonable alternative chosen algorithm. (3) The online emergency threat management performance. The algorithm is implemented using MatLab 2016b on a laptop computer with a 2.6 GHz Intel Core I5 processor.

5.1. Comparative simulation of multiple path exploration

We design three different scenarios, which are non-obstacle scenario, rounded obstacle crowed scenario, and irregular polygons crowed scenario, to perform comparative simulations. All the scenarios are 2D with $100^{*}65\ m^2$ space, and obstacles randomly generated.

For scenario 1, it is a non-obstacle environment, and we set the variable threshold as a set with value {3, 5, 7, 8, 11,13,15,18,20,25,28} for representation. As we know the length of EFA threshold affects the goal biased probability, which directly decide the area of the newly obtained Voronoi region of the neighbor branches, we design a set of comparative experiments to study the effects between EF length and RAPs. An intuitive result of the relationship between planned paths and EF length after 10,000 iterations are provided in **Figure 9(a)–(d)**. MGART can find 37 paths after 10,000 iterations if EF length is set as 3 step-length, and the number decreases to 22 if the EF length is set as 28. The reason is that the longer the EF length, the further the neighbor branches can obtain the goal biased resource. Thus, the neighbor branches need more steps to exploring toward the goal, that is, less paths will be achieved with better homology performance. As we can see that the paths in **Figure 9(d)** have a better homology performance than **Figure 9(a)**. The same EF length variation experiment is also deployed in scenario 2, and results are shown in **Figure 9(e)–(h)**. For RAPs, it is the same with the results in scene 1 that RAPs decrease with the increasing of the EF length. However, as the increasing of EF length enables more branches to explore toward the goal as well as increasing the homologous paths (see in **Figure 9(e)–(h)**), the number of the RAPs increasing with the increasing of the EF length. The statistic relation between the RAPs and the EF length is illustrated in **Figure 10**, which further proves the conclusion.

The EFA can be used to any SBAs by shifting the goal biased resource to achieve multiple RAPs for online switching. This paper compares the performance between MGART and MRRT* in three scenarios with 10,000 iterations. We compare the efficiency of generating a path, RAPs,

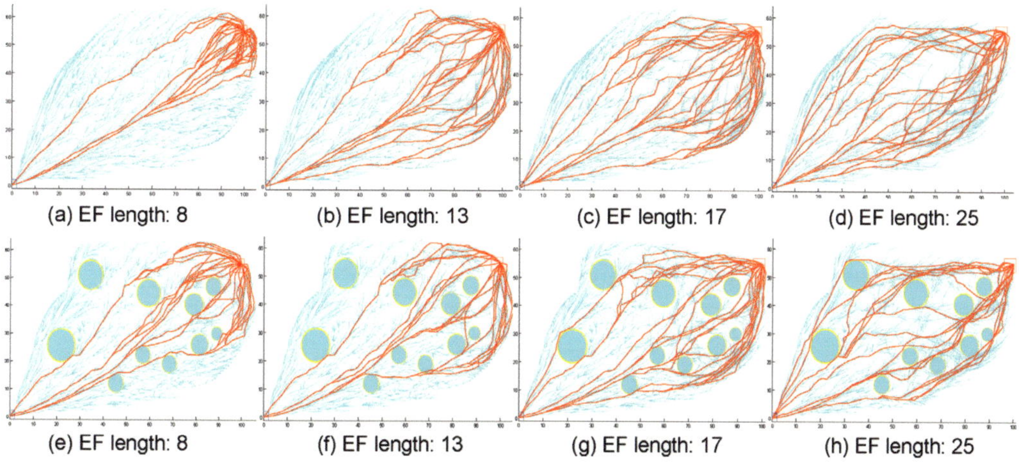

Figure 9. Illustration of alternative paths generated by MGART vary with representative backward EF length. (a)–(d) denotes the results in non-obstacles scenario, (e)–(h) denotes the results in obstacle crowed.

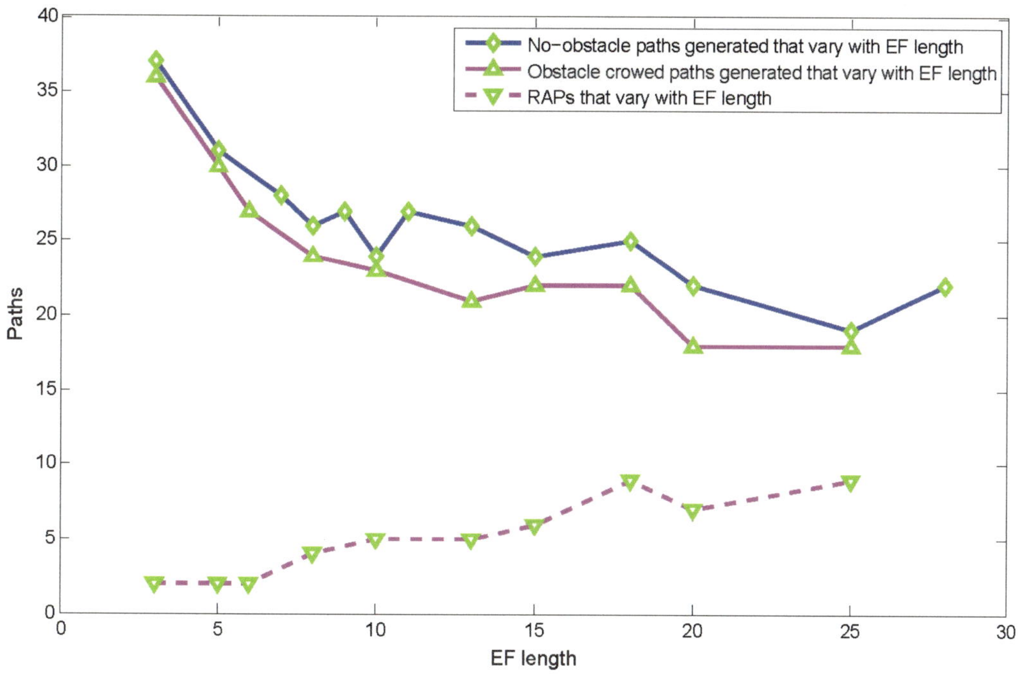

Figure 10. Relation between EF backward length and APs and RAPs with two scenarios. Here the solid diamond line denotes the relation of scene 1, and the triangle lines denotes the results of scene 2.

average time for finding a path, and average time for any RAP are compared in **Table 2**. GART has a better performance in both path exploration and RAP generation, such that MGART can find at least 3 times of the number of paths toward the goal than MRRT*. Because GART introduces the environmental information to speed up the exploring process, the results prove that MGART is more efficient in finding RAPs, which is almost 100% faster than MRRT*. For time performance, we can see that MGART also outperforms MRRT* with at least 3 times advantage.

Besides comparison of the time performance of finding online switching paths (that is RAPs), we also pay attention to the quality of the path generated. The average lengths and standard deviation of the length of all paths in each scene are illustrated in **Figure 11**. The average length of the paths that generated by MGART and MRRT* are illustrated in **Figure 11(a)**, we can see that MGART has a strong convergence performance than MRRT*. The standard deviation of the lengths is shown in **Figure 11(b)**, results demonstrate that MGART is more likely to find paths with smaller fluctuation as well as smaller cost.

5.2. Performance of reasonable alternative path chosen

We also test the path labeling algorithm, that is, the surrounding information pursuing using DTc and sensing envelop, which is used to obtain the reasonable alternative paths under Definition 5. It is should be noted that the under the definition, any two paths do not have the same information parameter, which enables fast switching when facing pop-up threat. As the path label method guarantees the unique labeling of all the paths, only the paths which stretch in a parallel way and within the same sensing envelop have the same labels.

The results of simulation after 10,000 iterations in scenario 2 and 3 are provided in **Table 3**. For each single path, the time needed for labeling the path mainly depends on the area, dimension,

Scenario	Algorithm	Paths after 10,000 iterations	RAPs after 10,000 iterations	Average time for a feasible path	Average time for a RAP
Scenario 1	MGART	48	/	2.081	/
	MMRRT*	12	/	7.825	/
Scenario 2	MGART	43	9	2.093	8.213
	MRRT*	11	5	7.169	15.772
Scenario 3	MGART	34	10	2.257	7.674
	MRRT*	11	5	6.789	14.935

Table 2. Detailed comparison of planning efficiency between MH-GART and MH-RRT* in all three scenarios.

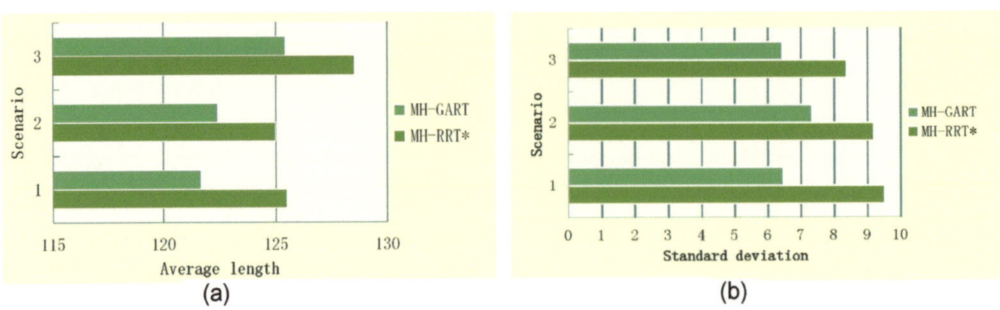

Figure 11. Comparison of (a) average length and (b) standard deviation of the APs generated by MGART and M-RRT* in three scenarios.

http://dx.doi.org/10.5772/intechopen.71410

the complexity of the configuration space. For our tested with area 100*60, the average time for acquiring the information for labeling 0.078 s (see in **Table 3**). The average time needs for RAP pursing of our cases is 0.139 s.

5.3. Experiments of emergency threat management

MGART can be used for 3D and 2D pop-up threat management, and the 3D environments can be easily segmented by DT. We evaluate the performance of our method in both 2D and 3D environments, and we also compared the time performance.

For 2D environments, we implement three tests with different number of dynamic threats. The RAP chosen algorithm works when robot realizes that the path will collide with the pop-up threat, that is, robot at position x_{UAV} detects the moving threat (see in **Figure 12(a)**). The simulation setting is illustrated in **Table 4**, where the robot speed is 10 m/s and the moving threat can be detected within 10 m detection range. Thus, the robot has less than 1 s to re-plan a path and executing to avoid the obstacle. RAP chosen algorithm first evaluates all its neighbor RAPs (the green lines) around the robot, and chooses the cost minimal and collision free path based on principle Eq. (17) (the dotted green path in **Figure 12(b)**). It is noted that **Figure 12(b)–(d)** are results of using MGART to avoid one, two, and three moving threats, respectively. The black parts along the navigation path denote the position where threat is detected by robot. We also execute test in 3D environment (see in **Figure 13**) with pup-up and moving threats. The on-line switching is supposed to be used for aerial robots in 3D, thus Dublin's Curves is used when switching from current position to safe path.

For all the experiments, we study the time efficiency of each switching to escape from current dangerous situation. For one moving threat avoiding (see in **Figure 12(b)**), the time needed to switch to other RAP is 0.0507 s, and the whole navigation duration is 13.14 s with 10 m/s

Scenario 2		Scenario 3	
Time for labeling (s)	**Time for RAPs (s)**	**Time for labeling (s)**	**Time for RAPs (s)**
0.0721	0.146	0.0842	0.132

Table 3. Time performance of proposed method in two scenarios.

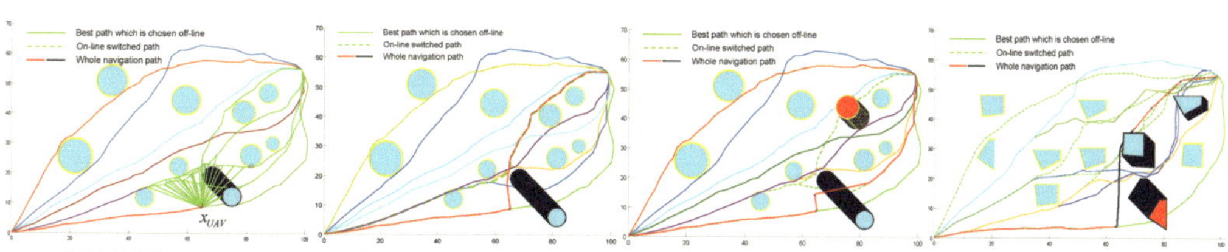

(a) Visible states (b) ETM with one obstacle (c) ETM with two obstacles (d) ETM with three obstacles

Figure 12. Tests of on-line switching to avoid dynamic threats using MGART in 2D scenarios. (a) Robot detects moving threat at position x_{UAV}, then it evaluates all its visible neighbor RAPs (the green lines) to choose the switching path. (b) Complete navigation of avoiding one moving threat, the red path is the navigation path. (c) Test of avoiding two moving threats, the black and red circles are threats. (d) Tests of avoiding three moving threats.

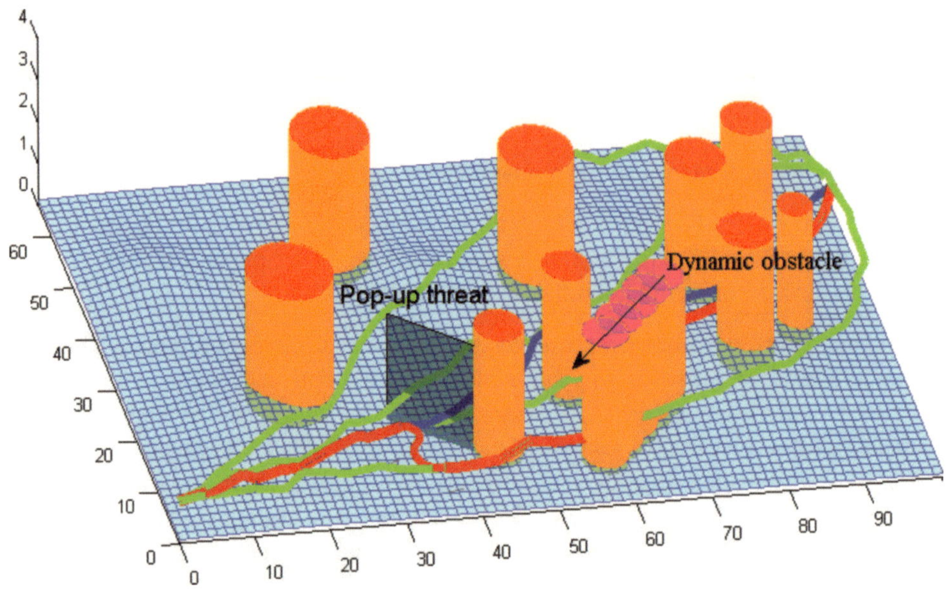

Figure 13. The experiment of using MGAT for 3D emergency threat management case, where pop-up threat and moving threat are exist in the environment.

Robot speed	10 m/s
Detection range	10 m
Speed of cyan obstacle (in **Figure 12(b)–(c)**)	1.4 m/s
Speed of red obstacles (in **Figure 12(c)–(d)**)	0.8 m/s
Speed of cyan obstacle (in **Figure 12(d)**)	0.8 m/s
Online switching range	20 m

Table 4. Setting of simulation for ETM.

speed. For two dynamic threats avoiding case (see in **Figure 12(c)**), the whole navigation time is 13.32 s, and the time spend to avoid the second threat is 0.0912 s. In scene 3, we designed a long duration for threat (see in **Figure 12(d)**). The two cyan threats disable the blue-path, thus robot has to switch for more times while tracking the dark path. The average time is no more than 0.15 s which can be decreased when implemented in robot's platform with C++ implementation, and the whole navigation time is 13.5 s.

6. Conclusion

The main contribution of this paper is that an online EMT planner is proposed, where pop-up threat and moving obstacle happen during tracking the pre-planned path. We propose a new multiple path planning approach called MGART, which is improved based on GART, by introducing an 'Extending Forbidden' algorithm to shift the goal biased probability to neighbor branches around goal reached branch. The algorithm is shown to inherit the merits of

GART and the ability of exploring in cluttered environments, and it guarantees asymptotically optimal and completeness. It is also shown that the algorithm can generate multiple paths without using variant cost principles, but only relying on the EFA threshold, thus it enables selection for online dynamical switching.

In the future, we would like to research on online visual positioning and environment perception topic, which is lack of discussion in this paper. We would like to enable cognitive sensing and autonomous for robots.

Acknowledgements

This work was supported by NSFC under grant No. 61528303.

Author details

Liang Yang[1,2,3], Yuqing He[1,3], Jizhong Xiao[1,2,3]*, Bing Li[2] and Zhaoming Liu[1,3]

*Address all correspondence to: jxiao@ccny.cuny.edu

1 State Key Laboratory of Robotics, Chinese Academy of Sciences, Shenyang, China

2 The City College/City University of New York, New York, USA

3 University of Chinese Academy of Sciences, Beijing, China

References

[1] Goerzen C, Kong Z, Mettler B. A survey of motion planning algorithms from the perspective of autonomous UAV guidance. Journal of Intelligent and Robotic Systems. 2010; **57**(1–4):65

[2] Yang L, Qi J, Song D, Xiao J, Han J, Xia Y. Survey of robot 3D path planning algorithms. Journal of Control Science and Engineering. 2016;**5**(2016):1-22

[3] Chen H, Chang K, Agate CS. UAV path planning with tangent-plus-Lyapunov vector field guidance and obstacle avoidance. IEEE Transactions on Aerospace and Electronic Systems. 2013;**49**(2):840-856

[4] Davoodi M, Panahi F, Mohades A, Hashemi SN. Multi-objective path planning in discrete space. Applied Soft Computing. 2013;**13**(1):709-720

[5] Rathbun D, Kragelund S, Pongpunwattana A, Capozzi B. An evolution based path planning algorithm for autonomous motion of a UAV through uncertain environments. In: Proceeding of IEEE Digital Avionics Systems Conference. Vol. 2; 2002. pp. 8D2-1-8D2-12

[6] Valenti RG, Jian Y-D, Ni K, Xiao J. An autonomous flyer photographer. In: Proc. of 2016 IEEE International Conference on Cyber Technology in Automation, Control, and Intelligent Systems (CYBER); 2016. pp. 273-278

[7] Coleman D, Sucan I, Chitta S, Correll N. Reducing the barrier to entry of complex robotic software: A moveit! case study. arXiv preprint. 2014;**1404**(3785):1-14

[8] Yang K, Gan SK, Sukkarieh S. A Gaussian process-based RRT planner for the exploration of an unknown and cluttered environment with a UAV. Advanced Robotics. 2013;**27**(6): 431-443

[9] Yoshida E, Kanehiro F. Reactive robot motion using path replanning and deformation. In: IEEE International Conference on Robotics and Automation (ICRA); 2011. pp. 5456-5462

[10] Lindemann SR, LaValle SM. Incrementally reducing dispersion by increasing Voronoi bias in RRTs. In: IEEE International Conference on Robotics and Automation. Vol. 4; 2004. pp. 3251-3257

[11] Kavraki L, Latombe J-C. Randomized preprocessing of configuration for fast path planning. In: Proc. of IEEE International Conference on Robotics and Automation; 1994 May 8. pp. 2138-2145

[12] LaValle SM, Kuffner JJ Jr. Randomized kinodynamic planning. The International Journal of Robotics Research. 2001;**20**(5):378-400

[13] Yang L, Xiao J, Qi J, Yang L, Wang L, Han J. GART: An environment-guided path planner for robots in crowded environments under kinodynamic constraints. International Journal of Advanced Robotic Systems. 2016;**13**(6):1-18

[14] Yang L, Song D, Xiao J, Han J, Yang L, Cao Y. Generation of dynamically feasible and collision free trajectory by applying six-order Bezier curve and local optimal reshaping. In: Proc. of IEEE/RSJ International Conference on Intelligent Robots and Systems (IROS); 2015 Sep 28. pp. 643-648

[15] Karaman S, Frazzoli E. Sampling-based algorithms for optimal motion planning. The International Journal of Robotics Research. 2011;**30**(7):846-894

[16] Urmson C, Simmons R. Approaches for heuristically biasing RRT growth. In: IEEE/RSJ International Conference on Intelligent Robots and Systems. Vol. 2; 2003, October. pp. 1178-1183

[17] Pan J, Manocha D. Fast probabilistic collision checking for sampling-based motion planning using locality-sensitive hashing. The International Journal of Robotics Research. 2016;**35**(12):1477-1496

[18] Yang L, Qi J, Jiang Z, Song D, Han J, Xiao J. Guiding attraction based random tree path planning under uncertainty: Dedicate for UAV. In: IEEE International Conference on Mechatronics and Automation (ICMA); 2014 Aug 3. pp. 1182-1187

[19] Levcopoulos C, Krznaric D. Quasi-greedy triangulations approximating the minimum weight triangulation. Journal of Algorithms. 1998;**27**(2):303-338

[20] Jaillet L, Siméon T. Path deformation roadmaps: Compact graphs with useful cycles for motion planning. The International Journal of Robotics Research. 2008;**27**(11–12):1175-1188